Food Molecular Microbiology

Food Biology Series

Food Molecular Microbiology

Editors

Spiros Paramithiotis

Laboratory of Food Quality and Hygiene
Departement of Food Science and Human Nutrition
Agricultural University of Athens
Athens, Greece

Jayanta Kumar Patra

Research Institute of Biotechnology & Medical Converged Science
Dongguk University
Goyang-si, Republic of Korea

CRC Press
Taylor & Francis Group
Boca Raton London New York

CRC Press is an imprint of the
Taylor & Francis Group, an **informa** business

A SCIENCE PUBLISHERS BOOK

CRC Press
Taylor & Francis Group
6000 Broken Sound Parkway NW, Suite 300
Boca Raton, FL 33487-2742

First issued in paperback 2020

© 2019 by Taylor & Francis Group, LLC
CRC Press is an imprint of Taylor & Francis Group, an Informa business

No claim to original U.S. Government works

ISBN-13: 978-1-138-08808-5 (hbk)
ISBN-13: 978-0-367-78026-5 (pbk)

Library of Congress Cataloging-in-Publication Data

Names: Paramithiotis, Spiros, editor.
Title: Food molecular microbiology / editor, Spiros Paramithiotis (Laboratory of Food Quality and Hygiene, Departement of Food Science and Human Nutrition, Agricultural University of Athens, Athens, Greece).
Description: Boca Raton, FL : CRC Press, 2019. | Series: Food biology series | "A science publishers book." | Includes bibliographical references and index.
Identifiers: LCCN 2018055950 | ISBN 9781138088085 (hardback)
Subjects: LCSH: Food--Microbiology.
Classification: LCC QR115 .F66268 2019 | DDC 579/.16--dc23
LC record available at https://lccn.loc.gov/2018055950

Visit the Taylor & Francis Web site at
http://www.taylorandfrancis.com

and the CRC Press Web site at
http://www.crcpress.com

Preface to the Series

Food is the essential source of nutrients (such as carbohydrates, proteins, fats, vitamins, and minerals) for all living organisms to sustain life. A large part of daily human efforts is concentrated on food production, processing, packaging and marketing, product development, preservation, storage, and ensuring food safety and quality. It is obvious therefore, our food supply chain can contain microorganisms that interact with the food, thereby interfering in the ecology of food substrates. The microbe-food interaction can be mostly beneficial (as in the case of many fermented foods such as cheese, butter, sausage, etc.) or in some cases, it is detrimental (spoilage of food, mycotoxin, etc.). The *Food Biology* series aims at bringing all these aspects of microbe-food interactions in form of topical volumes, covering food microbiology, food mycology, biochemistry, microbial ecology, food biotechnology and bio-processing, new food product developments with microbial interventions, food nutrification with nutraceuticals, food authenticity, food origin traceability, and food science and technology. Special emphasis is laid on new molecular techniques relevant to food biology research or to monitoring and assessing food safety and quality, multiple hurdle food preservation techniques, as well as new interventions in biotechnological applications in food processing and development.

The series is broadly broken up into food fermentation, food safety and hygiene, food authenticity and traceability, microbial interventions in food bio-processing and food additive development, sensory science, molecular diagnostic methods in detecting food borne pathogens and food policy, etc. Leading international authorities with background in academia, research, industry and government have been drawn into the series either as authors or as editors. The series will be a useful reference resource base in food microbiology, biochemistry, biotechnology, food science and technology for researchers, teachers, students and food science and technology practitioners.

Ramesh C Ray
Series Editor

Preface

Advances in the field of molecular biology, methodological approach and equipment have provided the required sophisticated tools that allowed the intensive study of food microbial communities from a molecular perspective. Information from genomic, transcriptomic, proteomic and metabolomic studies may be integrated through bioinformatic applications, improving our understanding regarding the interactions between biotic and abiotic factors and concomitantly the physiology of starter cultures, spoilage and pathogenic microbiota. Moreover, significant improvements in the speed, accuracy and reliability of food quality and safety assessment have taken place and at the same time the basis has been set for exciting future developments, including the exploitation of gene networks and applications of nanotechnology and systems biology.

Food molecular microbiology is a very active research area where a number of significant developments have been made that have altered the classical microbiological perspective. However, the building blocks of this field are yet to be assembled under one discipline. This is the first time that such an assembly is made, offering an integrated view of this field and facilitating our understanding as well as the identification of opportunities for future developments.

This book consists of twelve chapters. The first chapter offers an overview of food molecular microbiology and may be considered as an introduction to the topic. Tools and approaches currently used for the assessment of micro-ecosystems, microbial evolution and taxonomy as well as the recovery, detection and inactivation of foodborne viruses are analyzed in Chapters 2 to 4. The bioinformatics tools applied in contemporary food microbiology are presented in Chapter 5. Chapters 6 to 8 offer an overview of the 'omics approaches applied to microbial food safety and quality assessment as well as genomic, transcriptomic and proteomic insights into the physiology of *Campylobacter*, *Listeria monocytogenes* and *E. coli*. Chapters 9 and 10 are dedicated to the stress responses of lactic acid bacteria and yeasts. In Chapter 11 Gram negative food spoilers are analyzed from a genomic perspective. Finally, in Chapter 12, the applications of nanotechnology in food and agriculture are presented.

We would like to thank Dr. Ramesh C Ray, the editor of the 'Food Biology' book series for honoring us with the editing of this book, all contributing authors for devoting their time, effort and expertise as well as the Science Publishers/CRC Press for publishing this book.

Spiros Paramithiotis
Jayanta Kumar Patra

Contents

1

Food Molecular Microbiology
An Overview

Spiros Paramithiotis[1],* and *Jayanta Kumar Patra*[2]

1. Introduction

Food molecular microbiology may be defined as the discipline that aims to study, at the molecular level, the microorganisms that are able to grow on substrates considered for human consumption. The ultimate goal is to enable accurate and reliable prediction of microbial growth in foodstuff through the unraveling of the physical and transcriptional genomic organization, as well as the effect of food constituents and processing conditions used for food production and preservation. As a discipline, it has emerged following the advances in the field of molecular biology. Through the development of suitable tools, the assessment of the whole microbial entity was achieved. Thus, information from all levels of cellular organization, i.e., genomic, transcriptomic, proteomic and metabolomics, at both single cell and microecosystem level, was made accessible. Equally important were the bioinformatic tools that were developed in parallel, which enabled integration of these data and provided biologically meaningful conclusions.

Implementation of molecular tools have led to significant improvements throughout food microbiology, particularly in microbial taxonomy and evolution assessment, the study of microecosystems and in food safety.

The use of DNA in taxonomy assessment offered a phylogenetic insight into the dichotomously organized depiction of microdiversity and, at the same time, improved speed, throughput and in many cases discrimination. There is, however, still room for improvement since the genetic markers employed so far allow discrepancies with

[1] Department of Food Science and Human Nutrition, Agricultural University of Athens, Athens, Greece.
[2] Research Institute of Biotechnology & Medical Converged Science, Dongguk University, Goyang-si, Republic of Korea.
* Corresponding author: sdp@aua.gr

the classical taxonomic approach and lack adequate variability to offer the desired discrimination power.

The use of molecular approaches in the assessment of microecosystem composition allowed the study of the Viable but Non-Culturable (VNC) proportion of the microbiota, which in the case of food microbiology seems negligible. In addition, tools like Denaturing Gradient Gel Electrophoresis (DGGE) allow the depiction and monitoring of the total microbiota or the metabolically active proportion of an ecological niche, depending on the macromolecule used as a marker and the approach.

Molecular tools significantly improved food safety assessment in terms of epidemiological surveillance and understanding of the responses to food-associated stimuli. On the contrary, accurate and reliable detection of foodborne pathogens is still performed by classical microbiological approaches. Molecular tools have allowed accurate epidemiological surveillance through an array of techniques that are currently available, with advantages and limitations discussed in paragraph 4.2. Furthermore, the mechanisms of pathogenesis have been largely understood, at least regarding the major foodborne pathogens *L. monocytogenes*, *Salmonella* serovars and *E. coli*. In addition, the transcriptomic response to a variety of factors relative to the food industry is under study, both *in vitro* and in actual food systems.

Molecular approaches have offered novel insights that led to significant improvement of our understanding of food microbiology. Taking into consideration the speed of analysis, the amount of the generated data and the technology involved, molecular microbiology may seem more attractive than the 'old-fashioned' classical one (Gill 2017). This may lead to misunderstandings regarding the quality of the data obtained. It should be kept in mind that each approach generates both common and different types of data that need to be integrated in order to obtain biologically meaningful results. There are cases in which the data obtained from the classical approach may be more reliable than the respective obtained by the molecular one, and vice versa. For that purpose, these approaches should be regarded as complementary and not as antagonistic.

2. Study of Microecosystems

Assessment of microecosystem composition initially relied on the ability of the microorganisms to grow on the substrates used for their enumeration. However, a wide variety of bacteria, yeasts and fungi have been characterized as unculturable, either because of their entering into a VNC state, or simply due to their inability to grow in the standard media used in classical microbiology (Divol and Lonvaud-Funal 2005, Ghannoum et al. 2010, Steward 2012, Salma et al. 2013, Pinto et al. 2015). As a result, culture-independent approaches were extensively considered, either exclusively or in parallel with classical enumeration procedures in an attempt to gain an integrated view of a microecosystem (Table 1). Detection of microorganisms using such approaches may be based either on DNA or RNA. The use of the former has a major disadvantage, namely the ability of DNA to remain undegraded, at least for time relevant to food microbiology (Nielsen et al. 2007). As a result, analysis based on DNA may depict the history of a sample rather than the metabolically active microbiota. The use of ethidium monoazide (EMA) or propidium monoazide (PMA) constituted a major advancement. Both reagents, upon addition to a sample, intercalate with any accessible DNA

Table 1. Representative studies of microecosystem composition through culture-independent techniques.

Product	Microecosystem Assessment Approach	Comment	Reference
Fontina PDO cheese	RT-PCR-DGGE & pyrosequencing	Prevalence of the starter culture over autochthonous microbiota was verified	Dolci et al. (2014)
Fermented sausages	RT-PCR-DGGE & pyrosequencing, in parallel with culture-dependent	Possible PCR bias highlighted the need for parallel application of culture- dependent and –independent techniques	Greppi et al. (2015)
Barbera must fermentation	culture dependend in parallel with EMA-qPCR, RT-qPCR, EMA-PCR-DGGE, RT-PCR-DGGE	The study was focused on *S. cerevisiae*, *H. uvarum* and *St. bacillaris*. Statistically significant differences between the quantitative techniques (i.e., plating, EMA-qPCR & RT-qPCR) were reported. EMA-PCR-DGGE failed to detect *H. uvarum*	Wang et al. (2015)
Grana-like cheese	RT-PCR-DGGE & pyrosequencing, RT-qPCR	The competitiveness of the starter cultures was effectively assessed	Alessandria et al. (2016)
Spanish-style green table olive fermentation	RT-PCR-DGGE in parallel with culture-dependent	No direct comparison between the two approaches was possible	Benitez-Cabello et al. (2016)
Fabriano-like fermented sausages	RT-PCR-DGGE & Illumina sequencing, in parallel with culture-dependent	No direct comparison between the two approaches was possible	Cardinali et al. (2018)

S.: Saccharomyces; H.: Hanseniaspora; St.: Starmerella

molecule, i.e., molecules that reside either outside the cells or in the inside of membrane-compromised ones. Then, upon treatment with strong visible light, nitrene radicals are formed. This concomitantly leads to complexation with other organic molecules and ultimately these DNA molecules become unavailable for the subsequent PCR step (Zeng et al. 2016). However, this approach is characterized by significant limitations as well. EMA may penetrate metabolically active cells, depending on the treatment temperature; thus, underestimation of the viable microbiota may occur (Nocker and Camper 2006, Flekna et al. 2007). On the contrary, no such effect has been observed for PMA. In addition, the effectiveness of the treatment depends upon factors such as the population of the target microorganism, the dead-to-viable ratio and the amplicon size of the following PCR step. It has been reported that, when the cell density is high, the effectiveness of this approach may be compromised by the PMA itself, by affecting living cells as well (Elizaquivel et al. 2012, Zhu et al. 2012). In addition, a minimum population of 10^3 CFU/mL, as well as a ratio of dead to viable cells below 10^4 CFU/mL, have been reported as necessary for effective detection (Pan and Breidt 2007). The size of the amplicon is particularly important in qPCR applications as it may dictate the thermocycling conditions. Small amplicons, i.e., up to 200 bp, are compatible with fast protocols, while larger ones may require the separation of annealing and elongation steps and additional time, otherwise the efficiency of the reaction may not be optimal. In general, the longer the target genomic region, the higher the possibility of intercalation with PMA. Thus, for amplicon sizes below 300 bp, PMA treatment may not be effective. Indeed, this was verified by Pacholewicz et al. (2013), Banihashemi et al. (2012) and Luo et al. (2010). In the latter study, the ineffectiveness of PMA treatment for amplicons less than 190 bp was reported. In addition, Pacholewicz et al. (2013) reported a lack of concordance between classical microbiological counts of *Campylobacter* spp. on broiler chicken carcasses with the respective obtained by PMA-qPCR, which was attributed to the amplicon size of 287 bp. Similarly, Banihashemi et al. (2012) concluded that PMA treatment was effective when the size of the amplicon was more than 1.5 Kbp, whereas for amplicon size less than 200 bp interference by environmental DNA could not be excluded. The effectiveness of PMA treatment for large amplicons (1108 bp) was also reported by Schnetzinger et al. (2013). To this end, an alternative was provided by Soejima et al. (2016) and Soejima and Iwatsuki (2016) with the use of platinum and palladium compounds that chelate with DNA without the aforementioned disadvantages. This approach was effectively applied in milk for the discrimination of metabolically active from dead *Cronobacter sakazakii* and *E. coli* in the first case and members of the *Enterobacteriaceae* family in the second.

Metabolically active cells, i.e., cells that play a role in the development of the microecosystem under study, are better detected through the use of RNA (Arraiano et al. 1998). For that purpose, RNA is isolated, reverse transcribed and a locus relevant to the desired taxonomic level is amplified by PCR. Microecosystem composition assessment through this approach has gained specific attention over the last years (Livezey et al. 2013, Dolci et al. 2013, 2014, Greppi et al. 2015, Wang et al. 2015, Alessandria et al. 2016, Cravero et al. 2016). However, the major outcome of years of application remained unaltered: In food fermentations the unculturable proportion of a microecosystem is of minor, if any, significance. The situation may be different when the stress responses or growth under stressful conditions of pathogenic bacteria are considered. These were generally assessed through plate counts. However, there are some indications suggesting that the pathogen population may have not been accurately enumerated since a proportion may have entered a VNC state. Thus, verification by alternative methods, such as RT-qPCR may be necessary.

3. Evolution and Taxonomy Assessment

Phenotypic attributes, i.e., a set of observable properties including morphological attributes, as well as the ability to grow over a wide variety of substrates and conditions, was initially employed for taxonomic purposes. The microbiome was accordingly divided and speciation followed a dichotomous rationale. This division offered a glance at the diversity of the culturable proportion of the microbiome and at the same time was very useful in the assessment of microecosystem composition. On the contrary, it had limited usefulness in epidemiological surveillance.

The studies that revolutionized taxonomy by improving the organization of the microbiome and offering insight into the evolution of microorganisms were the ones by Zuckerlandl and Pauling (1965) and Fox et al. (1977). In the former, the macromolecules of living cells were classified according to their relevance to evolutionary history. It was concluded that DNA, RNA and polypeptide sequences were capable of revealing such information. Ribosomal DNA was the ideal molecule to perform such a task due to its omnipresence and the occurrence of both conserved and non-conserved regions. In prokaryotes, the use of 16S rRNA gene sequencing for that purpose was proposed by Fox et al. (1977). In eukaryotes, the genes encoding for the small (18S), large (28S) and 5.8S rRNAs were employed (Kurtzman 1994).

This 'new' approach, apart from the information on the evolutionary relationship between microorganisms, offered significant advantages in comparison to the 'classical' procedure, since speed, throughput, and in many cases discrimination ability, were improved. However, discrepancies between the two approaches often occurred. These were basically attributed to the type of information on which taxonomy was based and the fact that phylogenetic assessment does not obey dichotomous rules. From a technical perspective, inconsistencies were attributed to the molecular marker employed, namely the ribosomal DNA. The lack of adequate variability that characterizes ribosomal DNA does not allow reliable discrimination between closely related species. In addition, occurrence of multiple copies, not necessarily identical, the use of arbitrary cut-off values in species delineation and the dependence on bioinformatics algorithms constitute aspects that require specific attention. The latter has been the epicenter of intensive research, especially in the case of evolutionary assessments. A successful evolutionary assessment requires two successful steps, namely sequence alignment and phylogenetic tree construction. Sequence alignment is very challenging, since the recognition of homologous sites, especially in not conserved regions, may affect the result. Although several approaches have been described, the computational strength required compromises their use in meta- or whole- genome data. For that purpose, alignment-free approaches, termed k-mer-based ones, have been developed. These approaches offer an alternative that requires less computational strength but still lacks applicability, despite the improvements that have been proposed (Chan and Ragan 2013, Song et al. 2014, Gao et al. 2017, Sievers et al. 2017, Zhang and Alekseyenko 2017). Similarly, phylogenetic tree construction has been extensively studied and an extended variety of algorithms, divided into character-based and distance-based, are currently available (Horiike 2016, Bogusz and Whelan 2017) along with accuracy evaluation tests (Efron et al. 1996).

4. Food Safety Assessment

Implementation of molecular techniques in food safety assessment has significantly improved epidemiological surveillance, our understanding of pathogen physiology and mechanisms of pathogenicity but has so far failed to produce an alternative regarding pathogen detection.

4.1 Detection of Foodborne Pathogens

Detection of foodborne pathogens by classical microbiological techniques requires a series of steps, namely selective enrichment, cultivation on selective solid media, isolation of colonies exhibiting the typical phenotype and finally identification. This procedure may require more than a week, time that may exceed the shelf-life of the examined product.

Molecular techniques aimed at reducing the analysis time and at the same time improving specificity, selectivity, sensitivity and the overall reliability through the detection of markers specific to the desired taxonomic level, i.e., genus, species, serotype, strain. These genetic markers may either be a protein epitope or a nucleic acid sequence. The major challenges that still need to be addressed are the low numbers of the target microorganism in the presence of a dominant population several orders of magnitude higher and the effect of the food matrix.

When the genetic marker is a protein epitope, Enzyme-Linked ImmunoSorbent Assay (ELISA) is the approach most frequently used. Indeed, a wide range of ELISA protocols has been developed, from the relatively simple sandwich format to the more elaborated dipsticks and immunochromatographic strips and the fully automated systems commercially available (Bolton et al. 2000, Aschfalk and Mueller 2002, Jung et al. 2005, Kim et al. 2007, Kumar et al. 2011, Shukla et al. 2011, 2014, Cavaiuolo et al. 2013, Xu et al. 2013, Shen et al. 2014, Niu et al. 2014, Leem et al. 2014, Mata et al. 2016, Oueslati et al. 2016, Rodrigues et al. 2016, Zeleny et al. 2016). The major advantages of this approach are the relatively short detection time and the automation capacity; on the other hand, low sensitivity and cross-reactivity are the major disadvantages (Zhao et al. 2014).

DNA has been effectively employed for the detection of foodborne pathogens. The major challenge is to distinguish between metabolically active and deceased cells. This has been effectively addressed with the use of PMA, as already described in paragraph 2, and a variety of protocols for the detection of foodborne pathogens including *L. monocytogenes* (Pan et al. 2007), *Vibrio parahaemolyticus* (Zhu et al. 2012), *E. coli* (Elizaquivel et al. 2012, Luo et al. 2010, Schnetzinger et al. 2013), *Campylobacter* spp. (Pacholewicz et al. 2013, Banihashemi et al. 2012), *Salmonella* (Li and Chen 2013), etc., have been developed.

Fluorescence *In Situ* Hybridization (FISH) is another approach that, despite being characterized as very promising (De Long et al. 1989, Amann et al. 1990) and the protocols that have been developed (Kitaguchi et al. 2005, Schmid et al. 2005,

Fuchizawa et al. 2009, Laflamme et al. 2009, Angelidis et al. 2011, Oliveira et al. 2012, Almeida et al. 2013a,b) failed to meet widespread application, mostly due to the significant effect of the food matrix on the quality of the determination.

4.2 Epidemiological Surveillance

The aims of epidemiological studies are to detect the infection source and identify the transmission route as soon as possible, in order to restrict the extent of an outbreak and implement preventive measures. The use of molecular techniques has enabled such studies in relevant time and with adequate reliability, however there are several aspects that need special attention and will be discussed. The major issue is the need for epidemiological data, without them no technique can provide with reliable conclusions. In addition, the clonality of the pathogen under surveillance may affect the outcome, and interpretation of the results should be performed with great caution.

There are currently two classes of methods that have been effectively employed for outbreak investigation: the sequence-based ones and the genotypic profile-based ones. With the former, investigation is based on differences in the sequence of the whole genome or specific loci. Whole Genome Sequencing (WGS) may seem to be the ultimate approach, but the substantial effect of the sequencing platform and the bioinformatics approach applied on the quality of the result still compromises its use. Instead, Multi Locus Sequence Typing (MLST) has been widely considered. With MLST the nucleotide sequence of a number of loci, usually 5 to 10, the size of which usually does not exceed 600 bp, is compared. The discrimination capacity is adjusted by the selection of the number and the degree of conservation of the loci under examination. Selection of slowly evolving loci is suitable for long-term studies, while rapidly evolving ones are suitable for short-term studies. The isolates are distinguished into Sequence Types (STs) that are further grouped into Clonal Complexes (CCs). Conclusions of epidemiological nature are drawn on the basis of CCs and STs distribution within the samples. MLST protocols have been developed for all major foodborne pathogens (Paramithiotis et al. 2018). *L. monocytogenes* loci considered for that purpose include *abcZ, actA, betL, bglA, cat, clpP, dal, dapE, dat, gyrB, hlyA, inlB, inlC, ldh, lhkA, lisR, pgm, prfA, recA* and *sod* (Salcedo et al. 2003, Revazishvili et al. 2004, Zhang et al. 2004, Ragon et al. 2008); *Salmonella* loci include 16S rRNA, *aceK, aroC, atpD, dnaN, fimA, fimH, fliB, fliC, glnA, glpF, gyrB, hemD, hisD, icdA, manB, mdh, panB, pduF, pgm, purE, spaM, spaN, sseL, sucA* and *thrA* (Kotetishvili et al. 2002, Fakhr et al. 2005, Sukhnanand et al. 2005, Torpdahl et al. 2005, Tankouo-Sandjong et al. 2007, Liu et al. 2011b, Singh et al. 2012); *Campylobacter* spp. loci include *aspA, glnA, gltA, glyA, pgm, tkt,* and *uncA* (www.mlst.net), while *E. coli* O157:H7 include *adk, aspC, clpX, dinB, fadD, fumC, gyrB, icd, lysP, mdh, pabB, ploB, purA, putP, recA, trpA, trpB* and *uid* (Reid et al. 2000, Wirth et al. 2006, Jaureguy et al. 2008).

MLST has allowed the establishment of associations between CCs and ecological niches, providing us with an alternative insight. However, the applicability in outbreak investigation has been limited due to the limited discriminatory power and, therefore, the inability to cluster and separate outbreak-associated and -irrelevant isolates. Such a capacity may be attributed to the selection of the loci; therefore, improvement may be expected in the future.

As far as the methods that rely on the generation of genotypic profiles and their concomitant comparison are concerned, Pulsed-Field Gel Electrophoresis (PFGE) and Multi Locus Variable-number tandem repeat Analysis (MLVA) are the ones applied. In PFGE, the DNA of the isolates is cleaved with the use of specific restriction enzymes and the fragments are concomitantly separated with electrophoresis in pulsed electric field. Failure to separate isolates with the use of three restriction enzymes may lead to the conclusion that these strains are epidemiologically connected. These enzymes are *Asc*I, *Apa*I and *Xba*I for *L. monocytogenes*, *Xba*I, *Bln*I/*Avr*II and *Spe*I for *Salmonella*, *Sma*I and *Kpn*I for *Campylobacter* spp., *Xba*I and *Bln*I for *E. coli*, etc. The major advantages of this technique are the standardization potential and the capacity for data exchange. On the other hand, this technique is labor intensive, equipment and consumables are rather expensive and finally trained personnel is required for both execution and result interpretation.

With MLVA, the relatedness of the isolates is inferred through the comparison of the size of a number of loci containing a Variable Number of Tandem Repeat sequences (VNTRs). The discrimination capacity is adjusted, as in the case of MLST, through the selection of the number and degree of conservation of the loci. Several loci have been examined and several protocols have been proposed regarding the major foodborne pathogens (Paramithiotis et al. 2018). Although this approach provided discrimination comparable to the one obtained by PFGE, it has not gained widespread application mainly due to the fact that it is PCR-mediated and therefore difficult to standardize at a level required for data exchange between laboratories.

4.3 Physiology of Pathogenic Bacteria

Elucidation of pathogenesis and stress response mechanisms is of vital importance for the establishment of effective interventions. The molecular approaches that have been developed have greatly assisted in this endeavour, but a lot of study is still necessary. As an example, in the next paragraphs, the progress made regarding *L. monocytogenes* is discussed.

L. monocytogenes mode of pathogenesis has been extensively studied. Infection starts mostly with the ingestion of contaminated food. Then, the microorganism may enter the host either through the intestinal mucosa or through M cell transcytosis. In the first case, translocation is mainly mediated by InlA and InlB, two bacterial surface proteins that specifically recognize E-cadherin and Met, respectively, which reside on the surface of the host cells (Bonazzi et al. 2009). In the second case, internalization seems to be at least internalin independent (Corr et al. 2006, 2008, Jung et al. 2010). Irrespective the internalization mechanism, the pathogen is engulfed by a phagocytic vacuole, from which it escapes through the production of listeriolycin O (LLO), a cholesterol-binding, pore-forming toxin, encoded by *hly* (Kathariou et al. 1987, Cossart et al. 1989, Hamon et al. 2012). Then, via actin-based intra- and inter-cell motility, the pathogen moves to the neighboring cells, repeating the previous steps. Through this spread, the pathogen reaches the spleen and the liver and spreads throughout the host via the bloodstream.

In general, the factors that determine the outcome of a bacterial infection are: (a) the dose ingested, (b) the host immunological status and (c) the virulence potential of the ingested strain(s). Molecular approaches have greatly assisted in the study of the latter.

A series of genes that are very important for the infection cycle of *L. monocytogenes*, as described above, are physically gathered in a 9-kb gene cluster, termed Listeria Pathogenicity Island 1 (LIPI-1). LIPI-1 is functional in both pathogenic *Listeria* species, i.e., *L. monocytogenes* and *L. ivanovii*. On the contrary, it is absent from *L. innocua*, *L. welshimeri*, *L. marthii* and *L. grayi*, while a non-functional version has been detected in *L. seeligerii*.

The physical and transcriptional organization of LIPI-1 is presented in Fig. 1. *prs* and *orfX* are the borders of LIPI-1, which consists of *prfA*, *plcA*, *hly*, *mpl*, *actA* and *plcB*. *prfA* encodes for the transcriptional activator PrfA, which controls the expression of more than 140 genes throughout the genome of the pathogen. *plcA* and *plcB* encode for two phospholipases C, namely phosphatidylinositol phospholipase C (PI-PLC) and phosphatidylcholine phospholipase C (PC-PLC), respectively. The first has been successfully used as a marker for differentiation between non-pathogenic and pathogenic listeriae (Aurora et al. 2008), whereas both promote effective vacuole escape in collaboration with LLO (Schluter et al. 1998, Grundling et al. 2003). The latter is encoded by *hly*. *mpl* encodes a zinc-metalloprotease that is thermolysin-related and is involved in the maturation of pro-PlcB (Domann et al. 1991, Raveneau et al. 1992). Finally, *actA* encodes for ActA, a virulence factor with many functions including intra- and inter-cellular motility (Suarez et al. 2001, Birmingham et al. 2007).

LIPI-1 gene transcription is regulated mostly by PrfA, which binds to a 14 bp sequence, termed PrfA-box that resides -41 bp from the transcriptional start site. Binding efficiency, and concomitantly PrfA regulation, are affected by the symmetry of the PrfA box, the distance between the PrfA box and the -10 box as well as the sequence between the two boxes and downstream of the -10 box (Dickneite et al. 1998, Boeckmann et al. 2000, Lalic-Muelthaler et al. 2001, Luo et al. 2005).

The promoters designated as P_1prfA and P_2prfA (Fig. 1) lead to synthesis of monocistronic *prfA* mRNA. Bicistronic *plcA-prfA* message is initiated by the *plcA* promoter. P_1prfA is controlled by σ^A while P_2prfA by σ^A, σ^B and also contains a PrfA box (Mengaud et al. 1991, Nadon et al. 2002, Rauch et al. 2005, Schwab et al. 2005, Ollinger et al. 2009) that has been reported to affect *prfA* transcription negatively, thus creating a negative feedback loop. P_1prfA as well as the bicistronic message exhibit temperature dependence; below 37°C the total amount of PrfA is reduced and transcription carries on from P_2prfA that is not dependent on temperature (Chaturongakul et al. 2008). Other compounds that have been reported to affect *prfA* transcription, and concomitantly the expression of the genes controlled, include mono- and di-saccharides (Renzoni et al. 1997, Milenbachs et al. 1997, Milenbachs Lukowiak et al. 2004) and glycerol (Joseph et al. 2008). *hly* transcription is initiated by three promoter sites, namely P_1hly, P_2hly and P_3hly. The latter is PrfA-independent whereas the two first contain a PrfA box and are, therefore, PrfA-dependent (Domann et al. 1996). Type of carbohydrate as well as extracellular pH value have been reported to affect *hly* expression (Milenbachs et al. 1997, Behari and Youngman 1998). Regarding *mpl*, both PrfA - dependence and - independence has been demonstrated by Luo et al. (2004) through a GTP-dependent mechanism.

The expression of several other genes has been reported to promote *L. monocytogenes* virulence potential. Among them, *bsh*, *bilE*, *btlB*, *sigB*, *pva*, *prfA*,

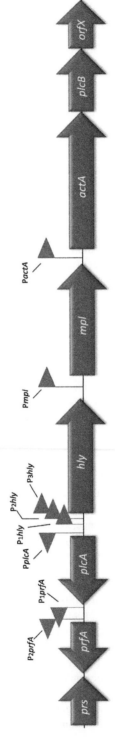

Fig. 1. Physical and transcriptional organization of the LIPI-1 in *L. monocytogenes*. The flags indicate the position of known promoters.

and *btlA* are involved in bile tolerance (Begley et al. 2005, Dowd et al. 2011, Zhang et al. 2011), *inlP* contribute to systemic virulence, particularly in the placenta (Faralla et al. 2016), *lap*, *dltA*, *inlJ*, *lapB* contribute to adhesion (Abachin et al. 2002, Linden et al. 2008, Sabet et al. 2008, Burkholder et al. 2009, Reis et al. 2010), *aut*, *iap*, *gtcA*, *lgt*, *lpeA*, *mprF* and *vip* contribute to invasion (Pilgrim et al. 2003, Reglier-Poupet et al. 2003, Cabanes et al. 2004, 2005, Thedieck et al. 2006, Cheng et al. 2008, Machata et al. 2008, Faith et al. 2009), *svpA* and *sipZ* to vacuole escape (Borezee et al. 2001, Bonnemain et al. 2004) and many more (Camejo et al. 2011).

The effect of various environmental stimuli on the *in vitro* transcriptomic responses of several *L. monocytogenes* strains has been assessed. Among them, the effect of carbohydrates (Milenbachs et al. 1997, Milenbachs Lukowiak et al. 1997, Renzoni et al. 1997), high hydrostatic pressure processing (Bowman et al. 2008), extracellular pH value (Behari and Youngman 1998), thermal, acidic and osmotic shock (Phan-Thanh and Gormon 1995, Ripio et al. 1998, Sleator et al. 1999, 2001a,b, Hanawa et al. 1999, 2002, Cotter et al. 2000, 2001,Gaillot et al. 2000, Duche et al. 2002, Brondsted et al. 2003, Nelson et al. 2004, Sleator and Hill 2005, van der Veen et al. 2007, Schmid et al. 2009, Durack et al. 2013, Milecka et al. 2015), essential oils (Hadjilouka et al. 2017) and various antimicrobials (Romanova et al. 2006, Elhanafi et al. 2010, Kastbjerg et al. 2010, van der Veen and Abee 2010, Stasiewicz et al. 2011, Dutta et al. 2013, Shi et al. 2013, Liu et al. 2014, Pleitner et al. 2014, Laursen et al. 2015, Hadjilouka et al. 2016a) has been thoroughly studied (NicAogain et al. 2016). On the other hand, the transcription of virulence genes during growth on food matrices has only been marginally studied. The transcription of virulence genes including *prfA*, *sigB*, *hly*, *actA*, *inlA*, *inlB*, *inlC*, *inlJ*, *plcA* and *plcB* during growth on fermented sausages, salmon, UHT milk, soft cheese, standard liver pate, minced meat, rocket and melon (Duodu et al. 2010, Olesen et al. 2010, Rantsiou et al. 2012a,b, Alessandria et al. 2013, Hadjilouka et al. 2016b) has been assessed and the effect of parameters like storage temperature, time, type of substrate and strain variability has been highlighted. In the majority of the cases, lack of a clear trend in the transcription of these genes was evident, suggesting the existence of still unexplored additional regulatory networks.

Conclusions-Future Perspectives

Food molecular microbiology is an exciting new discipline that has emerged due to the advances in the field of molecular biology. Through the development of molecular tools and approaches, information regarding the genomic, transcriptomic, proteomic and metabolomic organization of single cells and microecosystems has been derived, increasing our understanding of food microbiology in general. This is expected to carry on in the future as new tools offering new possibilities and higher throughput are constantly being developed, offering novel and improved insights.

References

Abachin, E., C. Poyart, E. Pellegrini, E. Milohanic, F. Fiedler, P. Berche and P. Trieu-Cuot. 2002. Formation of D-alanyl-lipoteichoic acid is required for adhesion and virulence of *Listeria monocytogenes*. Mol. Microbiol. 43: 1–14.

Alessandria, V., K. Rantsiou, P. Dolci, G. Zeppa and L. Cocolin. 2013. Comparison of gene expression of *Listeria monocytogenes* in vitro and in the soft cheese Crescenza. Int. J. Dairy Technol. 66: 83–89.

Alessandria, V., I. Ferrocino, F. De Filippis, M. Fontana, K. Rantsiou, D. Ercolini and L. Cocolin. 2016. Microbiota of an Italian Grana-like cheese during manufacture and ripening, unraveled by 16S rRNA-based approaches. Appl. Environ. Microbiol. 82: 3988–3995.

Almeida, C., J.M. Sousa, R. Rocha, L. Cerqueira, S. Fanning, N.F. Azevedo and M.J. Vieira. 2013b. Detection of *Escherichia coli* O157 by peptide nucleic acid fluorescence *in situ* hybridization (PNA-FISH) and comparison to a standard culture method. Appl. Environ. Microbiol. 79: 6293–6300.

Almeida, C., L. Cerqueira, N.F. Azevedo and M.J. Vieira. 2013a. Detection of *Salmonella enterica* serovar Enteritidis using Real Time PCR, immunocapture assay, PNA FISH and standard culture methods in different types of food samples. Int. J. Food Microbiol. 161: 16–22.

Amann, R.I., L. Krumholz and D.A. Stahl. 1990. Fluorescent-Oligonucleotide probing of whole cells for determinative, phylogenetic, and environmental studies in microbiology. J. Bacteriol. 172: 762–770.

Angelidis, A.S., I. Tirodimos, M. Bobos, M.S. Kalamaki, D.K. Papageorgiou and M. Arvanitidou. 2011. Detection of *Helicobacter pylori* in raw bovine milk by fluorescence *in situ* hybridization (FISH). Int. J. Food Microbiol. 151: 252–256.

Arraiano, C.M., S.D. Yancey and S.R. Kushner. 1998. Stabilization of discrete mRNA breakdown products in *ams*, *pnp* and *rnb* multiple mutants of *Escherichia coli* K-12. J. Bacteriol. 170: 4625–4633.

Aschfalk, A. and W. Mueller. 2002. *Clostridium perfringens* toxin types from wild caught atlantic cod (*Gadus morhua* L.) determined by PCR and ELISA. Can. J. Microbiol. 48: 365–368.

Aurora, R., A. Prakash, S. Prakash, D.B. Rawool and S.B. Barbuddhe. 2008. Comparison of PI-PLC based assays and PCR along with *in vivo* pathogenicity tests for rapid detection of pathogenic *Listeria monocytogenes*. Food Control 19: 641–647.

Banihashemi, A., M.I. Van Dyke and P.M. Huck. 2012. Long-amplicon propidium monoazide-PCR enumeration assay to detect viable *Campylobacter* and *Salmonella*. J. Appl. Microbiol. 113: 863–873.

Begley, M., R.D. Sleator, C.G. Gahan and C. Hill. 2005. Contribution of three bile-associated loci, *bsh*, *pva*, and *btlB*, to gastrointestinal persistence and bile tolerance of *Listeria monocytogenes*. Infect. Immun. 73: 894–904.

Behari, J. and P. Youngman. 1998. Regulation of *hly* expression in *Listeria monocytogenes* by carbon sources and pH occurs through separate mechanisms mediated by PrfA. Infect. Immun. 66: 3635–3642.

Benitez-Cabello, A., J. Bautista-Gallego, A. Garrido-Fernández, K. Rantsiou, L. Cocolin, R. Jiménez-Díaz and F.N. Arroyo-López. 2016. RT-PCR–DGGE analysis to elucidate the dominant bacterial species of industrial Spanish-style green table olive fermentations. Front. Microbiol. 7: 1291.

Birmingham, C.L., V. Canadien, E. Gouin, E.B. Troy, T. Yoshimori, P. Cossart, D.E. Higgins and J.H. Brumell. 2007. *Listeria monocytogenes* evades killing by autophagy during colonization of host cells. Autophagy 3: 442–451.

Boeckmann, R., C. Dickneite, W. Goebel and J. Bohne. 2000. PrfA mediates specific binding of RNA polymerase of *Listeria monocytogenes* to PrfA-dependent virulence gene promoters resulting in a transcriptionally active complex. Mol. Microbiol. 36: 487–497.

Bogusz, M. and S. Whelan. 2017. Phylogenetic tree estimation with and without alignment: New distance methods and benchmarking. Syst. Biol. 66: 218–231.

Bolton, F.J., E. Fritz and S. Poynton. 2000. Rapid enzyme-linked immunoassay for the detection of *Salmonella* in food and feed products: Performance testing program. J. AOAC Int. 83: 299–304.

Bonazzi, M., M. Lecuit and P. Cossart. 2009. *Listeria monocytogenes* internalin and E-cadherin: From bench to bedside. Cold Spring Harb. Perspect. Biol. 1: a003087.

Bonnemain, C., C. Raynaud, H. Reglier-Poupet, I. Dubail, C. Frehel, M.A. Lety, P. Berche and A. Charbit. 2004. Differential roles of multiple signal peptidases in the virulence of *Listeria monocytogenes*. Mol. Microbiol. 51: 1251–1266.

Borezee, E., E. Pellegrini, J.L. Beretti and P. Berche. 2001.SvpA, a novel surface virulence-associated protein required for intracellular survival of *Listeria monocytogenes*. Microbiology 147: 2913–2923.

Bowman, J.-P., C.-R. Bittencourt and T. Ross. 2008. Differential gene expression of *Listeria monocytogenes* during high hydrostatic pressure processing. Microbiology 154: 462–475.

Brondsted, L., B.H. Kallipolitis, H. Ingmer and S. Knochel. 2003. *kdpE* and a putative RsbQ homologue contribute to growth of *Listeria monocytogenes* at high osmolarity and low temperature. FEMS Microbiol. Lett. 219: 233–239.

Burkholder, K.M., K.P. Kim, K.K. Mishra, S. Medina, B.K. Hahm, H. Kim and A.K. Bhunia. 2009. Expression of LAP, a SecA2-dependent secretory protein, is induced under anaerobic environment. Microbes Infect. 11: 859–867.

Cabanes, D., O. Dussurget, P. Dehoux and P. Cossart. 2004. Auto, a surface associated autolysin of *Listeria monocytogenes* required for entry into eukaryotic cells and virulence. Mol. Microbiol. 51: 1601–1614.

Cabanes, D., S. Sousa, A. Cebria, M. Lecuit, F. Garcia-del Portillo and P. Cossart. 2005. Gp96 is a receptor for a novel *Listeria monocytogenes* virulence factor, Vip, a surface protein. EMBO J. 24: 2827–2838.

Camejo, A., F. Carvalho, O. Reis, E. Leitao, S. Sousa and D. Cabanes. 2011. The arsenal of virulence factors deployed by *Listeria monocytogenes* to promote its cell infection cycle. Virulence 2: 379–394.

Cardinali, F., V. Milanović, A. Osimani, L. Aquilanti, M. Taccari, C. Garofalo, S. Polverigiani, F. Clementi, E. Franciosi, K. Tuohy, M.L. Mercuri, M.S. Altissimi and M.N. Haouet. 2018. Microbial dynamics of model Fabriano-like fermented sausages as affected by starter cultures, nitrates and nitrites. Int. J. Food Microbiol. 278: 61–72.

Cavaiuolo, M., S. Paramithiotis, E.H. Drosinos and A. Ferrante. 2013. Development and optimization of an ELISA based method to detect *Listeria monocytogenes* and *Escherichia coli* O157 in fresh vegetables. Anal. Methods 5: 4622–4627.

Chan, C.X. and M.A. Ragan. 2013. Next-generation phylogenomics. Biol. Direct. 8:3.

Chaturongakul, S., S. Raengpradub, M. Wiedmann and K.J. Boor. 2008. Modulation of stress and virulence in *Listeria monocytogenes*. Trends Microbiol.16: 388–396.

Cheng, Y., N. Promadej, J.W. Kim and S. Kathariou. 2008. Teichoic acid glycosylation mediated by *gtcA* is required for phage adsorption and susceptibility of *Listeria monocytogenes* serotype 4b. Appl. Environ. Microbiol. 74: 1653–1655.

Cocolin, L., V. Alessandria, P. Dolci, R. Gorra and K. Rantsiou. 2013. Culture independent methods to assess the diversity and dynamics of microbiota during food fermentation. Int. J. Food Microbiol. 167: 29–43.

Corr, S., C. Hill and C.G. Gahan. 2006. An *in vitro* cell-culture model demonstrates internalin and hemolysin-independent translocation of *Listeria monocytogenes* across M cells. Microb. Pathog. 41: 241–250.

Corr, S.C., C.G. Gahan and C. Hill. 2008. M cells: origin, morphology and role in mucosal immunity and microbial pathogenesis. FEMS Immunol. Med. Microbiol. 52: 2–12.

Cossart, P., M.F. Vicente, J. Mengaud, F. Baquero, J.C. Perez-Diaz and P. Berche. 1989. Listeriolysin O is essential for virulence of *Listeria monocytogenes*: Direct evidence obtained by gene complementation. Infect. Immun. 57: 3629–636.

Cotter, P.D., C.G. Gahan and C. Hill. 2001. A glutamate decarboxylase system protects *Listeria monocytogenes* in gastric fluid. Mol. Microbiol. 40: 465–475.

Cotter, P.D., C.G.M. Gahan and C. Hill. 2000. Analysis of the role of the *Listeria monocytogenes* F0F1 ATPase operon in the acid tolerance response. Int. J. Food Microbiol. 60: 137–146.

Cravero, F., V. Englezos, K. Rantsiou, F. Torchio, S. Giacosa, S.R. Segade, V. Gerbi, L. Rolle and L. Cocolin. 2016. Ozone treatments of post harvested wine grapes: Impact on fermentative yeasts and wine chemical properties. Food Res. Int. 87: 134–141.

De Long, E.F., G.S. Wickham and N.R. Pace. 1989. Phylogenetic stains: Ribosomal RNA based probes for the identification of single cells. Science 243: 1360–1363.

Dickneite, C., R. Boeckmann, A. Spory, W. Goebel and Z. Sokolovic. 1998. Differential interaction of the transcription factor PrfA and the PrfA-activating factor (Paf) of *Listeria monocytogenes* with target sequences. Mol. Microbiol. 27: 915–928.

Divol, B. and A. Lonvaud-Funel. 2005. Evidence for viable but nonculturable yeasts in Botrytis-affected wine. J. Appl. Microbiol. 99: 85–93.

Dolci, P., S. Zenato, R. Pramotton, A. Barmaz, V. Alessandria, K. Rantsiou and L. Cocolin. 2013. Cheese surface microbiota complexity: RT-PCR-DGGE, a tool for a detailed picture? Int. J. Food Microbiol. 162: 8–12.

Dolci, P., F. De Filippis, A. La Storia, D. Ercolini and L. Cocolin. 2014. rRNA-based monitoring of the microbiota involved in Fontina PDO cheese production in relation to different stages of cow lactation. Int. J. Food Microbiol. 185: 127–135.

Domann, E., M. Leimeister-Wachter, W. Goebel and T. Chakraborty. 1991. Molecular cloning, sequencing, and identification of a metalloprotease gene from *Listeria monocytogenes* that is species specific and physically linked to the listeriolysin gene. Infect. Immun. 59: 65–72.

Domann, E., J. Wehland, K. Niebuhr, C. Haffner, M. Leimeister-Wachter and T. Chakraborty. 1996. Detection of PrfA-independent promoter responsible for listeriolysin gene expression in mutant *Listeria monocytogenes* strains lacking the PrfA regulator. Infect. Immun. 61: 3073–3075.

Dowd, G.C., S.A. Joyce, C. Hill and C.G.M. Gahan. 2011. Investigation of the mechanisms by which *Listeria monocytogenes* grows in porcine gallbladder bile. Infect Immun. 79: 369–379.

Duche, O., F. Tremoulet, P. Glaser and J. Labadie. 2002. Salt stress proteins induced in *Listeria monocytogenes*. Appl. Environ. Microbiol. 68: 1491–1498.

Duodu, S., A. Holst-Jensen, T. Skjerdal, J.-M. Cappelier, M.-F. Pilet and S. Loncarevic. 2010. Influence of storage temperature on gene expression and virulence potential of *Listeria monocytogenes* strains grown in a salmon matrix. Food Microbiol. 27: 795–801.

Durack, J., T. Ross and J.P. Bowman. 2013. Characterisation of the transcriptomes of genetically diverse *Listeria monocytogenes* exposed to hyperosmotic and low temperature conditions reveal global stress-adaptation mechanisms. PLoS ONE 8: e73603.

Dutta, V., D. Elhanafi and S. Kathariou. 2013. Conservation and distribution of the benzalkonium chloride resistance cassette *bcrABC* in *Listeria monocytogenes*. Appl. Environ. Microbiol. 79: 6067–6074.

Efron, B., E. Halloran and S. Holmes. 1996. Bootstrap confidence levels for phylogenetic trees. Proc. Natl. Acad. Sci. USA 93: 13429–13434.

Elhanafi, D., V. Dutta, and S. Kathariou. 2010. Genetic characterization of plasmid-associated benzalkonium chloride resistance determinants in a *Listeria monocytogenes* strain from the 1998–1999 outbreak. Appl. Environ. Microbiol. 76: 8231–8238.

Elizaquivel, P., G. Sanchez and R. Aznar. 2012. Application of propidium monoazide quantitative PCR for selective detection of live *Escherichia coli* O157:H7 in vegetables after inactivation by essential oils. Int. J. Food Microbiol. 159: 115–121.

Faith, N., S. Kathariou, Y. Cheng, N. Promadej, B.L. Neudeck, Q. Zhang, J. Luchansky and C. Czuprynski. 2009. The role of *L. monocytogenes* serotype 4b *gtcA* in gastrointestinal listeriosis in A/J mice. Foodborne Pathog. Dis. 6: 39–48.

Fakhr, M.K., L.K. Nolan and C.M. Logue. 2005. Multilocus sequence typing lacks the discriminatory ability of pulsed-field gel electrophoresis for typing *Salmonella enterica* serovar Typhimurium. J. Clin. Microbiol. 43: 2215–2219.

Faralla, C., G.A. Rizzuto, D.E. Lowe, B. Kim, C. Cooke, L.R. Shiow and A.I. Bakardjiev. 2016. InlP, a new virulence factor with strong placental tropism. Infect. Immun. 84: 3584–3596.

Flekna, G., P. Stefanic, M. Wagner, F.J. Smulders, S.S. Mozina and I. Hein. 2007. Insufficient differentiation of live and dead *Campylobacter jejuni* and *Listeria monocytogenes* cells by ethidium monoazide (EMA) compromises EMA/Real-Time PCR. Res. Microbiol. 158: 405–412.

Fox, G.E., K.R. Pechman and C.R. Woese. 1977. Comparative cataloging of 16S ribosomal ribonucleic acid - molecular approach to prokaryotic systematics. Int. J. Syst. Bacteriol. 27: 44–57.

Fuchizawa, I., S. Shimizu, M. Ootsubo, Y. Kawai and K. Yamazaki. 2009. Specific and rapid quantification of viable *Listeria monocytogenes* using fluorescence in situ hybridization in combination with filter cultivation. Microbes. Environ. 24: 273–275.

Gaillot, O., E. Pellegrini, S. Bregenholt, S. Nair and P. Berche. 2000. The ClpP serine protease is essential for the intracellular parasitism and virulence of *Listeria monocytogenes*. Mol. Microbiol. 35: 1286–1294.

Gao, X., H. Lin, K. Revanna and Q. Dong. 2017. A Bayesian taxonomic classification method for 16S rRNA gene sequences with improved species-level accuracy. BMC Bioinformatics 18: 247.

Gill, A. 2017. The importance of bacterial culture to Food Microbiology in the age of genomics. Front. Microbiol. 8: 777.

Ghannoum M.A., R.J. Jurevic, P.K. Mukherjee, F. Cui, M. Sikaroodi, A. Naqvi and P.M. Gillevet. 2010. Characterization of the oral fungal microbiome (mycobiome) in healthy individuals. PLoS Pathog. 6: e1000713.

Greppi, A., I. Ferrocino, A. La Storia, K. Rantsiou, D. Ercolini and L. Cocolin. 2015. Monitoring of the microbiota of fermented sausages by culture independent rRNA-based approaches. Int. J. Food Microbiol. 212: 67–75.

Grundling, A., M.D. Gonzalez and D.E. Higgins. 2003. Requirement of the *Listeria monocytogenes* broad-range phospholipase PC-PLC during infection of human epithelial cells. J. Bacteriol. 185: 6295–6307.

Hadjilouka, A., K. Nikolidakis, S. Paramithiotis and E.H. Drosinos. 2016a. Effect of co-culture with enterocinogenic *E. faecium* on *L. monocytogenes* key virulence gene expression. AIMS Microbiol. 2: 304–315.

Hadjilouka, A., C. Molfeta, O. Panagiotopoulou, S. Paramithiotis, M. Mataragas and E.H. Drosinos. 2016b. Expression of *Listeria monocytogenes* key virulence genes during growth in liquid medium, on rocket and melon at 4, 10 and 30ºC. Food Microbiol. 55: 7–15.

Hadjilouka, A., G. Mavrogiannis, A. Mallouchos, S. Paramithiotis, M. Mataragas and E.H. Drosinos. 2017. Effect of lemongrass essential oil on *Listeria monocytogenes* gene expression. LWT-Food Sci. Technol. 77: 510–516.

Hamon, M.A., D. Ribet, F. Stavru and P. Cossart. 2012. Listeriolysin O: The Swiss army knife of *Listeria*. Trends Microbiol. 20: 360–368.

Hanawa, T., M. Fukuda, H. Kawakami, H. Hirano, S. Kamiya and T. Yamamoto. 1999. The *Listeria monocytogenes* DnaK chaperone is required for stress tolerance and efficient phagocytosis with macrophages. Cell Stress Chaperones 4: 118–128.

Hanawa, T., S. Yamanishi, S. Murayama, T. Yamamoto and S. Kamiya. 2002. Participation of DnaK in expression of genes involved in virulence of *Listeria monocytogenes*. FEMS Microbiol. Lett. 214: 69–75.

Horiike, T., R. Minai, D. Miyata, Y.Nakamura and Y. Tateno. 2016. Ortholog-Finder: A tool for constructing an ortholog data set. Genome Biol. Evol. 8: 446–457.

Jaureguy, F., L. Landraud, V. Passet, L. Diancourt, E. Frapy, G. Guigon, E. Carbonnelle, O. Lortholary, O. Clermont, E. Denamur, B. Picard, X. Nassif and S. Brisse. 2008. Phylogenetic and genomic diversity of human bacteremic *Escherichia coli* strains. BMC Genomics 9: 560.

Joseph, B., S. Mertins, R. Stoll, J. Schaer, K.R. Umesha, Q. Luo, S. Mueller-Altrock and W. Goebel. 2008. Glycerol metabolism and PrfA activity in *Listeria monocytogenes*. J. Bacteriol. 190: 5412–5430.

Jung, B.Y., S.C. Jung and C.H. Kweon. 2005. Development of a rapid immunochromatographic strip for the detection of *Escherichia coli* O157. J. Food Prot. 68: 2140–2143.

Jung, C., J.P. Hugot and B. Frederick. 2010. Peyer's patches: The immune sensors of the intestine. Int. J. Inflam. 2010: 1–12.

Kastbjerg, V.-G., M.-H. Larsen, L. Gram and H. Ingmer. 2010. Influence of sublethal concentrations of common disinfectants on expression of virulence genes in *Listeria monocytogenes*. Appl. Environ. Microbiol. 76: 303–309.

Kathariou, S., P. Metz, H. Hof and W. Goebel. 1987. Tn916-induced mutations in the hemolysin determinant affecting virulence of *Listeria monocytogenes*. J. Bacteriol. 169: 1291–1297.

Kim, S.H., J.Y. Kim, W. Han, B.Y. Jung, P.D. Chuong and H. Joo. 2007. Development and evaluation of an immunochromatographic assay for *Listeria* spp. in pork and milk. Food Sci. Biotechnol. 16: 515–519.

Kitaguchi, A., N. Yamaguchi and M. Nasu. 2005. Enumeration of respiring *Pseudomonas* spp. in milk within 6 hours by fluorescence in situ hybridization following formazan reduction. Appl. Environ. Microbiol. 71: 2748–2752.

Kotetishvili, M., O.C. Stine, A. Kreger, J.G. Morris Jr. and A. Sulakvelidze. 2002. Multilocus sequence typing for characterization of clinical and environmental *Salmonella* strains. J. Clin. Microbiol. 40: 1626–1635.

Kumar, B.K., P. Raghunath, D. Devegowda, V.K. Deekshit, M.N. Venugopal, I. Karunasagar and I. Karunasagar. 2011. Development of monoclonal antibody based sandwich ELISA for the rapid detection of pathogenic *Vibrio parahaemolyticus* in seafood. Int. J. Food Microbiol. 145: 244–249.

Kurtzman, C.P. 1994. Molecular taxonomy of the yeasts. Yeast 10: 1727–1740.

Laflamme, C., L. Gendron, N. Filion, J. Ho and C. Duchaine. 2009. Rapid detection of germinating *Bacillus cereus* cells using fluorescent in situ hybridization. J. Rapid Methods Autom. Microbiol. 17: 80–102.

Lalic-Muelthaler, M., J. Bohne and W. Goebel. 2001. *In vitro* transcription of PrfA-dependent and -independent genes of *Listeria monocytogenes*. Mol. Microbiol. 42: 111–120.

Laursen, M.F., M.I. Bahl, T.R. Licht, L. Gram and G.M. Knudsen. 2015. A single exposure to a sublethal pediocin concentration initiates a resistance-associated temporal cell envelope and general stress response in *Listeria monocytogenes*. Environ. Microbiol. 17: 1134–1151.

Leem, H., S. Shukla, X. Song S. Heu and M. Kim. 2014. An efficient liposome-based immunochromatographic strip assay for the sensitive detection of *Salmonella* Typhimurium in pure culture. J. Food Saf. 34: 239–248.

Li, B. and J.Q. Chen. 2013. Development of a sensitive and specific qPCR assay in conjunction with propidium monoazide for enhanced detection of live *Salmonella* spp. in food. BMC Microbiol. 13: 273.

Linden, S.K., H. Bierne, C. Sabet, C.W. Png, T.H. Florin, M.A. McGuckin and P. Cossart. 2008. *Listeria monocytogenes* internalins bind to the human intestinal mucin MUC2. Arch. Microbiol. 190: 101–104.

Liu, X., U. Basu, P. Miller and L.M. McMullen. 2014. Stress response and adaptation of *Listeria monocytogenes* 08–5923 exposed to a sub-lethal dose of carnocyclin. Appl. Environ. Microbiol. 80: 3835–3841.

Livezey, K., S. Kaplan, M. Wisniewski and M.M. Becker. 2013. A new generation of food-borne pathogen detection based on ribosomal RNA. Annu. Rev. Food Sci. Technol. 4: 313–325.

Luo, J.-F., W.-T. Lin and Y. Guo. 2010. Method to detect only viable cells in microbial ecology. Appl. Microbiol. Biotechnol. 86: 377–384.

Luo, Q., M. Rauch, A.K. Marr, S. Mueller-Altrock and W. Goebel. 2004. *In vitro* transcription of the *Listeria monocytogenes* virulence genes *inlC* and *mpl* reveals overlapping PrfA-dependent and -independent promoters that are differentially activated by GTP. Mol. Microbiol. 52: 39–52.

Luo, Q., M. Herler, S. Mueller-Altrock and W. Goebel. 2005. Supportive and inhibitory elements of a putative PrfA-dependent promoter in *Listeria monocytogenes*. Mol. Microbiol. 55: 986–997.

Machata, S., S. Tchatalbachev, W. Mohamed, L. Jansch, T. Hain and T. Chakraborty. 2008. Lipoproteins of *Listeria monocytogenes* are critical for virulence and TLR2-mediated immune activation. J. Immunol. 181: 2028–2035.

Mata, G.M., E. Martins, S.G. Machado, M.S. Pinto, A.F. de Carvalho and M.C. Vanetti. 2016. Performance of two alternative methods for *Listeria* detection throughout Serro Minas cheese ripening. Braz. J. Microbiol. 47: 749–756.

Mengaud, J., S. Dramsi, E. Gouin, J.A. Vazquez-Boland, G. Milon and P. Cossart. 1991. Pleiotropic control of *Listeria monocytogenes* virulence factors by a gene that is autoregulated. Mol. Microbiol. 5: 2273–2283.

Milecka, D., A. Samluk, K. Wasiak and A. Krawczyk Balska. 2015. An essential role of a ferritin like protein in acid stress tolerance of *Listeria monocytogenes*. Arch. Microbiol. 197: 347–351.

Milenbachs Lukowiak, A., K.J. Mueller, N.E. Freitag and P. Youngman. 2004. Deregulation of *Listeria monocytogenes* virulence gene expression by two distinct and semi-independent pathways. Microbiol. 150: 321–333.

Milenbachs, A.A., D.P. Brown, M. Moors and P. Youngman. 1997. Carbon-source regulation of virulence gene expression in *Listeria monocytogenes*. Mol. Microbiol. 23: 1075–1085.

Nadon, C.A., B.M. Bowen, M. Wiedmann and K.J. Boor. 2002. σ^B contributes to PrfA-mediated virulence in *Listeria monocytogenes*. Infect. Immun. 70: 3948–3952.

Nelson, K.E., D.E. Fouts, E.F. Mongodin, J. Ravel, R.T. DeBoy, J.F. Kolonay, D.A. Rasko, S.V. Angiuoli, S.R. Gill, I.T. Paulsen, J. Peterson, O. White, W.C. Nelson, W. Nierman, M.J. Beanan, L.M. Brinkac, S.C. Daugherty, R.J. Dodson, A.S. Durkin, R. Madupu, D.H. Haft, J. Selengut, S. Van Aken, H. Khouri, N. Fedorova, H. Forberger, B. Tran, S. Kathariou, L.D. Wonderling, G.A. Uhlich, D.O. Bayles, J.B. Luchansky and C.M. Fraser. 2004. Whole genome comparisons of serotype 4b and 1/2a strains of the food-borne pathogen *Listeria monocytogenes* reveal new insights into the core genome components of this species. Nucleic Acids Res. 32: 2386–2395.

NicAogáin, K. and C.P. O'Byrne. 2016. The role of stress and stress adaptations in determining the fate of the bacterial pathogen *Listeria monocytogenes* in the food chain. Front. Microbiol. 7: 1865.

Nielsen, K.M., P.J. Johnsen, D. Bensasson and D. D'Affonchio. 2017. Release and persistence of extracellular DNA in the environment. Environ. Biosafety Res. 6: 37–53

Niu, K., X. Zheng, C. Huang, K. Xu, Y. Zhi, H. Shen and N. Jia. 2014. A colloidal gold nanoparticle-based immunochromatographic test strip for rapid and convenient detection of *Staphylococcus aureus*. J. Nanosci. Nanotechnol. 14: 5151–5156.

Nocker, A. and A.K. Camper. 2006. Selective removal of DNA from dead cells of mixed bacterial communities by use of ethidium monoazide. Appl. Environ. Microbiol. 72: 1997–2004.

Olesen, I., L. Thorsen and L. Jespersen. 2010. Relative transcription of *Listeria monocytogenes* virulence genes in liver pâtés with varying NaCl content. Int. J. Food Microbiol. 141: S60–S68.

Oliveira, M., M. Vieira-Pinto, P. Martins da Costa, C.L. Vilela, C. Martins and F. Bernardo. 2012. Occurrence of *Salmonella* spp. in samples from pigs slaughtered for consumption: A comparison between ISO 6579:2002 and 23S rRNA fluorescent in situ hybridization method. Food Res. Int. 45: 984–988.

Ollinger, J., B. Bowen, M. Wiedmann, K.J. Boor and T.M. Bergholz. 2009. listeria monocytogenes σ^B modulates PrfA-mediated virulence factor expression. Infect. Immun. 77: 2113–2124.

Oueslati, W., M.R. Rjeibi, M. Mhadhbi, M. Jbeli, S. Zrelli and A. Ettriqui. 2016. Prevalence, virulence and antibiotic susceptibility of *Salmonella* spp. strains, isolated from beef in Greater Tunis (Tunisia). Meat Sci. 119: 154–159.

Pacholewicz, E., A. Swart, L.J. Lipman, J.A. Wagenaar, A.H. Havelaar and B. Duim. 2013. Propidium monoazide does not fully inhibit the detection of dead *Campylobacter* on broiler chicken carcasses by qPCR. J. Microbiol. Methods 95: 32–38.

Pan, Y. and F. Breidt, Jr. 2007. Enumeration of viable *Listeria monocytogenes* cells by real-time PCR with propidium monoazide and ethidium monoazide in the presence of dead cells. Appl. Environ. Microbiol. 73: 8028–8031.

Paramithiotis, S., A. Hadjilouka and E.H. Drosinos. 2018. Molecular typing of major foodborne pathogens. pp. 421–472. *In*: A.M. Holban, and A.M. Grumezescu (eds.). Handbook of Food Bioengineering, vol. 15 Foodborne Diseases. Elsevier, Amsterdam, Netherlands.

Phan-Thanh, L. and T. Gormon. 1995. Analysis of heat and cold shock proteins in *Listeria* by two-dimensional electrophoresis. Electrophoresis 16: 444–450.

Pilgrim, S., A. Kolb-Maurer, I. Gentschev, W. Goebel and M. Kuhn. 2003. Deletion of the gene encoding p60 in *Listeria monocytogenes* leads to abnormal cell division and loss of actin-based motility. Infect. Immun. 71: 3473–3484.

Pinto, D., M.A. Santos and L. Chambel. 2015. Thirty years of viable but nonculturable state research: Unsolved molecular mechanisms. Crit. Rev. Microbiol. 41: 61–76.

Pleitner, A.M., V. Trinetta, M.T. Morgan, R.L. Linton and H.F. Oliver. 2014. Transcriptional and phenotypic responses of *Listeria monocytogenes* to chlorine dioxide. Appl. Environ. Microbiol. 80: 2951–2963.

Ragon, M., T. Wirth, F. Hollandt, R. Lavenir, M. Lecuit, A. Le Monnier and S. Brisse. 2008. A new perspective on *Listeria monocytogenes* evolution. PLoS Pathog. 4: e1000146.

Rantsiou, K., A. Greppi, M. Garosi, A. Acquadro, M. Mataragas and L. Cocolin. 2012a. Strain dependent expression of stress response and virulence genes of *Listeria monocytogenes* in meat juices as determined by microarray. Int. J. Food. Microbiol. 152: 116–122.

Rantsiou, K., M. Mataragas, V. Alessandria and L. Cocolin. 2012b. Expression of virulence genes of *Listeria monocytogenes* in food. J. Food Saf. 32: 161–168.

Rauch, M., Q. Luo, S. Muller-Altrock and W. Goebel. 2005. SigB-dependent *in vitro* transcription of *prfA* and some newly identified genes of *Listeria monocytogenes* whose expression is affected by PrfA *in vivo*. J. Bacteriol. 187: 800–804.

Raveneau, J., C. Geoffroy, J.L. Beretti, J.L. Gaillard J.E. Alouf and P. Berche. 1992. Reduced virulence of a *Listeria monocytogenes* phospholipase-deficient mutant obtained by transposon insertion into the zinc metalloprotease gene. Infect. Immun. 60: 916–921.

Reglier-Poupet, H., E. Pellegrini, A. Charbit and P. Berche. 2003. Identification of LpeA, a PsaA-like membrane protein that promotes cell entry by *Listeria monocytogenes*. Infect. Immun. 71: 474–482.

Reid, S.D., C.J. Herbelin, A.C. Bumbaugh, R.K. Selander and T.S. Whittam. 2000. Parallel evolution of virulence in pathogenic *Escherichia coli*. Nature 406: 64–67.

Reis, O., S. Sousa, A. Camejo, V. Villiers, E. Gouin, P. Cossart and D. Cabanes. 2010. LapB, a novel *Listeria monocytogenes* LPXTG surface adhesin, required for entry into eukaryotic cells and virulence. J. Infect. Dis. 202: 551–562.

Renzoni, A., A. Klarsfeld, S. Dramsi and P. Cossart. 1997. Evidence that PrfA, the pleiotropic activator of virulence genes in *Listeria monocytogenes*, can be present but inactive. Infect. Immun. 65: 1515–1518.

Revazishvili, T., M. Kotetishvili, O.C. Stine, A.S. Kreger, J.G. Morris, Jr. and A. Sulakvelidze. 2004. Comparative analysis of multilocus sequence typing and pulsed-field gel electrophoresis for characterizing *Listeria monocytogenes* strains isolated from environmental and clinical sources. J. Clin. Microbiol. 42: 276–285.

Ripio, M.T., J.A. Vazquez-Boland, Y. Vega, S. Nair and P. Berche. 1998. Evidence for expressional crosstalk between the central virulence regulator PrfA and the stress response mediator ClpC in *Listeria monocytogenes*. FEMS Microbiol. Lett. 158: 45–50.

Rodrigues, M.J., K. Martins, D. Garcia, S.M. Ferreira, S.C. Gonçalves, S. Mendes and M.F. Lemos. 2016. Using the mini-VIDAS® easy *Salmonella* protocol to assess contamination in transitional and coastal waters. Arch. Microbiol. 198: 483–487.

Romanova, N.A., P.F. Wolffs, L.Y. Brovko and M.W. Griffiths. 2006. Role of efflux pumps in adaptation and resistance of *Listeria monocytogenes* to benzalkonium chloride. Appl. Environ. Microbiol. 72: 3498–3503.

Sabet, C., A. Toledo-Arana, N. Personnic, M. Lecuit, S. Dubrac, O. Poupel, E. Gouin, M.A. Nahori, P. Cossart and H. Bierne. 2008. The *Listeria monocytogenes* virulence factor InlJ is specifically expressed *in vivo* and behaves as an adhesin. Infect. Immun. 76: 1368–1378.

Salcedo, C., L. Arreaza, B. Alcala, L. de la Fuente and J.A. Vazquez. 2003. Development of a multilocus sequence typing method for analysis of *Listeria monocytogenes* clones. J. Clin. Microbiol. 41: 757–762.

Salma, M., S. Rousseaux, A. Sequeira-Le Grand, B. Divol and H. Alexandre. 2013. Characterization of the viable but nonculturable (VBNC) state in *Saccharomyces cerevisiae*. PLoS One 8: e77600.

Schluter, D., E. Domann, C. Buck, T. Hain, H. Hof, T. Chakraborty and M. Deckert-Schluter. 1998. Phosphatidylcholine-specific phospholipase C from *Listeria monocytogenes* is an important virulence factor in murine cerebral listeriosis. Infect. Immun. 66: 5930–5938.

Schmid, B., J. Klumpp, E. Raimann, M.J. Loessner, R. Stephan and T. Tasara. 2009. Role of cold shock proteins in growth of *Listeria monocytogenes* under cold and osmotic stress conditions. Appl. Environ. Microbiol. 75: 1621–1627.

Schmid, M.W., A. Lehner, R. Stephan, K.H. Schleifer and H. Meier. 2005. Development and application of oligonucleotide probes for in situ detection of thermotolerant *Campylobacter* in chicken faecal and liver samples. Int. J. Food Microbiol. 105: 245–255.

Schnetzinger, F., Y. Pan and A. Nocker. 2013. Use of propidium monoazide and increased amplicon length reduce false-positive signals in quantitative PCR for bioburden analysis. Appl. Microbiol. Biotechnol. 97: 2153–2162.

Schwab, U., B. Bowen, C. Nadon, M. Wiedmann and K.J. Boor. 2005. The *Listeria monocytogenes* prfAP2 promoter is regulated by σ^B in a growth phase dependent manner. FEMS Microbiol. Lett. 245: 329–336.

Shen, Z., N. Hou, M. Jin, Z. Qiu, J. Wang, B. Zhang, X. Wang, J. Wang, D. Zhou and J. Li. 2014. A novel enzyme-linked immunosorbent assay for detection of *Escherichia coli* O157:H7 using immunomagnetic and beacon gold nanoparticles. Gut Pathog. 6: 14.

Shi, H., Q. Trinh, W. Xu, Y. Luo, W. Tian and K. Huang. 2013. The transcriptional response of virulence genes in *Listeria monocytogenes* during inactivation by nisin. Food Control 31: 519–524.

Shukla, S., H. Leem and M. Kim. 2011. Development of a liposome-based immunochromatographic strip assay for the detection of *Salmonella*. Anal. Bioanal. Chem. 401: 2581–2590.

Shukla, S., H. Leem, J.S. Lee and M. Kim. 2014. Immunochromatographic strip assay for the rapid and sensitive detection of *Salmonella* Typhimurium in artificially contaminated tomato samples. Can. J. Microbiol. 60: 399–406.

Sievers, A., K. Bosiek, M. Bisch, C. Dreessen, J. Riedel, P. Froß, M. Hausmann and G. Hildenbrand. 2017. K-mer content, correlation, and position analysis of genome DNA sequences for the identification of function and evolutionary features. Genes 8: 122.

Singh, P., S.L. Foley, R. Nayak and Y.M. Kwon. 2012. Multilocus sequence typing of *Salmonella* strains by high throughput sequencing of selectively amplified target genes. J. Microbiol. Methods 88: 127–133.

Sleator, R.D., C.G. Gahan, T. Abee and C. Hill. 1999. Identification and disruption of BetL, a secondary glycine betaine transport system linked to the salt tolerance of *Listeria monocytogenes* LO28. Appl. Environ. Microbiol. 65: 2078–2083.

Sleator, R.D., C.G.M. Gahan and C. Hill. 2001a. Identification and disruption of the *proBA* locus in *Listeria monocytogenes*: Role of proline biosythesis in salt tolerance and murine infection. Appl. Environ. Microbiol. 67: 2571–2577.

Sleator, R.D., C.G.M. Gahan and C. Hill. 2001b. Mutations in the Listerial *proB* gene leading to proline overproduction: Effects on salt tolerance and murine infection. Appl. Environ. Microbiol. 67: 4560–4565.

Sleator, R.D. and C. Hill. 2005. A novel role for the LisRK two-component regulatory system in *Listerial* osmotolerance. Clin. Microbiol. Infect. 11: 599–601.

Soejima, T. and K.J. Iwatsuki. 2016. Innovative use of palladium compounds to selectively detect live *Enterobacteriaceae* cells in milk by polymerase chain reaction. Appl. Environ. Microbiol. 82: 6930–6941.

Soejima, T., J. Minami, J.Z. Xiao and F. Abe. 2016. Innovative use of platinum compounds to selectively detect live microorganisms by polymerase chain reaction. Biotechnol. Bioeng. 113: 301–310.

Song Y., L. Li, Y. Ou, Z. Gao, E. Li, X. Li, W. Zhang, J. Wang, L. Xu, Y. Zhou, X. Ma, L. Liu, Z. Zhao, X. Huang, J. Fan, L. Dong, G. Chen, L. Ma, J. Yang, L. Chen, M. He, M. Li, X. Zhuang, K. Huang, K. Qiu, G. Yin, G. Guo, Q. Feng, P. Chen, Z. Wu, J. Wu, L. Ma, J. Zhao, L. Luo, M. Fu, B. Xu, B. Chen, Y. Li, T. Tong, M. Wang, Z. Liu, D. Lin, X. Zhang, H. Yang, J. Wang and Q. Zhan. 2014. Identification of genomic alterations in oesophageal squamous cell cancer. Nature 509: 91–95.

Stasiewicz, M.-J., M. Wiedmann and T-M. Bergholz. 2011. The transcriptional response of *Listeria monocytogenes* during adaptation to growth on lactate and diacetate includes synergistic changes that increase fermentative acetoin production. Appl. Environ. Microbiol. 77: 5294–5306.

Stewart, E.J. 2012. Growing unculturable bacteria. J. Bacteriol. 194: 4151–4160.

Suarez, M., B. Gonzalez-Zorn, Y. Vega, I. Chico-Calero and J.A. Vazquez-Boland. 2001. A role for ActA in epithelial cell invasion by *Listeria monocytogenes*. Cell. Microbiol. 3: 853–864.

Sukhnanand, S., S. Alcaine, L.D. Warnick, W.L. Su, J. Hof, M. Pat, J. Craver, P. McDonough, K.J. Boor and M. Wiedmann. 2005. DNA sequence-based subtyping and evolutionary analysis of selected *Salmonella enterica* serotypes. J. Clin. Microbiol. 43: 3688–3698.

Tankouo-Sandjong, B., A. Sessitsch, E. Liebana, C. Kornschober, F. Allerberger, H. Hachler and L. Bodrossy. 2007. MLST-v, multilocus sequence typing based on virulence genes, for molecular typing of *Salmonella enterica* subsp. *enterica* serovars. J. Microbiol. Methods 69: 23–36.

Thedieck, K., T. Hain, W. Mohamed, B.J. Tindall, M. Nimtz, T. Chakraborty, J. Wehland and L. Jänsch. 2006. The MprF protein is required for lysinylation of phospholipids in Listerial membranes and confers resistance to cationic antimicrobial peptides (CAMPs) on *Listeria monocytogenes*. Mol. Microbiol. 62:1325-1339,

Torpdahl, M., M.N. Skov, D. Sandvang and D.L. Baggesen. 2005. Genotypic characterization of *Salmonella* by multilocus sequence typing, pulsed-field gel electrophoresis and amplified fragment length polymorphism. J. Microbiol. Methods 63: 173–184.

van der Veen, S., T. Hain, J.A. Wouters, H. Hossain, W.M. de Vos, T. Abee, T. Chakraborty and M.H.J. Wells-Bennik. 2007. The heat-shock response of *Listeria monocytogenes* comprises genes involved in heat shock, cell division, cell wall synthesis, and the SOS response. Microbiology 153: 3593–3607.

van der Veen, S. and T. Abee. 2010. Importance of SigB for *Listeria monocytogenes* static and continuous-flow biofilm formation and disinfectant resistance. Appl. Environ. Microbiol. 76: 7854–7860.

Wang, C., B. Esteve-Zarzoso, L. Cocolin, A. Mas and K. Rantsiou. 2015. Viable and culturable populations of *Saccharomyces cerevisiae*, *Hanseniaspora uvarum* and *Starmerella bacillaris* (synonym *Candida zemplinina*) during Barbera must Fermentation. Food Res. Int. 78: 195–200.

Wirth, T., D. Falush, R. Lan, F. Colles, P. Mensa, L.H. Wieler, H. Karch, P.R. Reeves, M.C. Maiden, H. Ochman and M. Achtman. 2006. Sex and virulence in *Escherichia coli*: An evolutionary perspective. Mol. Microbiol. 60: 1136–1151.

Xu, D., X. Wu, B. Li, P. Li, X. Ming, T. Chen, H. Wei, and F. Xu. 2013. Rapid detection of *Campylobacter jejuni* using fluorescent microspheres as label for immune chromatographic strip test. Food Sci. Biotechnol. 22: 585–591.

Zeleny, R., Y. Nia, H. Schimmel, I. Mutel, J.A. Hennekinne, H. Emteborg, J. Charoud-Got and F. Auvray. 2016. Certified reference materials for testing of the presence/absence of *Staphylococcus aureus* enterotoxin A (SEA) in cheese. Anal. Bioanal. Chem. 408: 5457–5465.

Zeng, D., Z. Chen, Y. Jiang, F. Xue and B. Li. 2016. Advances and challenges in viability detection of foodborne pathogens. Front. Microbiol. 7: 1833.

Zhang, Q., Y. Feng, L. Deng, F. Feng, L. Wang, Q. Zhou and Q. Luo. 2011. SigB plays a major role in *Listeria monocytogenes* tolerance to bile stress. Int. J. Food Microbiol. 145: 238–243.

Zhang, W., B.M. Jayarao and S.J. Knabel. 2004. Multi-virulence-locus sequence typing of *Listeria monocytogenes*. Appl. Environ Microbiol. 70: 913–920.

Zhang, Y. and A.V. Alekseyenk. 2017. Phylogenic inference using alignment free methods for applications in microbial community surveys using 16s rRNA gene. PLoS One 12: e0187940.

Zhao, X., C.-W. Lin, J. Wang and D.H. Oh. 2014. Advances in rapid detection methods for foodborne pathogens. J. Microbiol. Biotechnol. 24: 297–312.

Zhu, R.G., T.P. Li, Y.F. Jia and L.F. Song. 2012. Quantitative study of viable *Vibrio parahaemolyticus* cells in raw seafood using propidium monoazide in combination with quantitative PCR. J. Microbiol. Methods 90: 262–266.

Zuckerkandl E. and L. Pauling. 1965. Molecules as documents of evolutionary history. J. Theor. Biol. 8: 357–366.

2

Molecular Tools for Food Micro-Ecosystems Assessment

Lorena Ruiz[1,*] and *Avelino Alvarez-Ordóñez*[2]

1. Introduction

Microorganisms have been extensively used for the fermentation of foods since ancient times of humanity, with the earliest records of artisanal and empirical food fermentations being conducted as early as 6000 BC (Buchholz and Collins 2013). Nowadays, food fermentations are conducted by deliberately adding starter microorganism's combinations to the raw food materials, under controlled conditions. Nevertheless, the main aim of food fermentation remains essentially unchanged. Through microorganisms' metabolic activity, fermentation processes modify the food matrix components, extending the shelf-life of the final products, as compared to the raw material, and providing characteristic taste, flavor, biochemical and specific properties to the fermented foods, overall enhancing the quality of the final food product (Hutkins 2006). In fact, recent evidence supports the enhanced health promoting effects of fermented food products, as compared to raw foods. Thus, these kinds of food products and the microorganisms responsible for their production are regaining popularity (Tamang et al. 2016, Marco et al. 2017).

On the opposite side, contamination of food products with either spoilage or pathogenic microorganisms is an area of global concern, as it can cause serious food borne intoxications and high economic burden for food producers. Thus, guaranteeing microbiological safety is critically important for both consumers and food producers.

[1] Departamento de Microbiología y Bioquímica de Productos Lácteos, Instituto de Productos Lácteos de Asturias (IPLA-CSIC), Villaviciosa, Asturias, Spain.
[2] Department of Food Hygiene and Technology and Institute of Food Science and Technology, University of León, León, Spain.
* Corresponding author: lorena.ruiz@ipla.csic.es

As a consequence, enforcement of monitoring and control systems, including full traceability, are established by national and international regulations and recommended by guidelines published by international organizations (FAO/WHO 2003).

In this context, substantial research efforts in the food safety field have focused on developing and implementing sensitive, rapid and cost-effective methods for monitoring microbiological risks along the food production chain. Furthermore, the increasing interest in fermented food products, due to their presumably health promoting characteristics and added value, makes it necessary to microbiologically monitor fermentation processes in order to guarantee the microbiological and biochemical composition of the final products, enhancing food quality (Capozzi and Spano 2011). Therefore, the characterization of microbial ecosystems present within food products, both throughout the food production chain and during their shelf-life, is critical in enhancing food quality and safety; and in facilitating surveillance (Kumar 2016). Characterizing microbial ecosystems within food fermentations is also essential when developing new fermented foods, as a means to understand at a molecular level the modifications that take place in the foodstuff during the fermentation process, aiding in the formulation of starters and functional strains combinations, in order to achieve reproducible fermentations and elaborate foodstuffs with increased added value.

Conventional microbiological analyses have long relied on culture-dependent methodologies, despite the limitations offered by such approaches, as they require long laboratory time to get analytical results—a minimum of 2–3 days for bacterial isolation and then a variable time required for further biochemical and genotyping tests to confirm microbial identification and establish genetic lineage—and offer a very limited view of the microbiological composition of foodstuffs. In addition, these approaches can lead to false negative results due to resolution limitations, as the particular characteristics of the microbial populations and culture conditions might impede the recovery of microorganisms found in low abundance, or with fastidious growth requirements. Besides, these approaches might be unable to detect microorganisms in a viable but not cultivable state, such as bacterial or fungal spores which might eventually germinate in the food product leading to food poisoning or rotting (Rastogi and Sani 2011). However, cultivation of microorganisms allows for the creation of collections of strains for further characterization, facilitating a historical tracking of strains evolution (e.g., antibiotic resistance genes acquisition), traceability of outbreak causing agents or even determining microbial fingerprints for the authenticity of foodstuffs origin and, thus, culture-dependent analyses are foreseen to continue to be an essential cornerstone in food microbiology analysis (Gill 2017).

Over the past years, with the rapid emergence and development of genomic-based tools for microbial communities assessment and multiple "-omic" technologies, research efforts in the food microbiology field have moved towards culture-independent approaches, which allow for rapid and accurate detection and monitoring of the microbial communities present in foodstuffs, as well as discerning the impact of their metabolic activities to the final technological and functional properties of foods. Overall, the shift from culture-dependent towards high-resolution culture-independent methodologies has led to a revolution in food microbiology studies which will eventually be implemented as standard tools for food quality and safety evaluation by food producers and/or regulatory agencies. In addition, the wide availability of

"-omic" tools that can be applied to food microbial ecology studies, holds a still undermined potential to assist risk-management and assessment activities in the food production chain that deserves further attention (Cocolin et al. 2017).

This book chapter presents an overview of the main molecular tools that have been used to characterize microbial communities in food products (Table 1, Fig. 1), highlighting their main applications, advances and limitations, as well as discussing new potential applications foreseen from recent technological developments in the field.

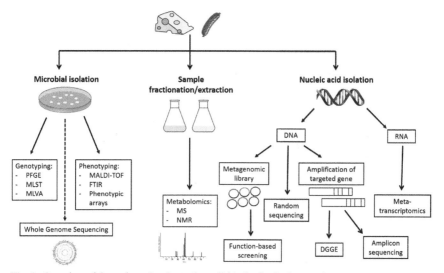

Fig. 1. Overview of the main molecular tools available for food micro-environments assessment.

2. Molecular Tools to Assist Culture-Dependent Approaches

The rapid expansion in the last decades of "-omics" studies, based on next generation sequencing (NGS) approaches, has boosted the acquisition of knowledge on the microbial diversity of a wide range of ecosystems, including foods and food-related environments. However, novel sequencing approaches alone cannot unveil the physiology and metabolism of specific microbial groups in complex ecosystems (Prakash et al. 2013). Indeed, cultivation of pure microorganisms isolated from such complex environments and their subsequent pheno- and geno-typic characterization are still relevant techniques which provide a plethora of invaluable pheno- and geno-typic data for micro-ecosystems assessment.

Nevertheless, researchers are only able to successfully cultivate a small fraction of the available microbial diversity (less than 10% of the existing microorganisms), while fastidious and recalcitrant microorganisms, which in most cases include the ecologically more relevant representatives, still remain "unculturable" (Giovannoni and Stingl 2007, Alain and Querellou 2009). As a consequence, research efforts are currently being dedicated to improve the culturability of previously unculturable microorganisms, which in some cases results in the identification of novel microbial species with relevant roles in particular ecosystems. Such research efforts include the formulation and optimization of novel media, the development of enrichment protocols for specific microbial groups, the use of simulated environments that

Table 1. Overview of the main applications and limitations of the molecular tools used for micro-environment assessment in foods.

Technique	Information Provided	PCR Biased	Application Examples	Limitations
Culture-dependent approaches				
WGS	Provides the most accurate and complete information on the genomic potential of culturable microorganisms.	No	• Allows further *in vitro* studies e.g.: Strains evolution and improvement; Design of starters based on functional traits of isolated strains.	• Imposed by our capability to isolate and culture microorganisms of interest. • Viability of certain microorganisms might be compromised in certain food matrixes.
Culture-independent approaches				
DGGE	Semi-quantitative assessment of microbial community composition.	Yes	• 16S rRNA or 18S rRNA based analyses of microbial communities' composition.	Imposed by: • Efficiency and universality of the primers used. • Dying and gel visualization technique.
Targeted metagenomics	Variable depending on the studied genes.	Yes	• 16S rRNA or 18S rRNA-based metataxonomic analyses. • Other selected targeted genes, e.g., distribution of biogenic amine synthetic genes.	Imposed by the efficiency and universality of the primers used, e.g. • Different 16S rRNA variable regions have different resolution of certain groups. • Limit of detection biased by the microbial community structure: due to the amplification step, highly abundant groups will appear overrepresented and low abundant groups will appear underrepresented or even undetected. • Other custom targeted metagenomics approaches require prior design and validation on mock communities with defined known composition.

Shotgun metagenomics	Microbial community composition & functional potential.	No	• Provides information on the genomic potential of the whole community present in a given sample.	• Requires high-sequencing depth in complex microbial communities. • Requires a minimum amount of high-quality DNA that might be difficult to achieve in certain food matrixes. • A mixture of sequences from food and microbial DNA will be obtained, what might increase the sequencing depth required for the analysis and, thus, bioinformatic pipelines need to be adapted in order to filter undesired sequences from the analyses.
Meta-transcriptomics	Microbial community composition & expression of functions at selected sampling points.	No	• Provides information on the groups that are metabolically active within the food matrix and, thus, that can have the most relevant effect and impact on the quality of the food products.	• RNA is generally more unstable and, thus, obtaining the desired concentrations of high-quality RNA might be limiting. • Requires higher sequencing depth than previous methodologies and might need high computing capacity in complex microbial communities.
Metabolomics	Metabolites profiling.	-	• In combination with results from the above-mentioned techniques, allows to identify microbial groups responsible for the production of metabolites of interest in the food matrix.	• Comprehensive metabolic profiling requires combination of multiple extraction and metabolomic profiling approaches. • Association with methods of assessment of microbial community composition and activity is essential in identifing microbial-based modifications in the metabolomic profiling of the foodstuff.

mimic the natural conditions prevailing in the studied ecosystems, the employment of oligotrophic media or the implementation of *in situ* cultivation strategies (Giovannoni and Stingl 2007, Cárdenas and Tiedje 2008, Alain and Querellou 2009). Indeed, data generated through NGS of relevant ecosystems have a huge potential to aid in the design of novel strategies aimed at facilitating the isolation and cultivation of previously unculturable microorganisms (Browne et al 2016, Lagier et al 2016).

2.1 Molecular Tools for Typing of Microbial Strains

Once a pure microbial culture is obtained, there are a wide range of molecular techniques available for completing the typing of isolates. The most widely used are described in the following lines.

Pulsed Field Gel Electrophoresis (PFGE) has been classically considered as the "gold standard" typing method for a vast number of microbial species, including several pathogenic microorganisms. PFGE is based on the restriction analysis of released microbial DNA and the separation of the fragments obtained through a gel electrophoresis in the presence of an electric field that periodically changes direction (PulsedNetUSA 2009). It is a molecular tool with a high discriminatory power that allows for the comparison of bacterial strains worldwide, if standardised methods are used. However, the process is lengthy, uses expensive equipment and requires highly trained staff (Jordan et al. 2015).

Ribotyping is also based on the cleavage of total genomic DNA by restriction enzymes, but strain differentiation is achieved through differences in location and number of rRNA gene sequences. Ribotyping analysis involves Southern blot transfer and hybridisation of labelled rRNA (ribosomal RNA) with a radiolabelled ribosomal operon probe (Schumann and Pukall 2013). Its discriminatory power has generally been reported to be weaker than that of PFGE or multilocus sequence typing (MLST), but its capability to be automated offers a distinct advantage over other methods (Jordan et al. 2015).

Multi Locus Sequence Typing (MLST), which is based on sequencing a set of housekeeping genes, is widely used as a typing technique, although it is nonetheless not equally developed for all microbial groups (Glaeser and Kämpfer 2015). Indeed, while it is a fully standardized method for some particular microbial species, no set of housekeeping genes has been globally agreed for others, with individual laboratories often choosing different sets of genes to sequence, which has made comparison between laboratories difficult. A global database of MLST sequences, available at https://pubmlst.org/databases.shtml, does however exist for several microbial species. This allows comparison of MLST sequences with previously available ones, which facilitates the identification of common MLST types and comparison of MLST types across geographically distinct locations (Jordan et al. 2015).

Multiple-Locus Variable Tandem Repeat Analysis (MLVA) is a typing tool developed for some particular microbial species which uses tandem repeat sequences in regions of genomic DNA to differentiate between strains (Nadon et al. 2013). It has less discriminatory power than other typing methods, but it is rapid, easy to perform and allows inter-laboratory comparison of isolates.

Phenotypic typing methods are commonly used in combination with molecular tools in order to elucidate physiological features of isolated microbial strains. High-throughput phenotypic arrays, such as the API system from Biomerieux or the OmniLog phenotype Microarrays, which uses a redox dye to measure cell metabolism simultaneously under up to 1,920 different growth conditions, are commonly used with this aim (Fox and Jordan 2014). Other molecular tools that provide information on the overall cellular composition, such as Fourier-Transformed Infrared (FTIR) spectroscopy and Matrix-assited laser desorption/ionization time-of-flight mass spectrometry (MALDI-TOF), are also frequently used for the typing and characterization of culturable microorganisms, since they allow for the generation of data on the main molecular constituents of the cell (proteins, lipids, carbohydrates and nucleic acids), facilitating the discrimination of strains based on their differential cellular/molecular composition (Alvarez-Ordóñez et al. 2011, Cheng et al. 2016).

More recently, whole genome sequencing (WGS) of microbial strains is becoming a highly powerful, very affordable and fast tool for microbial typing, achieving much higher resolution than previous technologies due to the continuous decline in the associated costs (Loman and Pallen 2015). WGS costs in US dollars (USD) can be as little as USDS 100 per bacterial genome, including sample preparation, library quality control and sequencing (Koser et al. 2012), and prices are going down year after year. A genome-wide gene-by-gene analysis tool, through extended multilocus sequence typing (eMLST) or through a "pan-genome approach" can then be implemented in order to infer relationships among isolates (Maiden et al. 2013).

3. Nucleic Acid-Based Culture-Independent Tools for Micro-Environments Assessment

3.1 Nucleic Acids Extraction: Impact on Culture-Independent Analyses

A common initial step in most culture-independent analyses requires total nucleic acid extraction from the corresponding sample. Most NGS-based studies have been designed and optimized for the analyses of highly complex microbial communities, such as those present in the gut microbiota of mammals, soils and water, and few studies to date have aimed at optimizing dedicated procedures for nucleic acid extractions from food matrixes. However, taking into account the conclusions drawn from studies conducted in other types of matrixes, it is reasonable to assume that the selection of the sample processing, DNA isolation and preservation procedures and the gene fragment(s) selected for taxonomic assignment will have a significant impact on the results of culture-independent analyses (Costea et al. 2017, Walden et al. 2017). Specific biochemical components present in the food matrix, like high concentrations of fat or fibers, might interfere with standard DNA isolation procedures. In addition, the presence of complex mixtures of eukaryotes and prokaryotes, with very different cell wall structures, might require the utilization of different cell lysis methods in order to get a good representation of nucleic acids from the whole microbial populations present in the sample. On the other hand, specific maturation processes occurring during the ripening of some food products involve extensive proteolysis what might imply a concomitant destruction of free nucleic acids present in the sample. Thus, before conducting culture-independent analysis in foodstuffs, it might be advisable to optimize nucleic acids extraction procedures using the specific food matrix under assessment.

3.2 PCR-Denaturing Gel Gradient Electrophoresis (DGGE)

DGGE is a molecular tool that allows the separation of small polymerase chain reaction (PCR) amplicons of the same length along polyacrylamide gels, according to differences in their DNA sequences. The methodology is based on the fact that different DNA sequences have different denaturing (melting) conditions. Thus, by using gradients of denaturing agents (usually urea and/or formamide) in the gels, sequence-based separation of DNA fragments is possible. PCR-DGGE utilization with 16S rRNA amplicons was introduced for microbial ecology studies back in 1993 (Muyzer 1999) and, for a couple of decades, has been the gold standard molecular fingerprinting methodology to study microbial ecosystems.

The typical workflow of PCR-DGGE analyses requires the DNA extraction from the foodstuffs, followed by PCR amplification of selected targeted gene fragments, usually 16S or 26S ribosomal RNA fragments, which are the most commonly used for the identification of prokaryotes and eukaryotes, respectively. Further electrophoretic separation of PCR products in polyacrylamide gels containing a gradient of a denaturing agent is conducted. Gel bands visualization allows comparisons of band patterns among samples. Further band excision from the gel and Sanger sequencing, either directly or following a cloning step, allows the identification of the microbial group represented by each of the observed bands.

As compared to culture-dependent analyses, DGGE has the advantage of detecting both alive and dead cells, culturable and unculturable species, allowing for a rapid comparison among communities. However, since it relies on electrophoretic images, resolution can be low. In addition, targeted sequences, oligonucleotides design and PCR amplification conditions, can introduce biases in the community fingerprinting. Besides, results from DGGE-based microbial community analyses are qualitative or semi-quantitative and might fail to detect low abundant groups within the community.

Despite their limitations, DGGE-based analyses have been the gold standard in microbial ecology studies and are still being frequently employed in food microbial ecology, having been applied to fermented meat products, milk derived foods, cocoa fermentation products (Quigley et al. 2011, Delgado et al. 2013, Greppi et al. 2015, Ferreira et al. 2015), as well as to other locally produced artisanal fermented products such as fermented rice (Ono et al. 2014) or spontaneous fermented milk products (Motato et al. 2017). Most frequently, these analyses are employed in order to monitor food fermentation and maturation processes.

3.3 Metagenomic Tools for Food Micro-Environments Assessment

Rapid evolution of NGS technologies has changed the way scientists approach microbial community studies, including research in the food microbiology field. Increasing and continuous improvements in the power and resolution of NGS platforms, accompanied by a concomitant decrease in associated costs, has boosted microbial ecology studies, enabling a deeper understanding of complex microbial communities through sequencing or expressing the whole set of DNA that can be extracted from a given environment (Mayo et al. 2014, Cao et al. 2017).

3.3.1. Functional metagenomics. Functional metagenomic approaches can overcome, in part, the existing barriers to harness yet unculturable bacteria (Coughlan et al. 2015). Functional metagenomics is based on the expression of total DNA from a sample (including DNA from uncultured microorganisms) in a routinely culturable host, usually *Escherichia coli* strains, followed by the execution of function-based screenings to identify desired activities conferred to the host by the inserted DNA. The advantage of functional metagenomics over sequencing-based metagenomic approaches is that the latter are limited to the study and identification of genes homologous to those that are already known and functionally annotated in the databases, while function-based screening of metagenomic expression libraries can unearth previously undescribed proteins and small molecules derived from not yet culturable microorganisms (Coughlan et al. 2015).

Functional metagenomic analyses involve the completion of the following sequential steps: isolation of total DNA from the relevant sample, cloning of the DNA into a suitable vector, expression of the vector in a host, and analysis of the obtained clones through phenotypic-based screenings. Methods for DNA isolation in metagenomic studies and their impact in the success of this sort of studies have been described above. Regarding the vector used for cloning, the use of large insert vectors capable of propagating in more than one host and accommodating large biosynthetic gene clusters or operons, such as cosmids or fosmids (Streit and Schmitz 2004) is advisable. The selection of an accurate host for expressing the metagenomic library is also a relevant factor to consider. *E. coli* is the most commonly used host, due to its accessibility as a molecular tool. Indeed, available fosmid commercial kits use *E. coli* as a host for the construction of metagenomic libraries. However, expressing DNA fragments isolated from native diverse microorganisms in a relatively domesticated host, such as *E. coli,* is often challenging (Banik and Brady 2010). For instance, foreign DNA may not be successfully transcribed and translated in *E. coli*, or protein folding into an active form might not occur. Therefore, great research efforts are being dedicated to the development of alternative hosts. Indeed, Craig et al. (2009) discovered two novel gene operons (one displaying antimicrobial activity and the second expressing a carotenoid biosynthetic gene cluster) through functional screening of a metagenomic library, constructed using *E. coli* as a host and then transferred to *Ralstonia metallidurans* for activity-based screenings. Interestingly, those two inserts showing activity in *R. metallidurans* did not confer the same metabolic abilities onto the *E. coli* host. Another study comparing six different Proteobacteria as hosts for the same metagenomic cosmid library showed that active clones recovered from the library in the frame of different functional screenings were expressed by different heterologous hosts with minimal overlap between hosts. This again shows the usefulness of Broad-Host Range vectors for overcoming host expression related barriers (Craig et al. 2010). Finally, efficient screening methods must be used to detect the presence of interesting genes conferring specific activities within the metagenomic library. As the probability of identifying a metagenomic clone, among possibly thousands of others, with a specific desired activity is low, high-throughput screening (HTS) protocols may improve the chances of obtaining an active clone, by allowing higher numbers of clones to be screened simultaneously (Uchiyama and Miyazaki 2009). In those HTS protocols, metagenomic clones may be grown on specific indicator media in order to allow visual identification of an active clone (Henne et al. 2000, Rondon et al. 2000). In other occasions, libraries are screened for the presence of zones of inhibition in soft agar overlay assays using indicator microorganisms, in order to identify antimicrobial agents produced by an active clone (Tannieres et al. 2013, Iqbal et al. 2014). Other

HTS procedures are based on the growth or survival of clones onto which the activity of interest has been conferred. These latter function-based screenings include for instance the ability to metabolize a given substrate (Entcheva et al. 2001), the ability to resist a potent antimicrobial agent (Donato et al. 2010) or the ability to grow in the presence of a lethal concentration of a heavy metal (Staley et al. 2015).

Within the Food Science field, functional metagenomic approaches have so far been followed in order to identify novel bio-catalysts for food processing reactions (e.g., enzymes for starch degradation, brewing, baking, synthesis of sugar and corn syrups, etc., which include amylases, lipases, esterases and β-galactosidases, among others), novel gene clusters responsible for the synthesis of bioactives (e.g., synthesis of biotin), or novel antimicrobials as biopreservative agents. Further details on potential applications of functional metagenomics in the food microbiology field may be found in the review article by Coughlan et al. (2015).

3.3.2. Amplicon sequencing approaches. Metagenomic tools based on high-throughput sequencing of exclusively selected gene markers, such as 16S rRNA or 18S rRNA genes, which allow for taxonomic assignment of prokaryotes and eukaryotes, respectively, have been the most widely exploited tools to study the taxonomic composition of microbial populations, having also been applied to food microbiology studies. Overall, the methodological process to conduct a gene marker metagenomic sequencing analysis of a food product implies: (i) total DNA isolation from the food matrix; (ii) PCR amplification of the marker gene(s); (iii) introduction of barcodes and sequencing platform adapters to prepare the sequencing libraries; (iv) NGS, generating millions of reads per sample; and (v) sequencing reads processing and analysis through bioinformatic tools.

Among the gene markers most frequently analyzed through these approaches, 16S rRNA and 18S rRNA genes have been the most widely applied to study the microbial community dynamics during food fermentation processes (De Filippis et al. 2017, Walsh et al. 2017). As an example, dynamics of bacterial communities during spontaneous or controlled fermentation of kimchi, kefir, wine or fermented dairy products, among others, have been studied through 16S rRNA gene metagenomic sequencing (Hsieh et al. 2012, Lusk et al. 2012, Escobar-Zepeda et al. 2016, Stefanini et al. 2016, Ceugniez et al. 2017, Lee et al. 2017). From these studies, several practical applications have arisen, aimed at improving and optimizing food production processes for instance through finding associations among bacterial communities' compositions and specific quality attributes in the final product, such as flavor, sensorial and nutritional properties. In this context, microbial communities composition has been employed in order to identify core bacteria producing aromatic compounds during vinegar fermentation (Wang et al. 2016, Wu et al. 2017). In addition to optimizing food production, metagenomic studies have also been applied as a diagnostic tool for microbial-related defects in cheese fermentation as is the case for pink discoloration defects which were attributed to *Thermus* bacteria (Quigley et al. 2016). It is also worth highlighting that identification of microbial communities associated with certain food ingredients has even proved valuable as a fingerprint to identify the geographical origin of foodstuffs and has, thus, even been proposed as an additional signature to certify the Protected Designation of Origin for some ingredients and foodstuffs (Capozzi and Spano 2011, Delcenserie et al. 2014, Mezzasalma et al. 2017). Thus, overall, discerning the microbial community structure in food matrixes has allowed for a better understanding of the role of microbial interactions in food production processes, providing valuable tools to improve food traceability, quality and safety (Wolfe et al. 2014, Dugat-Bony et al. 2015).

In addition to taxonomic markers, metagenomic studies have also been applied to other gene markers of interest in foodstuffs. For instance, some bacteria in cheese can lead to the accumulation of biogenic amines like histamine, tyramine, putrescine and cadaverine, whose consumption might be associated with detrimental effects, depending on the susceptibility of the individual. While traditional detection of biogenic amine producers has usually been conducted through bacterial isolation, in a recent work, O'Sullivan and colleagues have exploited the potential of NGS to detect biogenic amine producing genes in cheese samples, which allows not only identifying the presence of biogenic amine producing bacteria but also to identify the specific bacteria harboring the detected genes (O'Sullivan et al. 2015). A similar approach has been conducted by Zhu and colleagues to quantify antibiotic resistance genes in organic and conventionally produced lettuce (Zhu et al. 2017).

While 16S rRNA amplicon sequencing provides exclusively taxonomic information on the microbial community composition, the wide availability of complete bacterial genomes in sequence databases, has made it possible to create functional prediction tools based on taxonomic information. As such, both PICRUST and Tax4Fun are bioinformatic tools designed to infer metagenomes from taxonomic information in microbial communities (Langille et al. 2013, Aßhauer et al. 2015). It is important to highlight that these are just predictive tools, that at the moment are only available for bacterial communities, and still present important limitations since functional potential might strongly vary among strains and functional prediction is exclusively based on a closed collection of reference available genomes. Nevertheless, it is a tool available to complement 16S metagenomic surveys which has not been used into food microbiology surveys yet.

3.3.3. Whole-metagenome shotgun sequencing (WMS) approaches. WMS involves the fragmentation and subsequent sequencing, assembly and annotation of total genomic DNA isolated from a given sample (e.g., a food or a food processing environment), and allows us to gain information on its entire (prokaryotic and eukaryotic) gene content (Franzosa et al. 2015). In doing so, WMS outcompetes targeted amplicon sequencing in that it provides species- or even strain-level identification (Truong et al. 2017), and, in addition, it offers insights into the metabolic potential of microbial communities (Walsh et al. 2016, 2017), while amplicon sequencing only provides information on population structure at Family or Genus level and produces only very general predictions on the metabolic potential of the microbial community.

Some examples of WMS applications in the Food Science field are: (i) the detection of foodborne pathogens in food, as for example shown by Leonard et al. (2015), who detected *E. coli* in fresh spinach using this approach; (ii) the investigation of outbreaks or the transmission of microorganisms through food production chains, as for instance shown by Yang et al. (2016), who evaluated the effect of food processing on the microbiota of beef; (iii) the monitoring of microbial succession throughout fermentation of foods, as shown for several fermented foods, such as kimchi, kefir, cheese or cocoa (Jung et al. 2011, Wolfe et al. 2014, Illeghems et al. 2015, Walsh et al. 2016), which can facilitate the identification of the microbes that are most relevant during fermentation or that mainly contribute to the flavour, organoleptic or quality properties of foodstuffs (Walsh et al. 2016); (iv) or the identification of changes in microbial populations leading to food defects or spoilage, as shown in some investigations linking particular microbes to spoilage or defects for some fermented foods such as Chinese rice wine (Hong et al. 2016) and cheese (Quigley et al. 2016).

3.4 Studying Population Functionality Through Metatranscriptomics

Despite their greater power and resolution, WMS approaches, described in the previous section, still have some limitations, since while they predict the metabolic capabilities of microbial communities, providing information on all genes present in a given sample, this does not necessarily mean that such genes are transcribed and have a relevant role in the global micro-ecosystem. On the other hand, metatranscriptomic approaches, i.e., isolation of total RNA from a given sample followed by RNA sequencing, annotation and transcripts quantification, can measure the extent to which different genes are transcribed and, therefore, provide more solid data to elucidate the role of some particular microbes or genes in a given ecosystem. Nevertheless, metatranscriptomic methods have been scarcely used yet to study microbial populations in foods or food-related environments due to methodological constraints (described in Table 1). Anyway, some recent studies have shown the potential of this technology to characterize the microbial dynamics in some fermented foods, such as kimchi (Jung et al. 2013) or different types of cheese (Dugat-Bony et al. 2015, De Filippis et al. 2016, Monet et al. 2016).

4. Metabolomics as a Mean to Study Food Microbial Ecosystems

During food fermentation, the metabolic activity of microbial populations within the food matrix results in critical transformations of the raw material, which confer characteristic biochemical properties onto the final product. In this context, metabolomics approaches have demonstrated great potential in studying the capacity of different starters to confer the desired nutritional, sensorial, functional, and biochemical characteristics to the final products (Jiang and Bratcher 2016).

Metabolomics encompasses a wide range of tools aimed at studying the molecular fingerprint of a biological system through the identification and quantification of the collection of metabolites present at a particular time point. In this way, metabolomics gives a snapshot of the biological processes that are taking place in that particular system, in response to biological, genetic or environmental changes. Globally, metabolomics can follow a targeted approach, if only a predefined set of metabolites of interest are meant to be studied, or an untargeted approach, if as many metabolites as possible will be measured and compared between samples. Experimental design, sample preparation and metabolomics analytical tools need to be selected based on sample complexity and targeted metabolites to be analyzed. Some of the most commonly used methodologies, including examples of their application to food microbiology analyses, are summarized below.

Most samples generally contain a highly complex mixture of metabolites and, thus, the first step in sample preparation for metabolomics analyses will be the sample separation into simplified fractions. This can be achieved through gas chromatography (GC), high-performance liquid chromatography (HPLC) or capillary electrophoresis (CE). Following sample pre-separation, metabolites are separated from one another and identified by different means. This is usually done by employing mass spectrometry (MS) or nuclear magnetic resonance (NMR) spectroscopy analyses (Gowda and Djukovic 2014). Both types of analytical tools are complementary and their combination offers more accurate metabolomics assignments (Marshall and Powers 2017).

MS metabolomics is based on measuring the "mass to charge" ratio of ionized particles. Individual molecules result in specific "mass to charge" patterns that can be used for structural elucidation or identification via spectral matching to authentic compound data libraries, which serves to identify the molecules from which they have originated. MS is a highly sensitive tool which allows detection of pico- to fento-mole concentrations of many primary and secondary metabolites.

NMR is based on the magnetic properties of certain atoms' nuclei (^1H, ^{12}C, ^{31}P, ^{19}F) and their ability to absorb and emit energy. Under the presence of strong magnetic fields, nuclear spins are aroused and rapid changes in the magnetic fields allow for nuclei relaxation with a concomitant electromagnetic energy release. Besides, the molecular surroundings of the nucleus which is being analyzed affect its resonance frequency, thus making it possible to distinguish nuclei surrounded by different atoms in a given molecule. An undeniable advantage of NMR based metabolomics is that it requires little sample manipulation, is highly sensitive, non-destructive and can be virtually applied on crude intact samples, although spectrum analyses and interpretation requires great expertise (Kruk et al. 2017).

To date, metabolomics approaches have not been fully exploited in food microbiology research, however they offer an enormous potential in discerning the biological processes behind many food production and preservation methodologies, such as those taking place during food fermentation and maturation. In addition, they have been deemed as valuable tools for food safety analyses, as well as for nutritional, functional, sensorial and traceability studies of food production processes. For instance, metabolomics analyses have been used to study fermentation characteristics associated to different starter bacteria, including the differentiation in the production of functional metabolites such as γ-aminobutyric acid (GABA) or certain phenolic compounds (Hagi et al. 2016, Tomita et al. 2017). Similarly, metabolomics tools have also been used to evaluate the effect of environmental parameters on fermentation processes, as is the case of salt concentration (Kim et al. 2017) and to evaluate optimal microbial culture conditions aimed at maintaining traits of industrial interest in starter strains (Cretenet et al. 2014). Metabolomic profiles have been associated to quality attributes of the final product in the case of crab paste fermentation (Chen et al. 2016); and bacterial communities have been associated to particular flavor metabolites during the fermentation of aromatic vinegar (Wang et al. 2015).

On the other hand, it is worth remarking that metabolomics fingerprinting has also been deemed suitable for differentiating foodstuffs elaborated with different raw materials, or by using different fermentation processes or starters, as in the case of Italian white wines produced by different yeast starters (Mazzei et al. 2013). Indeed, metabolomics have been used to determine the geographical origin of foodstuffs, being thus suitable for the analysis of authenticity markers (Díaz et al. 2014, Jumhawan et al. 2016).

Finally, targeted metabolomics approaches have also been applied for some food safety applications, such as those aimed at the early detection of toxins, including mycotoxins in cereals (Rubert et al. 2017), marine toxins in shellfish (Orellana et al. 2014) and other contaminants (Inoue et al. 2016). Metabolomic footprinting of foods has also been used to identify early biomarkers, revealing the presence of spoilage bacteria, prior to any significant product degradation (Johanningsmeier and McFeeters 2015).

Conclusions and Future Prospects

The development of novel molecular tools to study and characterize food and food-related microbial ecosystems has allowed the blooming in the last few decades of ambitious investigations aimed at understanding the role that microbial communities colonizing food processing environments and present in foods play in microbial safety and in the nutritional, organoleptic and functional attributes of the final food products. Some such molecular tools allow the characterization of microbial populations in a culture-independent manner, obtaining information on the microbial richness and functionality of a given sample, by sequencing or expressing nucleic acids directly extracted from the target sample or by characterizing the metabolites they produce through physico-chemical analyses. However, in most cases, in order to assess micro-ecosystems linked to food and food production, it is still essential to isolate microbial strains in order to get detailed pheno- and geno-typic data on them. In this regard, whole genome sequencing is offering a very powerful tool which is revolutionizing our ability to get insights into the functional traits encoded by food associated bacteria, their evolutionary adaptation and their spread from farm to fork.

Metagenomic approaches are also valuable tools that allow for the culture-independent analysis of complex microbial communities, and that, therefore, have potential applications in food-related micro-environments assessment. They can provide access to all the genetic resources in a given environmental niche, which is essential for accessing the genomes of difficult-to-culture or non-culturable microorganisms. However, as Next Generation Sequencing technologies become more widely adopted, the key challenges of generating representative data sets and the development of harmonized protocols for sampling, nucleic acid extraction, library preparation and standardized bioinformatic pipelines to manage and interpret the generated data become increasingly pertinent.

In spite of the tremendous advances achieved in recent years, in terms of the ease of performing culture-independent surveys of complex microbial communities that have made possible their application to the food microbiology field, the establishment of network reconstruction analyses, capable of integrating the information obtained from multiple "-omic" datasets, is still required (Cocolin et al. 2017). These would allow the identification of key microbial groups or genes essential in achieving the desired characteristics in the final food products, allowing the translation of data gathered from food micro-ecosystems assessment into the development of new foodstuffs, with improved shelf life or nutritional, organoleptic or functional properties through the rationale exploitation of the microbial communities associated to food processing.

Nonetheless, despite the demonstrated potential of all these molecular tools in helping to improve food safety and functionality, its routine use by food companies do require for its transformation into cheaper, user-friendly approaches that could be used on site by non-specialized personnel (Hyeon et al. 2017). In this regard, prototypes of some miniaturized sequencers have recently become available and have demonstrated their potential to be used on site, generating results in real time (Benítez-Páez and Sanz 2017). In this context, the generation of open access databases specially dedicated to food microbiology ecosystems, updated in real time and freely accessible, can pave the way for a wider exploitation of these molecular tools (Taboada et al. 2017).

Acknowledgments

A. Alvarez-Ordóñez acknowledges the financial support by Fundación BBVA and Ministerio de Economía y Competitividad (AGL2016-78085-P). L. Ruiz is a post-doctoral researcher supported by the *Juan de la Cierva* Post-doctoral Trainee Program of the Spanish Ministry of Economy and Competitiveness (MINECO; IJCI-2015-23196).

References

Aßhauer, K.P., B. Wemheuer, R. Daniel and P. Meinicke. 2015. Tax4Fun: Predicting functional profiles from metagenomic 16S rRNA data. Bioinformatics 31(17): 2882–2884.

Alain, K. and J. Querellou. 2009. Cultivating the uncultured: Limits, advances and future challenges. Extremophiles 13: 583–594.

Alvarez-Ordóñez, A., D.J. Mouwen, M. López and M. Prieto. 2011. Fourier transform infrared spectroscopy as a tool to characterize molecular composition and stress response in foodborne pathogenic bacteria. J. Microbiol. Methods 84: 369–378.

Banik, J.J. and S.F. Brady. 2010. Recent application of metagenomic approaches toward the discovery of antimicrobials and other bioactive small molecules. Curr. Opin. Microbiol. 13: 603–609.

Benítez-Páez, A. and Y. Sanz. 2017. Multi-locus and long amplicon sequencing approach to study microbial diversity at species level using the MinION™ portable nanopore sequencer. Gigascience 6: 1–12.

Bokulich, N.A. and D.A. Mills. 2012. Next-generation approaches to the microbial ecology of food fermentations. BMB Rep. 45(7): 377–389.

Browne, H.P., S.C. Forster, B.O. Anonye, N. Kumar, B.A. Neville, M.D. Stares, D. Goulding and T.D. Lawley. 2016. Culturing of "unculturable" human microbiota reveals novel taxa and extensive sporulation. Nature 533(7604): 543–546.

Buchholz, K. and J. Collins. 2013. The roots—a short history of industrial microbiology and biotechnology. Appl. Microbiol. Biotechnol. 97: 3747–3762.

Cao, Y., S. Fanning, S. Proos, K. Jordan and S. Srikumar. 2017. A review on the applications of next generation sequencing technologies as applied to food-related microbiome studies. Front. Microbiol. 8: 1829.

Capozzi, V. and G. Spano. 2011. Food microbial biodiversity and microbes of protected origin. Front. Microbiol. 2(237): 1–3.

Cárdenas, E. and J.M. Tiedje. 2008. New tools for discovering and characterizing microbial diversity. Curr. Opin. Biotechnol. 19: 544–549.

Ceugniez, A., B. Taminiau, F. Coucheney, P. Jacques, V. Delcenserie and G. Daube. 2017. Use of a metagenetic approach to monitor the bacterial microbiota of 'Tomme d'Orchies' cheese during the ripening process. Int. J. Food Microbiol. 247: 65–69.

Chen, D., Y. Ye, J. Chen and X. Yan. 2016. Evolution of metabolomics profile of crab paste during fermentation. Food Chem. 192: 886–892.

Cheng, K., H. Chui, L. Domish, D. Herñandez and G. Wang. 2016. Recent development of mass spectrometry and proteomics applications in identification and typing of bacteria. Proteomics Clin. Appl. 10: 346–357.

Cocolin, L., M. Mataragas, F. Bourdichon, A. Doulgeraki, M-F. Pilet, B. Jagadeesan, K. Rantsiou and T. Phister. 2017. Next generation microbiological risk assessment meta-omics: The next need for integration. Int. J. Food Microbiol. https://doi.org/10.1016/j.ijfoodmicro.2017.11.008.

Costea, P.I., G. Zeller, S. Sunagawa, E. Pelletier, A. Alberti, F. Levenez, M. Tramontano, M. Driessen, R. Hercog, F.E. Jung, J.R. Kultima, M.R. Hayward, L.P. Coelho, E. Allen-Vercoe, L. Bertrand, M. Blaut, J.R.M. Brown, T. Carton, S. Cools-Portier, M. Daigneault, M. Derrien, A. Druesne, W.M. de Vos, B.B. Finlay, H.J. Flint, F. Guarner, M. Hattori, H. Heilig, R.A. Luna, J. van Hylckama Vlieg, J. Junick, I. Klymiuk, P. Langella, E. Le Chatelier, V. Mai, C. Manichanh, J.C. Martin, C. Mery, H. Morita, P.W. O'Toole, C. Orvain, K.R. Patil, J. Penders, S. Persson, N. Pons, M. Popova, A. Salonen, D. Saulnier, K.P. Scott, B. Singh, K. Slezak, P. Veiga, J. Versalovic, L. Zhao, E.G. Zoetendal, S.D. Ehrlich, J. Dore and P. Bork. 2017. Towards standards for human fecal sample processing in metagenomic studies. Nat. Biotechnol. 35(11): 1069–1076.

Coughlan, L.M., P.D. Cotter, C. Hill and A. Alvarez-Ordóñez. 2015. Biotechnological applications of functional metagenomics in the food and pharmaceutical industries. Front. Microbiol. 6: 672.

Craig, J.W., F.Y. Chang and S.F. Brady. 2009. Natural products from environmental DNA hosted in *Ralstonia metallidurans*. ACS Chem. Biol. 4: 23–28.

Craig, J.W., F.Y. Chang, J.H. Kim, S.C. Obiajulu and S.F. Brady. 2010. Expanding small-molecule functional metagenomics through parallel screening of broad-host-range cosmid environmental DNA libraries in diverse proteobacteria. Appl. Environ. Microbiol. 76: 1633–1641.

Cretenet, M., G. Le Gall, U. Wegmann, S. Even, C. Shearman, R. Stentz and S. Jeanson. 2014. Early adaptation to oxygen is key to the industrially important traits of *Lactococcus lactis* ssp. *cremoris* during milk fermentation. BMC Genomics 15 (1): 1054.

De Filippis, F., A. Genovese, P. Ferranti, J.A. Gilbert and D. Ercolini. 2016. Metatranscriptomics reveals temperature driven functional changes in microbiome impacting cheese maturation rate. Sci. Rep. 6: 21871.

De Filippis, F., E. Parente and D. Ercolini. 2017. Metagenomics insights into food fermentations. Microb. Biotechnol. 10(1): 91–102.

Delcenserie, V., B. Taminiau, L. Delhalle, C. Nezer, P. Doyen, S. Crevecoeur, D. Roussey, N. Korsak and G. Daube. 2014. Microbiota characterization of a Belgian protected designation of origin cheese, Herve cheese, using metagenomic analysis. J. Dairy Sci. 97(10): 6046–6056.

Delgado, S., C. Rachid, E. Fernández, T. Rychlik, A. Alegría, R.S. Peixoto and B. Mayo. 2013. Diversity of thermophilic bacteria in raw, pasteurized and selectively-cultured milk, as assessed by culturing, PCR-DGGE and pyrosequencing. Food Microbiol. 36(1):103–111.

Díaz, R., O.J. Pozo, J.V. Sancho and F. Hernández. 2014. Metabolomic approaches for orange origin discrimination by ultra-high performance liquid chromatography coupled to quadrupole time-of-flight mass spectrometry. Food Chem. 157: 84–93.

Donato, J.J., L.A. Moe, B.J. Converse, K.D. Smart, F.C. Berklein, P.S. McManus and J. Handelsman. 2010. Metagenomic analysis of Apple orchard soil reveals antibiotic resistance genes encoding predicted bifunctional proteins. Appl. Environ. Microbiol. 76: 4396–4401.

Dugat-Bony, E., C. Straub, A. Teissandier, D. Onésime, V. Loux, C. Monnet, F. Irlinger, S. Landaud, M.N. Leclercq-Perlat, P. Bento, S. Fraud, J.F. Gibrat, J. Aubert, F. Fer, E. Guédon, N. Pons, S. Kennedy, J.M. Beckerich, D. Swennen and P. Bonnarme. 2015. Overview of a surface-ripened cheese community functioning by meta-omics analyses. PloS One 10(4): e0124360.

Entcheva, P., W. Liebl, A. Johann, T. Hartsch and W.R. Streit. 2001. Direct cloning from enrichment cultures, are liable strategy for isolation of complete operons and genes from microbial consortia. Appl. Environ. Microbiol. 67: 89–99.

Escobar-Zepeda, A., A. Sanchez-Flores and C.Q. Baruch. 2016. Metagenomic analysis of a mexican ripened cheese reveals a unique complex microbiota. Food Microbiol. 57: 116–27.

Ferreira, A.C., E.L. Marques, J.C Dias and R.P. Rezende. 2015. DGGE and multivariate analysis of a yeast community in spontaneous cocoa fermentation process. Genet. Mol. Res. 14(144): 18465–18470.

Food and Agriculture Organization of the United Nations and World Health Organization. 2003. Assuring food safety and quality: Guidelines for strengthening national food control systems. Rome: ISBN 92-5-104918-1.

Fox, E.M. and K. Jordan. 2014. High-throughput characterization of *Listeria monocytogenes* using the OmniLog phenotypic microarray. Methods Mol. Biol. 1157: 103–108.

Franzosa, E.A., T. Hsu, A. Sirota-Madi, A. Shafquat, G. Abu-Ali, X.C. Morgan and C. Huttenhower. 2015. Sequencing and beyond: Integrating molecular 'omics' for microbial community profiling. Nat. Rev. Microbiol. 13: 360–372.

Gill, A. 2017. The importance of bacterial culture to food microbiology in the age of genomics. Front. Microbiol. 8: 777.

Giovannoni, S. and U. Stingl. 2007. The importance of culturing bacterioplankton in the "omics" age. Nat. Rev. Microbiol. 5: 820-826.

Glaeser, S.P. and P. Kämpfer. 2015. Multilocus sequence analysis (MLSA) in prokaryotic taxonomy. Syst. Appl. Microbiol. 38: 237–245.

Gowda, G.A. and D. Djukovic. 2014. Overview of mass spectrometry-based metabolomics: opportunities and challenges. Methods Mol. Biol. 1198: 3–12.

Greppi, A., I. Ferrocino, A. La Storia, K. Rantsiou, D. Ercolini and L. Cocolin. 2015. Monitoring of the microbiota of fermented sausages by culture independent rRNA-based approaches. Int. J. Food Microbiol. 212: 67–75.

Hagi, T., M. Kobayashi and M. Nomura. 2016. Metabolome analysis of milk fermented by γ-aminobutyric acid-producing *Lactococcus lactis*. J. Dairy Sci. 99(2): 994–1001.

Henne, A., R.A. Schmitz, M. Bomeke, G. Gottschalk and R. Daniel. 2000. Screening of environmental DNA libraries for the presence of genes conferring lipolytic activity on *Escherichia coli*. Appl. Environ. Microbiol. 66: 3113–3116.

Hong, X., J. Chen, L. Liu, H. Wu, H. Tan, G. Xie, Q. Xu, H. Zou, W. Yu, L. Wang and N Qin. 2016. Metagenomic sequencing reveals the relationship between microbiota composition and quality of Chinese rice wine. Sci. Rep. 6: 26621.

Hsieh, H.-H., S.-Y. Wang, T.-L. Chen, Y-L. Huang and M.-J. Chen. 2012. Effects of cow's and goat's milk as fermentation media on the microbial ecology of sugary kefir grains. Int. J. Food Microbiol. 157(1): 73–81.

Hutkins, R.W. 2006. Microbiology and Technology of Fermented Foods. Wiley-Blackwell, New Jersey, USA.

Hyeon, J.-Y., S. Li, D.A. Mann, S. Zhang, Z. Li, Y. Chen and X. Deng. 2017. Quasi-metagenomics and real time sequencing aided detection and subtyping of *Salmonella enterica* from food samples. Appl. Environ. Microbiol. In Press pii: AEM.02340-17.

Illeghems, K., S. Weckx and L. De Vuyst. 2015. Applying meta-pathway analyses through metagenomics to identify the functional properties of the major bacterial communities of a single spontaneous cocoa bean fermentation process sample. Food Microbiol. 50: 54–63.

Inoue, K., C. Tanada, T. Hosoya, S. Yoshida, T. Akiba, J.Z. Min, K. Todoroki, Y. Yamano, S. Kumazawa and T. Toyo'oka. 2016. Principal component analysis of molecularly based signals from infant formula contaminations using LC-MS and NMR in foodomics. J. Sci. Food Agricul. 96(11): 3876–3881.

Iqbal, H.A., J.W. Craig and S.F. Brady. 2014. Antibacterial enzymes from the functional screening of metagenomic libraries hosted in *Ralstonia metallidurans*. FEMS Microbiol. Lett. 354: 19–26.

Jiang, T. and C.L. Bratcher. 2016. Differentiation of commercial ground beef products and correlation between metabolites and sensory attributes: A metabolomic approach. Food Res. Int. 90: 298–306.

Johanningsmeier, S.D. and R.F. McFeeters. 2015. Metabolic footprinting of *Lactobacillus buchneri* strain LA1147 during anaerobic spoilage of fermented cucumbers. Int. J. Food Microbiol. 215: 40–48.

Jordan, K., D. Leong and A. Alvarez-Ordóñez. 2015. *Listeria monocytogenes* in the food processing environment. Springer Briefs in Food, Health and Nutrition. Springer International Publishing, Berlin, Germany.

Jumhawan, U., S.P. Putri, Yusianto, T. Bamba and E. Fukusaki. 2016. Quantification of coffee blends for authentication of asian palm civet coffee (kopi luwak) via metabolomics: A proof of concept. J. Biosci. Bioeng. 122(1): 79–84.

Jung, J.Y., S.H. Lee, J.M. Kim, M.S. Park, J-W. Bae, Y. Hahn, E.L. Madsen and C.O. Jeon. 2011. Metagenomic analysis of kimchi, a traditional Korean fermented food. Appl. Environ. Microbiol. 77: 2264–2274.

Jung, J.Y., S.H. Lee, H.M. Jin, Y. Hahn, E.L. Madsen and C.O. Jeon. 2013. Metatranscriptomic analysis of lactic acid bacterial gene expression during kimchi fermentation. Int. J. Food Microbiol. 163: 171–79.

Kim, D.W., B.-M. Kim, H.-J. Lee, G.-J. Jang, S.H. Song, J.-I. Lee, S-B. Lee, J.M. Shim, K.W. Lee, J.H. Kim, K.S. Ham, F. Chen and H.J. Kim. 2017. Effects of different salt treatments on the fermentation metabolites and bacterial profiles of kimchi. J. Food Sci. 82(5): 1124–1131.

Koser, C.U., M.J. Ellington, E.J. Cartwright, S.H. Gillespie, N.M. Brown, M. Farrington, M.T. Holden, G. Dougan, S.D. Bentley, J. Parkhill and S.J. Peacock. 2012. Routine use of microbial whole genome sequencing in diagnostic and public health microbiology. PLoS Pathog. 8: e1002824.

Kruk, J., M. Doskocz, E. Jodłowska, A. Zacharzewska, J. Łakomiec, K. Czaja and J. Kujawski. 2017. NMR techniques in metabolomic studies: A quick overview on examples of utilization. Appl. Magn. Reson. 48(1): 1–21.

Kumar, A. 2016. Role of microbes in food and industrial microbiology. J. Food Ind. Microbiol. 2(2): 1–2.

Lagier, J.-C., S. Khelaifia, M.T. Alou, S. Ndongo, N. Dione, P. Hugon, A. Caputo, F. Cadoret, S.I. Traore, E.H. Seck, G. Dubourg, G. Durand, G. Mourembou, E. Guilhot, A. Togo, S. Bellali, D. Bachar, N. Cassir, F. Bittar, J. Delerce, M. Mailhe, D. Ricaboni, M. Bilen, N.P. Dangui Nieko, N.M. Dia Badianoe, C. Valles, D. Mouelhi, K. Diop, M. Million, D. Musso, J. Abrahão, E.I. Azhar, F. Bibi, M. Yasir, A. Diallo, C. Sokhna, F. Djossou, V. Vitton, C. Robert, J.M. Rolain, B. La Scola, P.E. Fournier, A. Levasseur and D. Raoult. 2016. Culture of previously uncultured members of the human gut microbiota but culturomics. Nat. Microbiol. 1: 16203.

Langille, M.G.I., J. Zaneveld, J.G. Caporasso, D. McDonald, D. Knights, J.A. Reyes, J.C. Clemenmte, D.E. Burkepile, R.L. Vega Thurber, R. Knight, R.G. Beiko and C. Huttenhower. 2013. Predictive functional profiling of microbial communities using 16S rRNA marker gene sequences. Nat. Biotechnol. 31(9): 814–821.

Lee, M., J.H. Song, M.Y. Jung, S.H. Lee and J.Y. Chang. 2017. Large-scale targeted metagenomics analysis of bacterial ecological changes in 88 kimchi samples during fermentation. Food Microbiol. 66: 173–183.

Leonard, S.R., M.K. Mammel, D.W. Lacher and C.A. Elkins. 2015. Application of metagenomic sequencing to food safety: Detection of Shiga toxin–producing *Escherichia coli* on fresh bagged spinach. Appl. Environ. Microbiol. 81: 8183–8191.

Loman, N.J. and M.J. Pallen. 2015. Twenty years of bacterial genome sequencing. Nat. Rev. Microbiol. 13: 787–794.

Lusk, T.S., A.R. Ottesen, J.R. White, M.W. Allard, E.W Brown and J.A. Kase. 2012. Characterization of microflora in latin-style cheeses by next-generation sequencing technology. BMC Microbiol. 12(1): 254.

Maiden, M.C., M.J. Jansen van Rensburg, J.E. Bray, S.G. Earle, S.A. Ford, K.A. Jolley and N.D. McCarthy. 2013. MLST revisited: The gene-by-gene approach to bacterial genomics. Nat. Rev. Microbiol. 11: 728–736.

Marco, M.L., S. Heeney, S. Binda, C.J. Cifelli, P.D. Cotter, B. Foligné, M. Gänzle, R. Kort, G. Pasin, A. Pihlanto, E.J. Smid and R. Hutkins. 2017. Health benefits of fermented foods: Microbiota and beyond. Curr. Opin. Biotechnol. 44: 94–102.

Marshall, D.D. and R. Powers. 2017. Beyond the paradigm: Combining mass spectrometry and nuclear magnetic resonance for metabolomics. Prog. Nucl. Magn. Reson. Spectrosc. 100: 1–16.

Mayo, B., C.T. Rachid, A. Alegría, A.M. Leite, R.S. Peixoto and S. Delgado. 2014. Impact of next generation sequencing techniques in Food Microbiology. Curr. Genomics 15(4): 293–309.

Mazzei, P., R. Spaccini, N. Francesca, G. Moschetti and A. Piccolo. 2013. Metabolomic by [1] H NMR spectroscopy differentiates 'Fiano di Avellino' white wines obtained with different yeast strains. J. Agricult. Food Chem. 61(45): 10816–10822.

Mezzasalma, V., A. Sandionigi, I. Bruni, A. Bruno, G. Lovicu, M. Casiraghi and M. Labra. 2017. Grape microbiome as a reliable and persistent signature of field origin and environmental conditions in cannonau wine production. PloS One 12(9): e0184615.

Monnet, C., E. Dugat-Bony, D. Swennen, J.M. Beckerich, F. Irlinger, S. Fraud and P. Bonnarme. 2016. Investigation of the activity of the microorganisms in a Reblochon-style cheese by metatranscriptomic analysis. Front. Microbiol. 7: 536.

Motato, K.E., C. Milani, M. Ventura, F.E. Valencia, P. Ruas-Madiedo and S. Delgado. 2017. Bacterial diversity of the colombian fermented milk 'Suero Costeño' assessed by culturing and high-throughput sequencing and DGGE analysis of 16S rRNA gene amplicons. Food Microbiol. 68: 129–136.

Muyzer, G. 1999. DGGE/TGGE a method for identifying genes from natural ecosystems. Curr. Opin. Microbiol. 2(3): 317–322.

Nadon, C.A., E. Trees, L.K. Ng, E. Moller Nielsen, A. Reimer, N. Maxwell, K.A. Kubota and P. Gerner-Smidt. 2013. Development and application of MLVA methods as a tool for inter-laboratory surveillance. Euro Surveill. 18: 20565.

O'Sullivan, D.J., V. Fallico, O. O'Sullivan, P.L.H. McSweeney, J.J. Sheehan, P.D. Cotter and L. Giblin. 2015. High-throughput DNA sequencing to survey bacterial histidine and tyrosine decarboxylases in raw milk cheeses. BMC Microbiol. 15: 266.

Ono, H., S. Nishio, J. Tsurii, T. Kawamoto, K. Sonomoto and J. Nakayama. 2014. Monitoring of the microbiota profile in nukadoko, a naturally fermented rice bran bed for pickling vegetables. J. Biosci. Bioengin. 118(5): 520–525.

Orellana, G., J.V. Bussche, L. van Meulebroek, M. Vandegehuchte, C. Janssen and L. Vanhaecke. 2014. Validation of a confirmatory method for lipophilic marine toxins in shellfish using UHPLC-HR-Orbitrap MS. Anal. Bioanal. Chem. 406(22): 5303–5312.

Prakash, O., Y. Shouche, K. Jangid and J.E. Kostka. 2013. Microbial cultivation and the role of microbial resource centers in the omics era. Appl. Microbiol. Biotechnol. 97: 51–62.

PulseNetUSA, 2009. International Standard PulseNet protocol. https://www.cdc.gov/pulsenet/index.html. Accessed 10/12/2017.

Quigley, L., O. O'Sullivan, T.P. Beresford, R.P. Ross, G.F. Fitzgerald and P.D. Cotter. 2011. Molecular approaches to analysing the microbial composition of raw milk and raw milk cheese. Int. J. Food Microbiol. 150(2-3): 81–94.

Quigley, L., D.J. O'Sullivan, D. Daly, O. O'Sullivan, Z. Burdikova, R. Vana, T.P. Beresford, R.P. Ross, G.F. Fitzgerald, P.L. McSweeney, L. Giblin, J.J. Sheehan and P.D. Cotter. 2016. Thermus and the pink discoloration defect in cheese. mSystems 1(3): e00023–16.

Rastogi, G. and R.K. Sani. 2011. Molecular techniques to assess microbial community structure, function, and dynamics in the environment. pp. 29–57. *In*: I. Ahmad, F. Ahmad and J. Pichtel (eds.). Microbes and Microbial Technology. Springer, New York, USA.

Rondon, M.R., P.R. August, A.D. Bettermann, S.F. Brady, T.H. Grossman, M.R. Liles, K.A. Loiacono, B.A. Lynch, I.A. MacNeil, C. Minor, C.L. Tiong, M. Gilman, M.S. Osburne, J. Clardy, J. Handelsman and R.M. Goodman. 2000. Cloning the soil metagenome: A strategy for accessing the genetic and functional diversity of uncultured microorganisms. Appl. Environ. Microbiol. 66: 2541–2547.

Rubert, J., L. Righetti, M. Stranska-Zachariasova, Z. Dzuman, J. Chrpova, C. Dall'Asta and J. Hajslova. 2017. Untargeted metabolomics based on ultra-high-performance liquid chromatography–high-resolution mass spectrometry merged with chemometrics: A new predictable tool for an early detection of mycotoxins. Food Chem. 224: 423–431.

Schumann, P. and R. Pukall. 2013. The discriminatory power of ribotyping as automable technique for differentiation of bacteria. Syst. Appl. Microbiol. 36: 369–375.

Staley, C., D. Johnson, T.J. Gould, P. Wang, J. Phillips, J.B. Cotner and M.J. Sadowsky. 2015. Frequencies of heavy metal resistance are associated with land cover type in the Upper Mississippi River. Sci. Total Environ. 511: 461–468.

Stefanini, I., D. Albanese, A. Cavazza, E. Franciosi, C. De Filippo, C. Donati and D. Cavalieri. 2016. Dynamic changes in microbiota and mycobiota during spontaneous 'Vino Santo Trentino' fermentation. Microb. Biotechnol. 9(2): 195–208.

Streit, W.R. and R.A. Schmitz. 2004. Metagenomics—the key to the uncultured microbes. Curr. Opin. Microbiol. 7: 492–498.

Taboada, E.N., M.R. Graham, J.A. Carriço and G. Van Domselaar. 2017. Food safety in the age of next generation sequencing, bioinformatics, and open data access. Front. Microbiol. 8: 909.

Tamang, J.P., D-H. Shin, S.J. Jung and S.W. Chae. 2016. Functional properties of microorganisms in fermented foods. Front. Microbiol. 7: 578.

Tannieres, M., A. Beury-Cirou, A. Vigouroux, S. Mondy, F. Pellissier, Y. Dessaux and D. Faure. 2013. A metagenomic study highlights phylogenetic proximity of quorum-quenching and xenobiotic-degrading amidases of the AS-family. PLoS ONE 8: e65473.

Tomita, S., K. Saito, T. Nakamura, Y. Sekiyama and J. Kikuchi. 2017. Rapid discrimination of strain-dependent fermentation characteristics among *Lactobacillus* strains by NMR-based metabolomics of fermented vegetable juice. PLoS One 12(7): e0182229.

Truong, D.T., A. Tett, E. Pasolli, C. Huttenhower and N. Segata. 2017. Microbial strain-level population structure and genetic diversity from metagenomes. Genome Res. 27: 626–638.

Uchiyama, T. and K. Miyazaki. 2009. Functional metagenomics for enzyme discovery: Challenges to efficient screening. Curr. Opin. Biotechnol. 20: 616–622.

Walden, C., F. Carbonero and W. Zhang. 2017. Assessing impacts of DNA extraction methods on next generation sequencing of water and wastewater samples. J. Microbiol. Methods 141: 10–16.

Walsh, A.M., F. Crispie, K. Kilcawley, O. O´Sullivan, M.G. O´Sullivan, M.J. Claesson and P.D. Cotter. 2016. Microbial succession and flavor production in the fermented dairy beverage kefir. mSystems 1: 5.

Walsh, A.M., F. Crispie, M.J. Claesson and P.D. Cotter. 2017. Translating omics to food microbiology. Ann. Rev. Food Sci. Technol. 8(1): 113–134.

Wang, Z.M., Z.M. Lu, Y.J. Yu, G.Q. Li, J.S. Shi and Z.H. Xu. 2015. Batch-to-batch uniformity of bacterial community succession and flavor formation in the fermentation of Zhenjiang aromatic vinegar. Food Microbiol. 50: 64–69.

Wang, Z.M., Z.M. Lu, J.S. Shi and Z.H. Xu. 2016. Exploring flavour-producing core microbiota in multispecies solid-state fermentation of traditional chinese vinegar. Sci. Rep. 6: 26818.

Wolfe, B.E., J.E. Button, M. Santarelli and R.J. Dutton. 2014. Cheese rind communities provide tractable systems for in situ and *in vitro* studies of microbial diversity. Cell 158(2): 422–433.

Wu, L.H., Z.M. Lu, Z.J. Zhang, Z.M. Wang, Y.J. Yu, J.S. Shi and Z.H. Xu. 2017. Metagenomics reveals flavour metabolic network of cereal vinegar microbiota. Food Microbiol. 62: 23–31.

Yang, X., N.R. Noyes, E. Doster, J.N. Martin, L.M. Linke, R.J. Magnuson, H. Yang, I. Geronaras, D.R. Woerner, K.L. Jones, J. Ruiz, C. Boucher, P.S. Morley and K.E. Belk. 2016. Use of metagenomic shotgun sequencing technology to detect foodborne pathogens within the microbiome of the beef production chain. Appl. Environ. Microbiol. 82(8): 2433–2443.

Zhu, B., Q. Chen, S. Chen and Y.G. Zhu. 2017. Does organically produced lettuce harbor higher abundance of antibiotic resistance genes than conventionally produced? Environ. Int. 98: 152–159.

3

Molecular Tools for Evolution and Taxonomy Assessment

Stamatoula Bonatsou,[1,*] *Spiros Paramithiotis*[2] *and*
Eleftherios H. Drosinos[2]

1. Introduction

The roots of modern systematics and classification, as a depiction of the order and diversity in nature, are traced back to the work of Aristotle and his student, Theophrastus. The former, in his 'History of Animals', described comparatively more than 500 animal species, including invertebrates, fish, amphibians, reptiles, birds, insects, mammals and human. He presented aspects of their anatomy, physiology, habitats, breeding habits and ethos, and classified them accordingly. Theophrastus, in his 'Enquiry into Plants', described more than 500 plant species, divided into trees, shrubs, under-shrubs and herbs. These were described giving particular attention to their 'essential nature', i.e., the features that are characteristic for a plant and differentiate one from another. However, he recognized that several plants may diverge from their 'essential nature' depending on the cultivation conditions. Moreover, he presented additional classification systems based on properties such as size, the ability to bear fruits, etc. Moreover, special reference was made to tree cultivation, as well as the quality and the uses of timber. Finally, in his fourth book, he described trees and plants specific to regions conquered by Alexander the Great, since he had access to the notes and the observations made by adequately trained observers that were involved in the campaigns.

[1] Laboratory of Microbiology and Biotechnology of Foods, Department of Food Science and Human Nutrition, Agricultural University of Athens, Athens, Greece,
[2] Laboratory of Food Quality Control and Hygiene, Department of Food Science and Human Nutrition, Agricultural University of Athens, Athens, Greece.
* Corresponding author: m.bonatsou@aua.gr

Carl Linnaeus is considered as the father of modern taxonomy and his works 'Species Plantarum', published in 1753, and 'Systema Naturae' 10th edition, published in 1758, are thought to be the starting points of modern botanical and zoological nomenclature, respectively. Linnaeus proposed that the taxonomy of all living creatures be divided into three kingdoms, which in turn were further subdivided into classes, orders, genera and species. In addition, he used the binomial nomenclature, improving and simplifying classification.

Regarding microbiology, in 1872 Ferdinand Cohn was the first to apply the Linnaean principles to microbial taxonomy; he used morphological characteristics, growth requirements and pathogenic potential as the taxonomic criteria. These were constantly enriched during the 20th century and gradually growth potential ina wide variety of conditions and substrates, rather than growth requirements,was used to assign a taxonomic affiliation. It was not until the late 1960s that numerical taxonomy, in combination with several chemotaxonomic approaches, such as DNA-DNA hybridization, DNA base composition, cell-wall peptidoglycan types, cellular fatty acid composition, etc., emerged and, in some cases, significantly improved taxonomy. Advances in the field of molecular biology, along with technological improvements, allowed for the generation of data from different levels of a microbial entity, i.e., DNA, transcriptome, proteome, cellular composition and physiology, leading to the adoption of a polyphasic approach for resolving inconsistencies and contradictions that occur during classification. This approach is currently dominating taxonomy and 16S rRNA gene sequencing is considered as the backbone of modern bacterial taxonomy. As a result, Rosselló-Móra and Amann (2015) proposed that a species is '*a category that circumscribes monophyletic, and genomic and phenotypically coherent populations of individuals that can be clearly discriminated from other such entities by means of standardized parameters*' providing a contemporary insight into the species definition debate which has been ongoing since the dawn of classification (Vandamme et al. 1996, Ludwig and Klenk 2005, Schleifer 2009, Sestausa and Fournier 2013). Very often, the discrimination between species, and especially between closely related ones, may not be achieved by 16S-rRNA gene sequencing, and the use of additional markers is required. Therefore, approaches such as multilocus sequence analysis (MLSA) and average nucleotide identity (ANI) were developed. However, the boundaries between the species remain vague and depend upon arbitrary cut-off values, supporting the notion that speciation does not really occur in nature and that the microbiome forms a continuum, like a cloud, that may be denser at some places (Vandamme et al. 1996). After all, taxonomy is an attempt to organize living creatures, including microbiota, primarily for practical reasons.

2. Assessment of Evolutionary Relationships

Evolutionary assessment is based on the idea presented by Zuckerlandl and Pauling (1965). According to them, phylogenetic history may be inferred through the comparison of macromolecules. For that purpose, they classified the macromolecules of living cells according to their relevance to evolutionary history and concluded that this type of information may be retrieved by DNA, RNA and polypeptide sequences. Fox et al. (1977) were the first to apply 16S rRNA gene sequencing for both taxonomic and phylogenetic purposes.

Assessment of evolutionary relationships consists of two steps. Sequence alignment is the first critical step towards accurate phylogenetic depiction. The main challenge of sequencing alignment is recognizing the homologous sites, especially in genomic regions that are not conserved. Such a task requires the use of appropriate algorithms, as well as adequate computational strength. The latter in particular makes the handling of whole genome and metagenome data problematic. Development of alignment-free approaches, namely *k-mer*-based ones, provided an alternative to alignment-based ones. These approaches reduced the required computational strength but have, at least, questionable applicability in phylogenomics, since they suffer from the absence of positional information (Chan and Ragan 2013). In order to address this issue, several improvements have been proposed (Zhou et al. 2008, Song et al. 2014, Gao et al. 2017, Sievers et al. 2017, Zhang and Alekseyenko 2017).

Phylogenetic tree construction, on the basis of the results obtained from sequence alignments or alignment-free methods, and evaluation are the second critical step. A wide variety of algorithms, classified into distance-based and character-based ones, have been developed. The former, represented by Unweighted Pair Group Method with Arithmeticmean (UPGMA) and Neighbor-Joining (NJ), offer short calculation times, while the latter, represented by Maximum Likelihood method (ML), are more elaborated, since an evolution model may be included in the analysis (Horiike 2016, Bogusz and Whelan 2017). The accuracy of the constructed phylogenetic trees is evaluated by the bootstrapping method proposed by Falsenstein (1985). This approach provides a confidence assessment for each clade of a phylogenetic tree. An improvement of this approach was provided by Efron et al. (1996).

3. Molecular Methods for Microbial Taxonomy

Microbial taxonomy is the field of science concerning the systematic classification of microorganisms, providing insights into evolution, phenotype and relationships between microorganisms. Classification based on the phenotype could be a difficult and unreliable method, as microorganisms of related or same species may be diverse in morphological characteristics under different culture conditions (Roselló-Móra and Amann 2001, Staley 2006, Richter and Roselló-Móra 2009), while unrelated species may exhibit similar or even identical characteristics. Classification using the molecular approach may eliminate these difficulties and distinguish microorganisms by the identification and the definition of their similarities and differences (Alexander et al. 2015). Analysis of genotypic characteristics can be achieved by molecular techniques that target different sites of the genome to assess genus-, species- and strain-level relationships. Overall analysis of the DNA has shown a connection between the content of guanine and cytosine (% G+C) and the microorganisms of a related or distinct species. More particularly, the molar percentage seemed to be consistent for microorganisms belonging to the same genus. It has been proposed that taxonomy as a process includes three main stages, known as alpha, beta and gamma taxonomy, which include naming and identification (Zhi et al. 2012), assignment of species to a specific class according to their phenotypic and genotypic characteristics (Priest and Austin 1993) and classification based on intraspecific categories subspecies level, respectively (Zhi et al. 2012).

Analysis of DNA sequence is called fingerprinting or genotyping and reduces the complexity of DNA into simple patterns, which may be representative for a certain taxonomic level (i.e., genus, species, strain), depending on the approach employed. Fingerprinting includes the detection of variable loci of the DNA by electrophoresis of the DNA fragments after the hybridization of multilocus probes, known as molecular markers. DNA fingerprinting is an important tool for identification, gene mapping and the taxonomy.

3.1 DNA-DNA Hybridization

DNA-DNA hybridization (DDH) is a method based on the complementary property of the double stranded DNA (Alexander et al. 2015). Denaturated DNA of two microorganisms is mixed to form hybrid double-stranded DNA. The degree of the relevance of the two genomes is mainly based on the degree of hybrid reassociation or the thermal stability of the hybrids. When the hybridized sequences show a high degree of similarity, duplex formation is extensive. DDH technique can effectively distinguish between related and unrelated species,and a threshold more than 70% of genome hybridization and 5°C or less ΔT_m for the stability of heteroduplex molecules hasbeen proposed by Wayne et al. (1987) as recommended standards for microorganisms of the same species, although several studies have used stricter cut-off values. DDH techniques for the taxonomy of prokaryotes have been used as the gold standard for the genomic similarity analyses of pairwise sets of strains (Rosselló-Móra et al. 2011). Although, DDH method has been shown to exhibit several advantages, there are also important disadvantages. This technique requires larger amounts of high quality input DNA in comparison to PCR-based methods and, thus,can be time-consuming and labor-intensive (Alexander et al. 2015). Moreover, because there are several approaches which may lead to different results, DDH technique is often considered inaccurate and incapable of producing a cumulative and comprehensive strain database for comparison, when compared to genome sequencing. Due to hybridization, it was revealed that some regions of the genome,especially those which encode the genes of ribosomal RNA, are highly conserved among related species and new methods based on sequencing of specific targets of the genome have been proposed as alternatives to DNA-DNA hybridization.

Other genetic targets used for bacterial classification include methods based on variations in restriction endonuclease sites, genetic loci, repetitive DNA sequences and gene sequences.

3.2 Use of Restriction Endonucleases

Restriction endonucleases are enzymes that are able to recognize a specific sequence motif of the DNA, known as a restriction site, consisting of four to seven nucleotides, and cleave the DNA within or adjacent to this site and not randomly, therefore restricting the invasion by foreign DNA (Stephenson et al. 2016). The frequency of the restriction areas depends on the base composition of the DNA and % G+C content. These enzymes detect the functional groups of each base without annealing the double helix of the DNA. More specifically, restriction enzymes make cuts either on opposite strands,producing complementary single stranded ends, or straight across the helix (Nadin-Davis 2007). According to this taxonomic method, a restriction map can be created for a small DNA segment (usually less 10 Mbp) through the enzymatic digestion of DNA into smaller fragments of different sizes (less than 40 kbp), depending on the number and the location of recognition sites that exist on the specific DNA sequence and are separated by electrophoresis (Saraswathy and Ramalingam 2011). A DNA segment may be subjected to single or double digest, which may take place sequentially or simultaneously. Two widespread molecular techniques based on this approach are Restriction Fragment Length Polymorphism (RFLP) and Pulsed-Field Gel Electrophoresis (PFGE).

3.2.1. Restriction fragment length polymorphism (RFLP). RFLP is a commonly used method, based on restriction enzymes, that is used for the classification of both bacteria and yeasts (Nadin-Davis et al. 2007). According to RFLP analysis, the genotypic profile that occurs after enzymatic digestion of specific genomic regions of unidentified samples is compared to the ones of reference strains; identical profiles are necessary for taxonomic assignment at species level. The discriminant ability of this method increases with the increase of the employed restriction enzymes.

3.2.2. Pulsed-Field Gel Electrophoresis (PFGE). Regarding large DNA fragments, Pulsed-Field Gel Electrophoresis (PFGE) is proposed for their separation in an agarose gel under an electrical field with a pulsed current. This classification technique is based on the electrophoretic mobility of each different sized molecule under the changes of the electric field during electrophoresis. Isolates with identical genotypic profiles may be considered as identical at species level, isolates with one to three or six band differences are considered as closely or possible related, respectively, while isolates with seven or more band differences are considered as distinct strains. This method has been extensively used, especially for the subtyping of pathogenic microorganisms, due to its high discriminative power, as well as for the assessment of the biodiversity of several microecosystems (Paramithiotis et al. 2008, Doulgeraki et al. 2010, 2011). The use of reference strains may enhance the reproducibility of this technique. Apart from the advantages of PFGE analysis, there are several disadvantages. Although, faster protocols have been described, PFGE is rather time-consuming as one run often takes 30 to 40 hours, while the quality of the result depends on various factors, such as the quality and the concentration of the DNA, the agarose concentration and the conditions of the electrophoresis. Moreover, this method is not always efficiently discriminant within the same species, which may exhibit significant variation among their band profiles. Finally, PFGE analysis is very expensive and requires experienced personnel for both the execution of the protocol and the interpretation of the result.

3.3 Repetitive DNA Amplification

DNA sequencing has revealed a great number of repetitive DNA-sequence motifs, which are considered to play a significant structural and functional role. The repetitive DNA sequences show specificity for a species and are widely distributed in a taxonomic family or a genus (Rao et al.2010).These sequences are highly polymorphic, constitute 30-90% of the total genome and are present in both prokaryotic and eukaryotic microorganisms. Because of the fact that they exhibit variability in the sequence and the copy number over evolution, they can be used as markers for taxonomic and phylogenetic studies (Smith and Flavell 1974). Repetitive DNA can be divided into two main classes, namely (a) tandem repetitive sequences (satellite DNA), which make up about 30% of the total genome and are localized in the centromere or the subtelomeric regions of the chromosomes, and (b) interspersed repeats (transposons and retrotransposons), found throughout the genome. These interspersed repeats have entered the genome via RNA during evolution. Comparison of the sequence, frequency and localization of these repetitive DNA sequences, as well as their role in transcription or gene expression, can provide significant information regarding their nature and role between related, less related and distinct species. These repeats have the ability to transfer horizontally between species and they play an important role in the variability, gene expression under stress conditions, insertion in the genome and the centromere definition (Biscotti et al. 2015). Repetitive DNA occurs randomly by unequal crossovers during cell division and it is suggested that the cell possesses a mechanism for sensing these repeats and targets them (Twyman 2009). Features of the organization of repetitive sequences in eukaryotic genomes, and their distribution in natural populations, reflect the evolutionary forces acting on DNA (Charlesworth et al. 1994). These sequences are transferred horizontally between species. Several studies have shown that interspersed repeats, derived predominantly from the past amplification of transposable elements, constitute this repetitive DNA. Analysis of the genome revealed that transposable elements, as well as tandem sequence,show rapid evolution in comparison with genes, which are more conservative and some of them even detectable across many different species (Waterston 2002).Moreover, these elements seem to catalyze a lineage-specific genome evolution (Marino-Ramirez et al. 2005, Mikkelsenet al. 2007, Wang et al. 2007, Bourque et al. 2008, Feschotte 2008) and throughout the spontaneous mutation determine the structure of the chromosomes and the gene sequence in the nucleus of the eukaryotic cell (Kidwell and Lisch 2001, Eichler and Sankoff 2003, Kazazian 2004, Feschotte and Pritham 2007, Belancio et al. 2008, Feschotte 2008). According to Wicker et al. (2007), two transposable elements belong to the same family, when the consensus sequences exhibit more than 80% similarity to each other over at least 80% of their length and at least 80 bp of sequence. However, only a few transposable elements are conserved among the species and, because of this, identification and classification of these entities is becoming complex and difficult (Feschotte et al. 2009).

3.4 Single Nucleotide Polymorphism (SNP) and Single Nucleotide Variation (SNV)

Single nucleotide polymorphisms (SNPs) are polymorphisms caused by point mutations that may occur in coding or noncoding regions of the genome. This change may refer to a replacement of one nucleotide with another and/or deletion or addition of a single nucleotide (Ben-Ari and Lavi 2012). The single amino acid replacements caused by SNPs can be adaptive and lead to novel biological species and prove the significance of the functional changes occurred by a single mutation (Perutz 1983). SNP's might be selected outside of exons or the direct cause of a genetic mutation, depending on their application (Butler 2012). SNP's are widely used in several DNA sequencing approaches for the identification of microorganisms (Jin et al. 2016). Their use as predominant markers in molecular classification is common because of two main reasons, namely, their abundance in the genome and the availability of high-throughput technologies for their genotyping, which have permitted the identification and characterization of relevant phenotypes. However,these technologies require quite expensive machinery and, due to the fact that there are usually only two alleles at a SNP locus, SNPs exhibit low level of polymorphism (Daiger et al. 2013).

3.5 Ribosomal Gene Sequencing

Ribosomes are complex structures that convert the information found in mRNA into the chain of the amino acids needed for the synthesis of the proteins in cells. Both prokaryotic and eukaryotic ribosomes contain rRNA molecules, as well as multiple ribosomal proteins. Prokaryotic ribosome is composed of three forms of RNA, more specifically, 16S, 23S and 5S rRNA and approximately 50 ribosomal proteins, 34 of which are conserved for all the taxonomic domains (Lecompte et al. 2002). Analysis of rRNA sequences has shown that prokaryotes can be divided into the lineages Archaea and Bacteria (Woese and Fox 1977).

3.5.1. 16S rRNA gene sequencing. The use of 16S rRNA gene sequencing is by far the most common approach for the determination of the phylogenetic relationships and the taxonomy of prokaryotic microorganisms for several reasons. 16S rRNA gene is a universal target for bacterial identification because of its presenceas an operon or a multigene family in almost all bacteria. A schematic for 16S rRNA is presented in Fig. 1. Moreover, 16S rRNA gene sequence is conserved and its function has not changed over time. The molecule is large enough with 1500bp for analysis,and contains 50 functional domains, which is an important factor as selected changes in one domain do not change significantly the sequences in other domains (Woese 1987). The presence of nine hypervariable regions (V1–V9, HVR's), flanked by conserved regions which are used for the amplification of the variable regions, make this ribosomal gene a suitable marker for the taxonomic identification (Janda and Abbott 2007, Chaudhary et al. 2015). A great number of 16S rRNA sequences are deposited in databases.However, classification using 16S rRNA gene sequences remains a challenge for taxonomists as the tools available for 16S sequencing are not sufficient for classification at species level and the results obtained are not reliable enough. (Gao et al. 2017). One of the main limitations of 16S rRNA gene sequencing is the presence of multiple nucleotide variations within the rRNA operons in a single genome and changes in relationships among taxa can be observed because of horizontal gene transfer. Moreover, because of the fact that 16S rRNA gene is rather conserved, it is not reliable for species level identification. Isolates are considered as belonging to related species when a 16S rRNA gene sequence identity higher than 99% is obtained, otherwise they belong to distinct species. Identity of 98.7% may indicate novel species. However, a study showed that strains which shared less than 50% DNA similarity by DNA-DNA hybridizationdisplayed16S rRNA gene sequences with a similarity of 99% to 100%.

16S rRNA gene

◫ Conserved areas   ◰ Variable regions

Fig. 1. A schematic for 16S rRNA (1500 base pair gene), located on the small ribosomal subunit (30S). Conserved and variable regions are indicated with different colors.

Sequence similarity can be further assessed by the construction of phylogenetic trees that can provide estimation for the evolutionary distance among the isolate sequence (Petti 2007). The available sequencing technology can provide 500bp of sequence data and, therefore, sequencing of the total genome requires several sequences which are rather expensive and time-consuming for common laboratories. However, the sequencing of the entire 16S rRNA gene has shown a region of high heterogeneity in the first 500 bases of the 5'end, indicating that sequencing of this region is sufficient enough for the identification and further sequencing of the entire gene is unnecessary (Tang et al. 1998). The full gene and 500 base (5' end) gene identity agree at genus level and a percentage of 93.1% of the species assignment was common for the two approaches (Tang et al. 1998). Sequencing of 16S rRNA gene depends on the reaction conditions, namely reagents and equipment;there are, however, specific procedures. Sequencing from a single-culture leads to clear sequence without overlapping peaks on the electrophoregram. This messy sequence profile could be the result of a mixed culture or even a single culture when the isolate contains several copies of the ribosomal gene and each copy has a different sequence. More specifically, a bacterial genome may contain up to 15 copies of the rRNA operon, and each of these copies usually exhibits sequence diversity less than 1.5%, but sequence heterogeneity close to 6% has also been observed (Yarza et al. 2008, 2010). Gene-based sequencing is more reliable and accurate than phenotypic identification or DNA-DNA hybridization but requires expensive equipment and guidelines for the correct deposition of the obtained sequences to the common databases for comparisons. Therefore, to avoid misinterpretation of data, taxonomists propose a multifactor approach for bacterial identification rather than 16S rRNA gene sequencing alone. 16S rRNA sequencing requires specific reagents and suitable instrument for amplification and sequencing, as well as the development of software for editing of sequences and databases with known sequences of strains for comparisons. Complete unambiguous nucleotide sequences and the application of the correct genus and species name according to specific guidelines can minimize mistakes during deposition and comparisons in the database.

3.5.2. Internal Transcribed Spacer Regions (SSU) and 26S rRNA gene sequence (LSU). For the identification and classification of yeasts and fungi, the gene targets used are the variable ITS1 and ITS2 regions, located between the conserved areas of 18S, 5.8S and 28S rRNA genes in the small subunit of the ribosomes (Fig. 2). However, because of the limitations of the ITS regions for yeast identification, the 600-nucleotide variable region (D1/D2) at the 5' end of 26S rRNA gene in the large subunit of eukaryotic ribosomes is commonly used (Petti 2007, Kurtzman and Robnett 1997). 26S rRNA gene is composed of 7 domains with conserved areas. Unpaired regions exhibit sequence homology, while paired regions contain base changes, most of which are present in domain II (D2), while domain VI (D5) is highly conserved (Veldman et al. 1981). Sequence of D2 region showed high resolution of heterothallic sibling species for different genera (Peterson and Kurtzman 1991). However, another study showed that strains with complementarity less than 40% nuclear DNA could not be indistinguishable by the differences in the sequences of the D1/D2 region (Kurtzman and Robnett 1997). Disadvantages during yeast and fungi classification with ribosomal gene sequences are common with 16S rRNA gene sequence for bacteria.

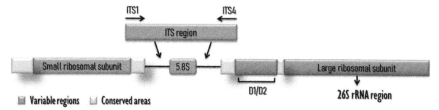

Fig. 2. ITS regions between the genes for small and large rDNA subunits. Conserved and variable regions are indicated with different colors. D1/D2 is located on 26S rRNA gene of the LSU.

3.5.3. Operational Taxonomic Units (OTUs). The standard pipeline for 16S and 26S rRNA gene analysis is clustering sequences within a percent sequence similarity threshold (typically 97% according to study showed that most isolated showed 97% similarity for 16S rRNA (Konstantinidis and Tiedge 2005)) into 'Operational Taxonomic Units' (OTUs). From the OTU clusters, a single sequence characteristic for each OTU is selected as a representative and is annotated with the method of 16S classification, then annotation is applied to the remaining sequences of this OTU (Wang et al. 2007, Chaudhary et al. 2015). The main advantage of OTU clustering is the reduction of the 16S reads from millions to thousands of OTUs, and therefore a faster analysis. However, using percent sequence similarity to define OTUs can lead to the overestimation of the evolutionary similarity between pairs of sequences (Rosenberg 2005, Ogden et al. 2006). A previous study showed that using OTU clustering can lead to a query sequence, which can be under-classified (when the representative sequence is classified at genus level and the query sequence is classified at the family level), over-classified (when the representative sequence is classified at family level and the query sequence is classified at the genus level), or even-conflicting (when the representative sequence and the query sequence are classified in different genera) (Nguyen et al. 2016). Moreover, the 97% threshold is ambiguous in terms of species classification, as different species may exhibit up to 99% 16S sequence similarity (Fox et al. 1992). Sequencing techniques based on 16S and 26S rRNA are known for the limited short read lengths, the errors occurred during sequencing, the differences in sequences obtained for the various regions of the ribosomal units and the difficulties for assessing OTUs (Quince et al. 2009, 2011, Youssef et al. 2009). Furthermore, prevalence of horizontal gene transfer and the difficulty in defining bacterial species are the main disadvantages of rRNA gene-based techniques and,therefore, alternative molecular methods are developed for higher resolution, such as multilocus sequence typing (MLST), using 5–10 housekeeping genes and shotgun metagenomic sequencing, which can identify all genes of a strain present in a sample (Poretsky et al. 2014).

3.6 Multilocus Sequence Analysis (MLSA)

Multilocus sequence analysis (MLSA), first introduced by Maiden et al. in 1998, is widely used for the resolution of the phylogenetic relationships between microorganisms of the same genus or family. MLSA uses the partial sequencing of specific genes which encode proteins with conserved functions (they are considered more stable with respect to genetic mutations) by generating and comparing phylogenetic trees of each individual gene. Nevertheless, despite the fact that MLSA has become an accepted and widely used method in prokaryotic taxonomy, no generally accepted recommendations have been devised to date for either the whole area of microbial taxonomy or for taxa-specific applications of individual MLSA schemes. As a molecular method, MLSA exhibits higher resolution than 16S rRNA gene-based approaches. However, because there are no generally accepted guidelines for the individual MLSA schemes that occur by each gene sequence and, thus, there are variations among the type and the number of the genes, comparing of the sequences is becoming difficult and leads to conflicting phylogenies which should be reviewed critically and re-evaluated. MLSA is based on the theory of electrophoresis of multiple enzyme loci of the genome (Maiden 2006). A change in a single nucleotide may lead to the assignment of a new unique allele of the respective gene. The selection of loci and the number of loci for this analysis depends on the taxonomic group and the clustering process that will be performed for the comparison of the phylogenetic trees, which show the relationships of the species and support the identification and the taxonomy of the microorganisms. Phylogenetic trees based on 16S rRNA gene include the overall phylogenetic diversity and the relationships between species and, although this method is frequently unable to discriminate the isolates at species or even genus level, it is common and widely used as the basic molecular method for taxonomic purposes. This drawback of this method is overcome by the use of several genetic markers, such as MLSA, which can efficiently resolute the microorganisms at genus and species levels, but it still needs important improvements for a generally accepted application. More particularly, not only studies over the genome of all known species within genera but also, user-friendly hardware and software should be established in order to allow sequencing analysis and comparisons by all researchers, even users with no statistical background. Moreover, the most significant prerequisite forMLSA optimization is the clear definition of the criteria that the analysis should require in order to avoid misinterpretation of the resulting sequence (Glaeser and Kämpfer, 2015). If draft genome is used to extend MLSA analysis and phenotype is included on the information of each isolate, MLSA can successfully replace DNA-DNA hybridization analysis (Kämpfer and Glaeser 2012, Cody et al. 2013).

3.7 *Metagenomic Sequencing*

The early metagenomic approaches were based on the complete sequencing of 16S rRNA gene using the Sanger method, an approach that was insufficient in covering the complexity of bacterial diversity (Chaudhary et al. 2015). Metagenomics can describe and classify microorganisms without the bias of PCR amplification of single genes (Sentausa and Fournier 2013), and provide a taxonomic assignment based on contig dynamics, or even the reconstruction of the complete 16S rRNA gene (Poretsky et al. 2014). Metagenomic sequence may identify more phyla and genera than amplicon sequence. However, the confidence of this approach remains a challenge. Whole genome shotgun sequencing (WGS) can generate whole prokaryotic genome sequences in a very short time (Sentausa and Fournier 2013), without targeting and amplifying a specific gene (Poretsky et al. 2014). WGS uses additional loci for higher resolution of intraspecies differences and provides an extent ribosomal sequence comparison from the rRNA operon to the ribosomal protein genes (Jolley et al. 2012, Yutin et al. 2012). Whole-genome sequence-based approaches provide the opportunity to develop the current classification techniques to a functional genomic taxonomy system (Reis-Filho 2009).

3.7.1. Next generation sequencing. Next generation sequencing or "massively parallel sequencing" can overcome the limitation of Sanger sequencing method, as it promotes millions of parallel sequencing reactions by attaching the DNA molecules to a surface or beads existing on the sequencing platforms to be sequenced (Reis-Filho 2009). NGS reads are produced from DNA fragments which are ligated to both their endsby specific adaptor oligos. The platforms used are creating libraries using random fragmentation of DNA molecule into millions of shorter read lengths fragments (35–250 bp), followed by ligation, during which DNA strands are joined. Many copies of a particular library fragment occur and their amplification is required so that the received signal is strong enough to be detected, counted and quantified accurately, in order to allow the identification of mutations in the genome. NGS allows sequencing from single DNA molecules and can also be applied to sequencing RNA. The quality of the sequences is not well understood yet. These short-reads sequences may be applied to *de novo* assembly and genome resequencing (Mardis et al. 2008). NGS requires only one or two instruments and little DNA input to produce the library. Next generation platforms may sequence uncultured and unpurified samples and explore microbial diversity by comparing the metagenomic sequences.

3.7.2. Pyrosequencing. Pyrosequencing is based on the chain of enzymatic reactions during DNA synthesis as a result of nucleotide incorporated by polymerase. More particularly, pyrophosphate is converted to adenosine triphosphate (ATP), which is in turn converted to light due to luciferin oxidation by luciferase (Ronaghi et al. 1996, Simner et al. 2015). A previous study evaluated the use of pyrosequencing for the analysis of bacilli isolates in comparison to phenotypic analysis and Sanger sequencing. The results showed that pyrosequencing identified 98% of the total isolates in comparison with 16S rRNA gene sequencing, which was not able to differentiate some isolates at species level (Bao et al. 2010). Generally, pyrosequencing method does not exhibit as high a discriminative power as Sanger sequencing but it is less expensive and faster than chain termination sequencing because ig does not require electrophoresis or any other fragment separation procedure. The main advantage of this taxonomic approach is the fact that it enables a great number of sequences to be obtained in a single run, throughout massively parallel sequencing (El-Bondkly et al. 2014).

Conclusions and Future Perspectives

Despite the depth of the novel sequence methods, such as NGS, WGS or MLSA and their discrimination power, validation with traditional sequencing methods, namely 16S and 26S rRNA gene sequence, is required for the identification and classification of the microbial diversity. Classification of microbial species is a difficult and very demanding process which requires the application of several genetic markers to be sequenced and compared in order to fully understand the phylogenetic relationships and the evolution of the species. Specific guidelines for the deposition of the sequences are required for the correct and unambiguous characterization and classification of the species. Next generation sequencing can allow the massive parallel sequencing and lead to millions of short-read sequences, while MLSA allows the complete sequence of the ribosomal genes by the use of different genetic markers for the various regions of the small and the large subunits of the ribosomes.

References

Alexander, D.C., P.N. Levett and C.Y. Turenne. 2015. Molecular taxonomy. pp. 369–379. *In*: Y.W. Tang, M. Sussman, D. Liu, I. Poxton and J. Schartzman (eds.). Molecular Medical Microbiology (2nd Edition). Academic Press, Amsterdam.

Bao, J.R., R.N. Master, D.A. Schwab and R.B. Clark. 2010. Identification of acid-fast bacilli using pyrosequencing analysis. Diagn. Microbiol. Infect. 67(3): 234–238.

Belancio, V.P., D.J. Hedges and P. Deininger 2008. Mammalian non-LTR retrotransposons: For better or worse, in sickness and in health. Genome Res. 18: 343–358.

Ben-Ari, G. and U. Lavi. 2012. Marker-assisted selection in plant breeding. Plant Biotechnology and Agriculture. Prospects for the 21st Century. Academic Press. 11: 163–184.

Biscotti, M.A., E. Olmo and J.S. Heslop-Harrison. 2015. Repetitive DNA in eukaryotic genomes. Chromosome Res. 3: 415–420.

Bogusz, M. and S. Whelan. 2017. Phylogenetic tree estimation with and without alignment: New distance methods and benchmarking. Syst. Biol. 66(2): 218–231.

Bourque, G. 2008. Evolution of the mammalian transcription factor binding repertoire via transposable elements. Genome Res. 11: 1752–1762.

Butler, M. 2012. Advanced topics in forensic DNA typing. Methodology 12: 347–369.

Chan, C.X. and M.A. Ragan. 2013. Next-generation phylogenomics. Biol. Direct. 8:3

Charlesworth, B., P. Sniegowski and W. Stephan. 1994. The evolutionary dynamics of repetitive DNA in eukaryotes. Nature. 371: 215–220.

Chaudhary, N., A.K. Sharma, P. Agarwal, A. Gupta and V.K. Sharma. 2015. 16S classifier: A tool for fast and accurate taxonomic classification of 16S rRNA hypervariable regions in metagenomic datasets. PLoS One. 10(2): e0116106.

Cody, A.J., N.D. McCarthy, M. Jansen van Rensburg, T. Isinkaye, S.D. Bentley, J. Parkhill, K.E. Dingle, I.C. Bowler, K.A. Jolley and M.C Maiden. 2013. Real-time genomic epidemiological evaluation of human *Campylobacter* isolates by use of whole-genome multilocus sequence typing. J. Clin. Microbiol. 51(8): 2526–2534.

Daiger, S.P., L.S. Sullivan and S.J. Bowne. 2013. Genetic mechanisms of retinal disease. Retina (5th edition). 31: 624–634.

Doulgeraki, A.I., S. Paramithiotis, D.M. Kagkli and G.J.E. Nychas. 2010. Lactic acid bacteria population dynamics during minced beef storage under aerobic or modified atmosphere packaging conditions. Food Microbiol. 27: 1028–1034.

Doulgeraki, A.I., S. Paramithiotis and G.J.E. Nychas. 2011. Characterization of the Enterobacteriaceae community that developed during storage of minced beef under aerobic or modified atmosphere packaging conditions. Int. J. Food Microbiol. 145: 77–83.

Efron, B., E. Halloran and S. Holmes. 1996. Bootstrap confidence levels for phylogenetic trees. Proc. Natl. Acad. Sci. USA 93: 13429–13434.

Eichler, E.E. and D. Sankoff. 2003. Structural dynamics of eukaryotic chromosome evolution. Science 301: 793–797.

El-Bondkly, A.M. 2014. Sequence analysis of industrially important genes from Trichoderma. pp. 377–392. *In*: V.K. Gupta, M. Schmoll, A. Herrera-Estrella, R.S. Upadhyay, I. Druzhinina and M.G. Tuohy (eds.). Biotechnology and Biology of Trichoderma, Elsevier, Amsterdam.

Felsenstein, J. 1985. Confidence limits on phylogenies: An approach using the bootstrap. Evolution 39: 783–791.

Feschotte, C. 2008. Transposable elements and the evolution of regulatory networks. Nat. Rev. Genet. 9: 397–405.

Feschotte, C. and E.J. Pritham. 2007. DNA transposons and the evolution of eukaryotic genomes. Annu. Rev. Genet. 41: 331–368.

Feschotte, C., U. Keswani, R. Ranganathan, M.L. Guibotsy and D. Levine. 2009. Exploring repetitive DNA landscapes using REPCLASS, a tool that automates the classification of transposable elements in eukaryotic genomes. Genome Biol. Evol. 42(1): 205–220.

Fox, G.E., L.J. Magrum, W.E. Balch, R.S. Wolfe and C.R. Woese. 1977. Classification of methanogenic bacteria by 16S ribosomal RNA characterization. Proc. Natl. Acad. Sci. USA 74(10): 4537–41.

Fox, G.E., J.D. Wisotzkey and P.Jr. Jurtshuk. 1992. How close is close: 16S rRNA sequence identity may not be sufficient to guarantee species identity. Int. J. Syst. Bacteriol. 42: 166–170.

Gao, X., H. Lin, K. Revanna and Q. Dong. 2017. A Bayesian taxonomic classification method for 16S rRNA gene sequences with improved species-level accuracy. BMC Bioinformatics 18: 247.

Glaeser, S.P. and P. Kämpfer. 2015. Multilocus sequence analysis (MLSA) in prokaryotic taxonomy. Syst. Appl. Microbiol. 38(4): 237–245.

Horiike, T., R. Minai, D. Miyata, Y. Nakamura and Y. Tateno. 2016. Ortholog-Finder: A tool for constructing an ortholog data set. Genome Biol. Evol. 8(2): 446–457.

Janda, J.M. and S.L. Abbott. 2007. 16S rRNA gene sequencing for bacterial identification in the diagnostic laboratory: Pluses, perils, and pitfalls. J. Clin. Microbiol. 45(9): 2761–2764.

Jin, Y., S. Liu, Z. Yuan, Y. Yang, S. Tan and Z. Liu. 2016. Catfish genomic studies: Progress and perspectives. pp. 73–104. *In*: S. MacKenzie and S. Jentoft (eds). Genomics in Aquaculture, Academic Press, Amsterdam.

Jolley, K.A., C.M. Bliss, J.S. Bennett, H.B. Bratcher, C. Brehony, F.M. Colles, H. Wimalarathna, O.B. Harrison, S.K. Sheppard, A.J. Cody and M.C. Maiden. 2012. Ribosomal multilocus sequence typing: Universal characterization of bacteria from domain to strain. Microbiology 158(4): 1005–1015.

Kämpfer, P. and S.P. Glaeser. 2012. Prokaryotic taxonomy in the sequencing era--the polyphasic approach revisited. Environ. Microbiol. 14(2): 291–317.

Kazazian, H.H. 2004. Mobile elements: Drivers of genome evolution. Science 303: 1626–1632.

Kidwell, M.G and D.R. Lisch. 2001. Perspective: Transposable elements, parasitic DNA, and genome evolution. Evolution 55(1): 1–24.

Konstantinidis, K.T. and J.M. Tiedje. 2005. Genomic insights that advance the species definition for prokaryotes. Proc. Natl. Acad. Sci. USA 102(7): 2567–2572.

Kurtzman, C.P. and C.J. Robnett. 1997. Identification of clinically important ascomycetous yeasts based on nucleotide divergence in the 5' end of the large-subunit (26S) ribosomal DNA gene. J. Clin. Microbiol. 35(5): 1216–1223.

Lecompte, O., R. Ripp, J.C. Thierry, D. Moras and O. Poch. 2002. Comparative analysis of ribosomal proteins in complete genomes: an example of reductive evolution at the domain scale. Nucleic Acids Res. 30(24): 5382–5390.

Ludwig, W. and H.P. Klenk. 2005. Overview: A Phylogenetic Backbone and Taxonomic framework for procaryotic systematics. Bergey's Manual of Systematic Bacteriology. pp. 49–65.

Maiden, M.C., J.A. Bygraves, E. Feil, G. Morelli, E. Russell, R. Urwin, Q. Zhang, J. Zhou, K. Zurth, D.A. Caugant, I.M. Feavers, M. Achtman and B.G. Spratt. 1998. Multilocus sequence typing: A portable approach to the identification of clones within populations of pathogenic microorganisms. Proc. Natl. Acad. Sci. USA 95(6): 3140–3145.

Maiden, M.C. 2006. Multilocus sequence typing of bacteria. Annu. Rev. Microbiol. 60: 561–588.

Mardis, E.R. 2008. The impact of next-generation sequencing technology on genetics. Trends Genet. 24(3): 133–141.

Marino-Ramirez, L., K.C. Lewis, D. Landsman and I.K. Jordan. 2005. Transposable elements donate lineage-specific regulatory sequences to host genomes. Cytogenet. Genome Res. 110: 333–341.

Mikkelsen, T.S, M.J. Wakefield, B. Aken, C.T. Amemiya, J.L. Chang, S. Duke, M. Garber, A.J. Gentles, L. Goodstadt, A. Heger, J. Jurka, M. Kamal, E. Mauceli, S.M. Searle, T. Sharpe, M.L. Baker, M.A. Batzer, P.V. Benos, K. Belov, M. Clamp, A. Cook, J. Cuff, R. Das, L. Davidow, J.E. Deakin, M.J. Fazzari, J.L. Glass, M. Grabherr, J.M. Greally, W. Gu, T.A. Hore, G.A. Huttley, M. Kleber, R.L. Jirtle, E. Koina, J.T. Lee, S. Mahony, M.A. Marra, R.D. Miller, R.D. Nicholls, M. Oda, A.T. Papenfuss, Z.E. Parra, D.D. Pollock, D.A. Ray, J.E. Schein, T.P. Speed, K. Thompson, J.L. VandeBerg, C.M. Wade, J.A. Walker, P.D. Waters, C. Webber, J.R. Weidman, X. Xie and M.C. Zody. 2007. Genome of the marsupial Monodelphis domestica reveals innovation in non-coding sequences. Nature. 447(7141): 167–177.

Nadin-Davis, S.A. 2007. Molecular epidemiology. pp. 69–122. *In*: A.C. Jackson and W.H. Wunner (eds.). Rabies (2nd edition). Academic Press, Amsterdam.

Nguyen, N.P., T. Warnow, M. Pop and B. White. 2016. A perspective on 16S rRNA operational taxonomic unit using sequence similarity. NPJ Biofilms Microbiomes. 2: 16004.

Ogden, T.H. and M.S. Rosenberg. 2006. Multiple sequence alignment accuracy and phylogenetic inference. Syst. Biol. 55(2): 314–328.

Paramithiotis, S., D.M. Kagkli, V.A. Blana, G.J. Nychas and E.H. Drosinos. 2008. Identification and characterization of *Enterococcus* spp. in Greek spontaneous sausage fermentation. J. Food Prot. 71(6): 1244–1247.

Perutz, M.F. 1983. Species adaptation in a protein molecule. Mol. Biol. Evol. 1(1): 1–28.

Peterson, S.W. and C.P. Kurtzman. 1991. Ribosomal RNA sequence divergence among sibling species of yeasts. Syst. Appl. Microbiol. 14(2): 124–129.

Petti, C.A. 2007. Detection and Identification of microorganisms by gene amplification and sequencing. Clin. Infect. Dis. 44(8): 1108–1114.

Poretsky, R., R.M. Rodriguez-R, C. Luo, D. Tsementzi and K. Konstantinidis. 2014. Strengths and limitations of 16S rRNA gene amplicon sequencing in revealing temporal microbial community dynamics. PLoS One. 9(4): e93827.

Priest, F. and B. Austin. 1993. Modern bacterial taxonomy (2nd edition). Chapman & Hall, London.

Quince, C., A. Lanzen, R.J. Davenport and P.J. Turnbaugh. 2011. Removing noise from pyrosequenced amplicons. BMC Bioinformatics 12: 38.

Quince, C., A. Lanzén, T.P. Curtis, R.J. Davenport, N. Hall, I.M. Head, L.F. Read and W.T Sloan. 2009. Accurate determination of microbial diversity from 454 pyrosequencing data. Nat. Methods 6(9): 639–641.

Rao, S.R., S. Trivedi, D. Emmanuel, K. Merita and M. Hynniewta. 2010. DNA repetitive sequences-types, distribution and function: A review. J. Cell Mol. Biol. 7(2) & 8(1):1–11.

Reis-Filho, J.S. 2009. Next-generation sequencing. Breast Cancer Res. 11(3): S12.

Richter, M. and R. Rosselló-Móra. 2009. Shifting the genomic gold standard for the prokaryotic species definition. Proc. Natl. Acad. Sci. USA 106(45): 19126–19131.

Ronaghi, M., S. Karamohamed, B. Pettersson, M. Uhlén and P. Nyrén.1996. Real-time DNA sequencing using detection of pyrophosphate release. Anal. Biochem. 242(1): 84–89.

Rosenberg, M.S. 2005. Evolutionary distance estimation and fidelity of pair wise sequence alignment. BMC Bioinformatics 6: 102.

Rosselló-Móra, R. and R. Amann. 2001. The species concept for prokaryotes. FEMS Microbiol. Rev. 25(1): 39–67.

Rosselló-Móra, R., M. Urdiain and A. Lopéz-Lopéz. 2011. DNA-DNA hybridrization. Meth. Microbiol. 38: 325–347.

Rosselló-Móra, R. and R. Amann. 2015. Past and future species definitions for Bacteria and Archaea. Syst. Appl. Microbiol. 38(4): 209–216.

Saraswathy, N. and P. Ramalingam. 2011. Concepts and techniques in genomics and proteomics. Woodhead Publishing, Amsterdam.

Schleifer K.H. 2009. Classification of Bacteria and Archaea: Past, present and future. Syst. Appl. Microbiol. 32(8): 533–542.

Sentausa, E. and P.E. Fournier. 2013. Advantages and limitations of genomics in prokaryotic taxonomy. Clin. Microbiol. Inf. 19(9): 790–795.

Sievers, A., K. Bosiek, M. Bisch, C. Dreessen, J. Riedel, P. Froß, M. Hausmann and G. Hildenbrand. 2017. K-mer content, correlation, and position analysis of genome DNA sequences for the identification of function and evolutionary features. Genes 8: 122.

Simner, P.J., R. Khare and N.L. Wengenack. 2015. Rapidly Growing Mycobacteria. pp. 1679–1690. *In*: Y.W. Tang, M. Sussman, D. Liu, I. Poxton and J. Schartzman (eds.). Molecular Medical Microbiology (2nd Edition). Academic Press, Amsterdam.

Smith, D.B. and R.B. Flavell. 1974. The relatedness and evolution of repeated nucleotide sequences in the DNA of some *Graminae* species. Biochem. Genet. 12: 243–256.

Song Y., L. Li, Y. Ou, Z. Gao, E. Li, X. Li, W. Zhang, J. Wang, L. Xu, Y. Zhou, X. Ma, L. Liu, Z. Zhao, X. Huang, J. Fan, L. Dong, G. Chen, L. Ma, J. Yang, L. Chen, M. He, M. Li, X. Zhuang, K. Huang, K. Qiu, G. Yin, G. Guo, Q. Feng, P. Chen, Z. Wu, J. Wu, L. Ma, J. Zhao, L. Luo, M. Fu, B. Xu, B. Chen, Y. Li, T. Tong, M. Wang, Z. Liu, D. Lin, X. Zhang, H. Yang, J. Wang and Q. Zhan. 2014. Identification of genomic alterations in esophageal squamous cell cancer. Nature 509(7498): 91–95.

Staley, J.T. 2006. The bacterial species dilemma and the genomic–phylogenetic species concept Philos. Trans. R. Soc. B. Biol. Sci. 361(1475): 1899–1909.

Stephenson, F.H. 2016. Recombinant DNA. pp. 321–373. Calculations for Molecular Biology and Biotechnology (3rd Edition). Academic Press. Amsterdam.

Tang, Y.W., N.M. Ellis, M.K. Hopkins, D.H. Smith, D.E. Dodge and D.H. Persing.1998. Comparison of phenotypic and genotypic techniques for identification of unusual aerobic pathogenic Gram-negative bacilli. J. Clin. Microbiol. 36(12): 3674–3679.

Twyman, R.M. 2009. Single-Nucleotide Polymorphism (SNP) Analysis. pp. 871–875. *In*: L.R. Squire (eds.). Encyclopedia of Neuroscience. Academic Press, Amsterdam.

Vandamme P., B. Pot, M. Gillis, P. de Vos, K. Kersters and J. Swings. 1996. Polyphasic taxonomy, a consensus approach to bacterial systematics. Microbiol. Rev. 60(2): 407–438.

Veldman, G.M., J. Klootwijk, V.C. de Regt, R.J. Planta, C. Branlant, A. Krol and J.P. Ebel. 1981. The primary and secondary structure of yeast 26S rRNA. Nucleic Acids Res. 9(24): 6935–6952.

Wang, T., J. Zeng, C.B. Lowe, R.G. Sellers, S.R. Salama, M. Yang, S.M. Burgess, R.K. Brachmann and D. Haussler. 2007. Species-specific endogenous retroviruses shape the transcriptional network of the human tumor suppressor protein p53. Proc. Natl. Acad. Sci. USA 104 (47): 18613–18618.

Waterston, R.H. K. Lindblad-Toh, E. Birney, J. Rogers, J.F. Abril, P. Agarwal, R. Agarwala, R. Ainscough, M. Alexandersson, P. An, S.E. Antonarakis, J. Attwood, R. Baertsch, J. Bailey, K. Barlow, S. Beck, E. Berry, B. Birren, T. Bloom, P. Bork, M. Botcherby, N. Bray, M.R. Brent, D.G. Brown, S.D. Brown, C. Bult, J. Burton, J. Butler, R.D. Campbell, P. Carninci, S. Cawley, F. Chiaromonte, A.T. Chinwalla, D.M. Church, M. Clamp, C. Clee, F.S. Collins, L.L. Cook, R.R. Copley, A. Coulson, O. Couronne, J. Cuff, V. Curwen, T. Cutts, M. Daly, R. David, J. Davies, K.D. Delehaunty, J. Deri, E.T. Dermitzakis, C. Dewey, N.J. Dickens, M. Diekhans, S. Dodge, I. Dubchak, D.M. Dunn, S.R. Eddy, L. Elnitski, R.D. Emes, P. Eswara, E. Eyras, A. Felsenfeld, G.A. Fewell, P. Flicek, K. Foley, W.N. Frankel, L.A. Fulton, R.S. Fulton, T.S. Furey, D. Gage, R.A. Gibbs, G. Glusman, S. Gnerre, N. Goldman, L. Goodstadt, D. Grafham, T.A. Graves, E.D. Green, S. Gregory, R. Guigó, M. Guyer, R.C. Hardison, D. Haussler, Y. Hayashizaki, L.W. Hillier, A. Hinrichs, W. Hlavina, T. Holzer, F. Hsu, A. Hua, T. Hubbard, A. Hunt, I. Jackson, D.B. Jaffe, L.S. Johnson, M. Jones, T.A. Jones, A. Joy, M. Kamal, E.K. Karlsson, D. Karolchik, A. Kasprzyk, J. Kawai, E. Keibler, C. Kells, W.J. Kent, A. Kirby, D.L. Kolbe, I. Korf, R.S. Kucherlapati, E.J. Kulbokas, D. Kulp, T. Landers, J.P. Leger, S. Leonard, I. Letunic, R. Levine, J. Li, M. Li, C. Lloyd, S. Lucas, B. Ma, D.R. Maglott, E.R. Mardis, L. Matthews, E. Mauceli, J.H. Mayer, M. McCarthy, W.R. McCombie, S. McLaren, K. McLay, J.D. McPherson, J. Meldrim, B. Meredith, J.P. Mesirov, W. Miller, T.L. Miner, E. Mongin, K.T. Montgomery, M. Morgan, R. Mott, J.C. Mullikin, D.M. Muzny, W.E. Nash, J.O. Nelson, M.N. Nhan, R. Nicol, Z. Ning, C. Nusbaum, M.J. O'Connor, Y. Okazaki, K. Oliver, E. Overton-Larty, L. Pachter, G. Parra, K.H. Pepin, J. Peterson, P. Pevzner, R. Plumb, C.S. Pohl, A. Poliakov, T.C. Ponce, C.P. Ponting, S. Potter, M. Quail, A. Reymond, B.A. Roe, K.M. Roskin, E.M. Rubin, A.G. Rust, R. Santos, V. Sapojnikov, B. Schultz, J. Schultz, M.S. Schwartz, S. Schwartz, C. Scott, S. Seaman, S. Searle, T. Sharpe, A. Sheridan, R.Shownkeen, S. Sims, J.B. Singer, G. Slater, A. Smit, D.R. Smith, B. Spencer, A. Stabenau, N. Stange-Thomann, C. Sugnet, M. Suyama, G. Tesler, J. Thompson, D. Torrents, E. Trevaskis, J. Tromp, C. Ucla, A. Ureta-Vidal, J.P. Vinson, A.C. Von Niederhausern, C.M. Wade, M. Wall, R.J. Weber, R.B. Weiss, M.C. Wendl, A.P. West, K. Wetterstrand, R. Wheeler, S. Whelan, J. Wierzbowski, D. Willey, S. Williams, R.K. Wilson, E. Winter, K.C. Worley, D. Wyman, S. Yang, S.P. Yang, E.M. Zdobnov, M.C. Zody and E.S. Lander. 2002. Initial sequencing and comparative analysis of the mouse genome. Nature 420(6915): 520–562.

Wayne, L.G., D.J. Brenner, R.R. Colwell, P.A.D. Grimont, O. Kandler, L. Krichevsky, L. Moore, H., W.E.C. Moore, R.G.E. Murray, E. Stackebrandt, M.P. Starr and H.G Trüper. 1987. Report of the ad hoc committee on reconciliation of approaches to bacterial systematics. Int. J. Syst. Bacteriol. 37: 463–464.

Wicker, T., F. Sabot, A. Hua-Van, J.L. Bennetzen, P. Capy, B. Chalhoub, A. Flavell, P. Leroy, M. Morgante, O. Panaud, E. Paux, P. SanMiguel and A.H. Schulman. 2007. A unified classification system for eukaryotic transposable elements. Nat. Rev. Genet. 8(12): 973–982.

Woese, C.R. and G.E. Fox. 1977. Phylogenetic structure of the prokaryotic domain: The primary kingdoms. Proc. Natl. Acad. Sci. 74(11): 5088–5090.

Woese, C.R. 1987. Bacterial evolution. Microbial Rev. 51(2): 221–271.

Yarza, P., M. Richter and J. Peplies. 2008. The All-Species Living Tree project: A 16S rRNA-based phylogenetic tree of all sequenced type strains. Syst. Appl. Microbiol. 31(4): 241–250.

Yarza, P., W. Ludwig, J. Euzéby, R. Amann, K.H. Schleifer, F.O. Glöckner and R. Rosselló-Móra. 2010. Update of the All-Species Living Tree Project based on 16S and 23S rRNA sequence analyses. Syst. Appl. Microbiol. 33(6): 291–299.

Youssef, N., C.S. Sheik, L.R. Krumholz, F.Z. Najar, B.A. Roe and M.S. Elshahed. 2009. Comparison of species richness estimates obtained using nearly complete fragments and simulated pyrosequencing-generated fragments in 16S rRNA gene-based environmental surveys. Appl. Environ. Microbiol. 75(16): 5227–5236.

Yutin, N., P. Puigbò, E.V. Koonin and Y.I. Wolf. 2012. Phylogenomics of prokaryotic ribosomal proteins. PLoS One. 7(5): e36972

Zhang, Y. and A.V. Alekseyenk. 2017. Phylogenic inference using alignment free methods for applications in microbial community surveys using 16s rRNA gene. PLoS One 12(11): e0187940.

Zhi, X.Y., W. Zhao, W.J. Li and G.P. Zhao. 2012. Prokaryotic systematics in the genomics era. Anton. Leeuw. 101(1): 21–34.

Zhou, Z.J., F.M. Shi and Y. Huang. 2008. The complete mitogenome of the Chinese bush cricket, *Gampsocleis gratiosa* (Orthoptera: Tettigonioidea). J. Genet. Genomics 35: 341–348

Zuckerkandl, E. and L. Pauling. 1965. Molecules as documents of evolutionary history. J. Theor. Biol. 8(2): 357–366.

4

Tools and Techniques for Recovery, Detection, and Inactivation of Foodborne Viruses

Sushil Kumar Sahu

1. What are Food Borne Viruses

Food borne viruses are too small in size (nearly 25 to 30 nm) to be seen under a light microscope. A typical food virus consists of a nucleic acid molecule (DNA or RNA), a protein coat and, in some cases, an envelope of lipids. Food borne viruses are inert and incapable of performing any cellular activities on their own. Hence, they cannot multiply in food, water or the environment. Their obligate parasitic nature allows them to replicate and produce progenies only in a living host cell. These viruses usually have a low infectivity dose, i.e., as little as 10 to 100 particles can cause infection, resulting in a significant threat to public health. This, combined with the high numbers shed in vomiting and stool, leads to outbreaks in a short time. The following are the most important food borne viruses.

1.1 Human Norovirus

Norovirus (NoV) belongs to the family Caliciviridae. It was discovered in a stool sample by electron microscopy (Kapikian et al. 1972). They are 26 to 35 nm in size, non-enveloped and spherical in shape. NoV contains a positive-sense RNA genome of approximately 7.7 kilobases. Five genetic groups of NoV (Gr-I, Gr-II, Gr-III, Gr-IV and Gr-V) occur in nature. Among them, Gr-I and Gr-II contain all human NoV. Gr-V

Department of Pharmacology & Molecular Sciences, Johns Hopkins University, School of Medicine, Baltimore, MD 21205, USA.
E-mail: sahu.sushil@gmail.com

includes murine NoV, which is the laboratory substitute for human NoV. The rapid evolution of Gr-II NoV resulted in new strains (Siebenga et al. 2008). In 2006–2007, during the NoV outbreak in the United States, two new variants were recognized and named as Laurens and Minerva. NoV epidemics from 2001–2007 were caused by the global spread of GII.4 strains, which evolved under the pressure of population immunity (Siebenga et al. 2009). Human NoV enters into a host cell by receptor-mediated endocytosis. Replication of the virus occurs in the cytoplasm of host cells (de Graaf et al. 2016). These viruses have a low infectious dose, i.e., a minimum of 10 particles has the capacity to cause infection in humans (Caul 1996). In a study for quantifying NoV shedding in human stool by qPCR, 9.5×10^9 copies/g stool was reported (Atmar et al. 2008). NoV causes an inflammation of the stomach and intestines, known as gastroenteritis, and has an incubation period of 24 to 48 hours (Koopmans et al. 2002). Symptoms of NoV infections include diarrhea, nausea, vomiting and abdominal pain. NoV are the cause of more than 50% of all food borne disease outbreaks worldwide. They affect all age groups and are difficult to control, as evidenced by the frequent outbreaks on cruise ships (Patterson et al. 1997). After NoV infection, antibodies are deployed against viral antigens but do not last more than a year (Parrino et al. 1977). Hence, there is a lack of lifelong immunity against NoV, unlike the lifelong immunity against hepatitis A virus (Karst et al. 2003).

1.2 Hepatitis A Virus

Hepatitis A virus (HAV) belongs to the family Picornaviridae. It is 27 to 32 nm in size, non-enveloped and icosahedral in shape (Cristina and Costa-Mattioli 2007). It contains a positive-sense RNA genome of about 7.5 kilobases, which has a long open reading frame flanked by 5' and 3' non-translated regions. HAV enters into the host cell by receptor-mediated endocytosis. Replication of the virus takes place in the host cytoplasm. Codon bias occurs in HAV and is distinct from its host. It also has an internal ribosome entry site (Whetter et al. 1994). However, the region which code for the viral capsid is highly conserved and restricts antigenic variability (Aragones, Bosch, and Pinto 2008). A single serotype of HAV exists in nature (Cristina and Costa-Mattioli 2007) and lifelong immunity occurs after natural infection or by immunization (Lemon, Jansen, and Brown 1992). Vaccine for HAV is an inactivated preparation of a cell culture-adapted virus (Irving et al. 2012). The virus is usually transmitted from one person to another through the fecal-oral route (Brundage and Fitzpatrick 2006). Eating of shellfish cultivated in contaminated water is a major source of infection (Lees 2000). Spread by blood is very rare. The minimum infectious dose of the virus is 10 to 100 particles, as reported by the Center for Food Safety and Applied Nutrition (Hirneisen et al. 2009). The incubation period of virus infection ranges from 15 to 50 days. Symptoms of infection include fever, headache, fatigue, nausea and abdominal pain, which may lead to hepatitis after 2 to 3 weeks.

1.3 Other Food Borne Viruses

Hepatitis E virus (Yugo and Meng 2013) is recognized as a food borne virus. It possesses an RNA genome. Transmission of Hepatitis E virus occurs mostly by the fecal-oral route. The clinical signs and symptoms of the associated patients are similar to those of other viral hepatitis. Hepatitis E virus is self-limiting and severe hepatitis is rarely observed. This virus can be prevented by hand washing, proper sanitary conditions and heating foods appropriately.

Rotavirus (RoV) and adenovirus (AdV) are associated with many outbreaks of food borne illness, but less frequently than NoV and HAV. The genetic composition and organization of these two viruses differ from the other enteric viruses. RoV is composed of a double-stranded RNA and AdV is composed of a DNA genome. According to the Centers for Disease Control and Prevention (CDC), rotaviruses were the causative agents of nearly one-third of diarrhea-associated cases in children younger than 5 years of age (Hirneisen et al. 2009). AdV usually cause respiratory problems. However, serotypes 40 and 41 are associated with gastroenteritis as per CDC (Hirneisen et al. 2009).

Aichi virus (AiV) is another emerging food borne virus. It is a positive-sense single-stranded RNA virus. It was discovered in the stool specimen of a patient with oyster-associated gastroenteritis in Aichi, Japan (Yamashita et al. 1991). Oysters are the most common vehicle of AiV transmission (Yamashita et al. 2000). It has been identified in many parts of the world, including Pakistani children, Japanese travelers from Southeast Asia (Yamashita et al. 1995), patient samples from Germany and Brazil (Oh et al. 2006) as well as Finnish children (Kaikkonen et al. 2010). Two genotypes of AiV (A and B) were reported by using the methodology of reverse-transcriptase PCR and phylogenetic analysis (Yamashita et al. 2000). However, most of the gastroenteritis cases caused by Aichi viruses are self-limiting.

Tools and techniques to work with the above food borne viruses are shown in Fig. 1. The following major steps are included:

- Recovery of viral particles from food
- Qualitative and quantitative detection of viruses
- Inactivation of viruses

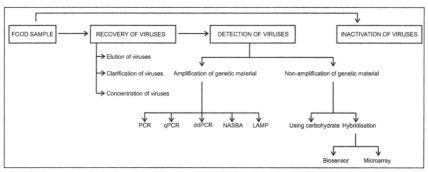

Fig. 1. Approaches including different steps for recovery, detection and inactivation of food borne viruses.

2. Virus Recovery from Koods

2.1 Types of Food Infected with Virus

The most common foods involved in the outbreaks of viral diseases are fresh produce (e.g., green vegetable salads, lettuce, cabbage, raspberries, strawberries, pomegranate seeds and raw frozen fruit mix) and shellfish, after being contaminated by virus-containing food and water or virus-infected food handlers. It is not easy to detect viruses in food matrices very quickly because they are usually found in low copy numbers and cannot replicate in host-free environments. Viruses, therefore, must be concentrated from large quantities of food sample. However, the complexity of food matrices makes it difficult to design a universal method of virus concentration that can be used to recover viruses from various food types, so development of efficient virus recovery methods is of great interest to the food safety community.

An ideal method for virus recovery from food samples should have the following criteria:

- The method should be universal so that small copies of viruses from large amounts of food can be recovered.
- The final extract should not be cytotoxic so that it can be used for cell cultures in order to reproduce more viruses.
- The final sample volumes should be small, concentrated with virus and free of PCR inhibitors so that they can be detected by molecular methods.

2.2 Food Sampling

A food sample obtained for detection of viruses has to be randomly selected and should be representative of the whole batch of food. There are no universal criteria for the amount of the food sample (e.g., fruit, vegetable) that should be tested but around 10 to 100 g of sample is usually required for this purpose.

2.3 Approach for Recovery of Viruses

Earlier techniques for detecting human enteric viruses in foods were based on inoculation of the food extracts into cell cultures that were susceptible to the virus of interest, e.g., HAV in Huh7-A-I cell line (Konduru and Kaplan 2006) and NoV culture in B cells (Jones et al. 2015). However, this is not easy and most of the time these viruses and other enteric viruses cannot grow in cell culture. Two methods employing porcine mucin-coated magnetic beads and ethidium monoazide dyes have been developed and show great potential for virus detection (Dancho et al. 2012, Moreno et al. 2015). Use of qPCR involving the amplification of conserved regions of the virus genome has become a popular way to detect viruses in foods (Salihah et al. 2016). The type of method used for virus recovery from foods depends on the food composition and characteristics. In this regard, the foods can be of the following categories:

- Foods composed of carbohydrate and water, mainly fruits and vegetables.
- Fat and protein-rich foods, such as ready-to-eat products.
- Food prepared from shellfish.

The approach of viral particle recovery involves the following steps:

- Elution of virus particles from the food matrix using an appropriate buffer.
- Clarification of the virus-containing eluate.
- Concentration of the recovered virus particles in a small final sample volume. A secondary concentration step can be added in order to obtain the virus particles in a more concentrated form.

2.3.1. Elution of viral particles from food matrices. This technique is referred to as 'acidic adsorption-elution technique' and involves the following major steps:

- Adsorption of positive-charge amino groups on the viral particles to the negative-charge carboxyl groups on a chromatographic column of activated carbon by electrostatic and hydrophobic interactions.
- Elution of viruses by adding a basic buffer (pH > 7) with different ionic strength.
- Collect different fraction of eluate as per the change in charge based alkaline pH environment.

This technique is used for the extraction of food borne viruses from a broad range of carbohydrate, fat and protein-based food materials, as well as from shellfish. However, the pH range in the basic buffer may change depending upon the type of food. For acidic fruits (e.g., strawberries), the elution buffers need be more basic than the elution buffers used for neutral foods (e.g., lettuce). This is because the acidic pH of the food can reduce the pH of the eluent solution, which may decrease the pH below neutrality with a reduction in the elution capacity. Usually an alkaline Tris-based buffer is employed. The presence of Tris in the elution buffer provides an appropriate medium when the eluate is to be used for PCR. On the other hand, Tris may decrease viral infectivity, which should be avoided if growth of viruses in a human packaging cell is planned. To overcome the viral infectivity loss caused by Tris-containing eluents, the eluents are often supplemented with $MgCl_2$, since enteric viruses are more resistant to inactivation in the presence of $MgCl_2$ (Dubois et al. 2002). To enhance the recovery of viral particles from foods, the elution buffers may be used in combination with some supplements, such as soya protein, beef extract, fetal bovine serum and peptone (Dubois et al. 2002, Fino and Kniel 2008, Sanchez et al. 2012). They reduce non-specific virus adsorption to the food matrix during the elution step. In addition, they facilitate flocculation of viruses on PEG molecules in the next step of concentration (Dubois et al. 2002, Stals et al. 2012). The contact between the sample and the elution buffers may not be enough for the viral particle to be released into the elution buffer. Some simple methods have been used for virus elution with mechanical assistance (e.g., pipetting, shaking, blending, stomaching, pulsifying, stirring and vortexing). This helps to decrease the time needed for elution and enhance its efficiency.

2.3.2. Clarification of the virus eluate. The virus elution step results in an alkaline eluate solution that contains viruses along with food impurities with a size larger than that of the virus particles. The retention of these impurities in the viral eluates causes hindering of the next concentration step by clogging the filter membranes used in the concentration step, decreases the efficiency of the viral DNA/RNA extraction step, and inhibits the PCR reaction due to the presence of PCR inhibitors.

Hence, it is important to clarify the virus eluate before starting the concentration. In general, the clarification step is performed under alkaline conditions in order to prevent the re-adsorption of viruses to the fruit or vegetable matter. The most commonly-used method for clarification of viral eluate is the use of a low-speed centrifugation step, which is enough for separating the fruit and vegetable matter in a pellet while viruses remain suspended in the supernatant. The centrifugation speed is usually 1000 rpm for 15 min at 4°C.

The second method for clarification of virus eluate is filtration through different kinds of filters, such as syringe filters of 0.2 or 0.45 μm porosity, glass wool, Whatman or cellulose triacetate filters and the filter compartment of certain plastic homogenization bags. It is always recommended to use both centrifugation and filtration methods for clarification.

Many clarification protocols perform a specific pretreatment of the filters or add some supplements to the viral eluate in order to increase the efficiency of the clarification step. Some relevant examples are given:

- Glass wool filters were pretreated with Eagle's basal maintenance medium and 2% fetal calf serum to prevent loss of virus by adsorption to the filter itself (Ward et al. 1982).
- After obtaining the alkaline viral eluate, the pH was adjusted to about 7.0 prior to filtration for preventing the adsorption of virus particles to the filter itself (Cheong et al. 2009).
- Cat-Floc™ (polydimethyldiallyl ammonium chloride, MW 500,000) is a cationic or polyelectrolyte flocculent, which is added as a supplement to the viral eluate to facilitate separation of the solid food during centrifugation by improving flocculation (Le Guyader et al. 2004).
- Bentonite is another flocculent used for enhancing the separation of the remaining food matter during centrifugation.
- Before centrifugation, Freon (1, 1, 2-trichloro-1, 2, 2-trifluoroethane) is added to the virus eluate as an organic solvent to remove PCR inhibitors (such as lipids). Freon is able to extract lipids and lipid bilayers without extracting proteinaceous (polar) material such as non-enveloped viruses (Stals et al. 2012).
- Vertrel® XF (1,1,1,2,3,4,4,5,5,5-decafluoropentane) is a newly developed, environmentally friendly Freon substitute (Love et al. 2008).
- Magnesium chloride is added to the elution buffer in order to minimize a reduction in viral infectivity during the elution and clarification processes (Dubois et al. 2002).
- Pectinase is added to remove residual pectin during clarification, preventing jelly formation during neutralization of the eluate. Without the addition of pectinase, jellies containing juices from soft fruits such as raspberries and frozen fruits can form in eluates (Dubois et al. 2002).
- A mixed organic solvent (chloroformbutanol; 1:1 v/v) is often added to the virus eluate in order to remove PCR inhibitors and cytotoxic substances (Love et al. 2008).

2.3.3. Concentration step. Concentrating the virus eluate is needed in order to improve virus recovery and remove PCR inhibitors. Methods for this purpose employ various techniques, as follows:

2.3.3.1. Organic flocculation. This method is based on the adsorption of eluted, positively charged virus particles on protein flocculate under acidic conditions. The pH of the virus eluate is adjusted to acidic values prior to adding the flocculent. The flocculate, with adsorbed virus, is separated by centrifugation, and the pellet containing the virus is eluted in a small volume of a suitable alkaline buffer. The alkalinity of the elution buffer is important in detaching the virus from the flocculate by changing the net charge on its surface to negative. NoV has been eluted from strawberry samples that were concentrated by the above technique using skimmed milk as an organic flocculent, followed by re-suspension of the pellet in PBS at pH 7.5 (Melgaco et al. 2016).

2.3.3.2. Polyethylene glycol precipitation. This method uses polyethylene glycol (PEG) as an aqueous precipitant for precipitating the eluted viruses. PEG is known to reduce the solubility of macromolecules. After separation of the precipitated virus by centrifugation, the virus pellet is re-suspended in a suitable buffer. In that way, precipitation of viruses at neutral pH and at high ionic concentrations without precipitation of other organic material is achieved. Therefore, it is important to adjust the pH of the virus eluate to about 7.0 before adding PEG. Different molecular weights of PEG are available. There was no significant difference in recovery of NoV from strawberries or raspberries when PEG molecules from 6000–20,000 Da were used. The PEG is usually added in final concentrations of 6–30% w/v of the virus eluate. Sodium chloride (0.3 Molar) is added in order to provide a high ionic strength needed for PEG precipitation. In some studies, higher concentrations of NaCl have been used, e.g., 1.2 to 1.5 Molar (Dubois et al. 2002).

2.3.3.3. Ultracentrifugation. In this procedure, very high centrifugal speed in the vacuum is used to sediment the viral particles from clarified virus eluates (Summa et al. 2012). After sedimentation of virus particles, the supernatant is decanted, and the pelleted virus is re-suspended in a small volume of suitable buffer. The centrifugal speed of 25,000 g for 2 hours at 4°C is sufficient to pellet most of the viruses from the supernatant. However, additional purification of virus eluates is required before being subjected to ultracentrifugation protocols because debris and other components originating from the food samples can sediment simultaneously with virus particles and interfere with their detection/isolation. Usually, purification by a conventional centrifugation (1000 rpm for 15 minutes) is followed by 0.22 to 0.45 μm filtration.

2.3.3.4. Ultrafiltration. This method concentrates viral particles by entrapping them from the eluate on the basis of their molecular weight, rather than by particle charge. The filter membranes have pore sizes of nearly 100 kDa, allowing only the passage of liquids and particles with less than 100 kDa molecular mass. Viruses have a molecular mass over this limit, so they are captured by the filters. When ultrafiltration is used for virus concentration, the eluate needs to undergo an efficient clarification and purification process to avoid clogging of the filter. The reported advantage of ultrafiltration procedures with the use of spin columns or microconcentrators is the ability to concentrate 1–80 mL of eluate to final volumes of 25–200 μL. An ultrafiltration-based technique was used to concentrate virus eluted from fresh strawberries, frozen raspberries, frozen blueberries and fresh raspberries, resulting in recoveries from 1.7 to 19.6%. It has been reported that virus recovery by ultrafiltration can be increased somewhat by treating the filters with bovine serum albumin or by sonication of the purified virus eluate (Butot et al. 2007, Jones et al. 2009).

2.3.3.5. Filter absorption-elution. In this technique, negatively charged filters are used to capture the positively charged viral particles from the eluate after adjusting the pH to about 3.5. The acidic pH is necessary for keeping the viral particles positively charged. After capturing the virus on the filter, a small volume of an alkaline elution buffer is used for eluting the virus by changing its charge. Various filters have been used, such as glass fiber, melanin-impregnated paper, epoxy pleated cartridge filters, Zeta plus 60 S filters and 0.45 μm negatively charged membrane filters. This technique requires an efficient clarification process for the eluate prior to filtration in order to avoid clogging of the filter (Ward et al. 1982).

2.3.3.6. Concentration by amorphous calcium phosphate. Amorphous calcium phosphate (ACP) particles are composed of hydroxyapatite precursors, spherically shaped and have multiple pores. The phosphate ions, calcium ions and hydroxide ions in ACP particles are uniformly distributed. The ACP particles interact with electrically charged substances due to the phenomenon of ion exchange or electrostatic attraction. Unlike adsorption on filters, the alkaline virus eluate does not need to be neutralized or adjusted to an acidic pH since the negatively charged virus particles in the alkaline eluate are replacing the negatively charged ions covering the ACP particles. ACP particles are added to the eluate and stirred for 1 h at room temperature and then collected by centrifugation. The pellet of ACP-virus particles is suspended in a small volume of a suitable buffer. This technique was used successfully for recovery of NoV from cabbage and lettuce, and resulted in 12–57% recovery (Shinohara et al. 2013).

2.3.3.7. Concentration by using magnetic beads coated with virus-specific ligands. This is a very specific method for concentration of certain viruses from the eluate. For the concentration of NoV, magnetic beads coated with either histo blood group antigens types A, B, H (2) and H (3) or type III porcine gastric mucin are used to bind NoV particles after elution from food samples. Subsequently, NoV particles are eluted from the magnetic beads using an appropriate buffer. Another type of magnetic bead covered with monoclonal (K3-2F2) antibodies against HAV has been used for their concentration. The concentration methods using magnetic beads coated with virus-specific ligands were used successfully for concentrating NoV and HAV from lettuce, green onions, strawberries and ham. GI and GII NoV genogroups were recovered from fruit and vegetable salad, blueberries and tomatoes with virus recoveries of 6–30%. Specific virus extraction from various food types and efficient removal of PCR inhibitors are the main advantages of this technique. However, questions could be raised about the long-term use of monoclonal antibodies due to the immunogenetic drift of these food-borne viruses (Summa et al. 2012, Stals et al. 2012, Pan et al. 2012).

3. Virus Detection

There are multiple methods of detection of food borne viruses that vary depending on the type of food matrices.

3.1 Non-Amplification Methods

3.1.1. Probe hybridization. In these assays, single-stranded RNA or DNA probes of variable length (usually 100–1000 bases long) and complementary to a viral genomic sequence are linked to a reporter (radioisotope, enzyme or chemiluminescent agent) and then hybridized with the ssDNA (Southern blotting) or RNA (Northern blotting) target immobilized on a membrane. Detection of a signal from the reporter after the hybridization reaction indicates the presence of the target nucleic acid. The following hybridization formats can be used: Solid-phase hybridization, liquid hybridization and *in situ* hybridization.

In solid-phase hybridization, the target nucleic acid is fixed to a nylon or nitrocellulose membrane and a solution containing the labeled probe is applied. After the hybridization reaction, the bound probe is detected by fluorescence, radioactivity or colorimetric development. In liquid-phase hybridization, both target and probe are in the solution at the time of hybridization. The probe signal is detected by measuring fluorescence change.

In situ hybridization uses a labeled complementary DNA, RNA or modified nucleic acid strand called probe. It localizes a targeted nucleic acid sequence in a cell or tissue sample.

Ligands that can be used to pull viruses out of a complex food matrix are:

• Biosensors
• Nucleic acid aptamers
• Histo blood group antigens

3.1.1.1 Biosensors. A typical biosensor consists of three components: A sensor platform functionalized with a bioprobe, a transduction platform that generates a measurable signal when the analyte is captured and an amplifier component which amplifies and processes the signal to give a quantitative estimate of the analyte. It can be directly applied for the detection of a pathogen in food samples generated by mincing and homogenization in the presence of detergents and/or proteolytic enzymes. The choice of a sample processing method depends on the type and complexity of the food materials. Although the use of biosensors for monitoring food and water samples has not been commercialized yet, several recent reports have shown the tremendous potential of this technique (Singh et al. 2013).

An electrochemical biosensor, using concanavalin A as the capture agent, was used for the detection of NoV in turbid food extracts (Hong et al. 2015). In this study, a nanostructure gold electrode, which had been treated with concanavalin A, was incubated with different concentrations of NoV GII.4 seeded in lettuce. The electrode was subsequently immersed in a solution of primary and secondary antibodies, followed by electrochemical measurement for NoV sensing. The system was able to detect norovirus in a concentration range of 102 to 106 copies/mL (Hong et al. 2015).

3.1.1.2 Nucleic acid aptamers. Aptamers are synthetic nucleic acids that fold into three-dimensional unique conformations. They are capable of binding a target with significant affinity and specificity. Nucleic acid aptamers are engineered through repeated rounds of *in vitro* selection, referred to as systematic evolution of ligands by exponential enrichment. Aptamers can bind to various molecular targets, such as: Nucleic acids, proteins in cells and tissues and have been employed successfully for detection of NoV (Escudero-Abarca et al. 2014).

3.1.1.3. Histo Blood Group Antigens. Interactions of carbohydrates with proteins or other carbohydrates play a key role in binding, entry and intracellular processes after an encounter with a pathogen. The involvement of carbohydrate moieties in NoV binding to the human gastrointestinal cell is known. Histo blood group antigens are complex carbohydrates linked to glycolipids or glycoproteins that are present as free antigens on various biological surfaces derived from plants and animals and serve as binding ligands to capture NoV. These carbohydrates have been used successfully to capture and detect NoV on lettuce, clams, mussels and oysters (Esseili et al. 2012).

3.1.1.4. Quantum Dots. The quantum dots (QDs) have unique optical properties that are advantageous for the development of novel chemical sensors and biosensors. QDs are fluorescent probes and have advantages over organic dyes in terms of brightness, resistance to photobleaching and detection of multicolor targets. These properties enable the use of QDs as optical labels for faster and easier molecular detection of biological targets. By using antibodies against a panel of different antigens, QDs have the capacity to simultaneously detect multiple viruses using the fluorescence intensity of the multicolored QDs, which are excited at a single wavelength and emit a different wavelength. The use of QDs for the qualitative detection of NoV from fresh lettuce sample has been reported (Lee et al. 2013).

3.1.1.5. Microarrays. Microarrays are based on the principle of hybridization between two complementary DNA strands. Typically, hundreds of samples are immobilized on a solid support, such as glass slides, silicon chips or nylon membranes. The size of the sample spot is less than 200 µM in diameter. The spots can be DNA, cDNA or oligonucleotides. To generate the labeled ssDNA template, viral RNA is first converted into cDNA which itself can be fluorescently labeled (e.g., Cy3 and Cy5) or can serve as a template to generate PCR products followed by post-PCR fluorescent labeling. The fluorescently labeled cDNA or PCR products are then denatured in order to obtain labeled ssDNA. Each molecule of the labeled ssDNA will bind to its complementary target sequence spotted on a solid support matrix and generate a signal. The total strength of the signal is directly correlated to the number of nucleic acid molecules bound to the probe. A laser excites each spot and the fluorescent emission is gathered through a photo-multiplier. This method has been used successfully to identify NoV, HAV, rotavirus and adenovirus (Vinje and Koopmans 2000, Kostrzynska and Bachand 2006, Ayodeji et al. 2009).

3.2 Target-Specific Amplification Methods

Most of the molecular methods for detecting food borne viruses are based on amplification of a partial region of the viral genome. Hence, polymerase chain reaction (PCR), reverse transcription PCR (RT-PCR), quantitative PCR (qPCR) and droplet PCR digital (ddPCR) methods have become the gold standard for detection of food borne viruses.

3.2.1. Polymerase chain reaction. Polymerase chain reaction (PCR) is a method used to amplify a specific DNA sequence generating millions of copies of that sequence (e.g., 2^{40} copies from 1 copy in 40 cycle of PCR). The reaction mixture for PCR contains: DNA polymerase (e.g., Taq polymerase), dNTPs (dATP, dTTP, dGTP and dCTP), target-specific oligonucleotide primers (forward and reverse primers) and an appropriate buffer for optimum activity and stability of the DNA polymerase.

There are three basic steps that are repeated through a variable number of cycles: Heat denaturation ($\sim$95°C), primer annealing (55–60°C) and primer extension ($\sim$72°C). During heat denaturation, double-stranded DNA will denature into two separate strands. The reaction mixtures are then cooled to a temperature that allows the primers to anneal to the target sequences of the separated DNA strands. During the primer extension step, the DNA polymerase forms a new strand by extending the bound primers with nucleotides, creating a complimentary copy of the target DNA sequence. When repeated, this cycle of denaturation, annealing and extension increases the number of target DNA sequences exponentially.

3.2.2. Reverse transcription PCR. Reverse transcription PCR (RT-PCR) is a modification of the PCR reaction that allows amplification of an RNA template. In the initial step, complementary DNA (cDNA) is synthesized, which is then amplified by normal PCR, using cDNA as the template. The cDNA synthesis step requires a DNA polymerase with reverse transcriptase activity, dNTPs (dATP, dTTP, dGTP and dCTP), an oligonucleotide primer and an appropriate buffer. The oligonucleotide primer should be template-specific, random hexamers or oligo-dT if the genomic region to be amplified is near the polyadenylated 5' end of the genome, such as NoV and HAV. In one-step RT-PCR, all reagents necessary for both cDNA synthesis and PCR amplification are added at the same time. In two-step RT-PCR, the cDNA is synthesized followed by PCR amplification. In nested and hemi-nested PCR, serial amplification of a target sequence was done using two different oligonucleotide primer pairs, of which the second set is internal to the first set. Nested PCR increases the specificity of the assay because both primer pairs must amplify the target sequence. The initial amplification is performed using an outer primer pair and 20 to 30 cycles of amplification. A second round of amplification is then performed using primers that anneal to a region internal to the initial two primers. In hemi-nested PCR, one of the primers used in the second round of amplification is the same as that used in the first round of amplification and the second primer anneals to a region on the opposite strand that is nested between the initial two primers (Haqqi et al. 1988). Multiplex PCR assays use two or more primer pairs to amplify different target sequences in a single tube (Chamberlain et al. 1988). This strategy allows for the evaluation of a sample for more than one virus at a time, or multiple genotypes of the test virus at the same time.

3.2.3. Quantitative PCR. In quantitative PCR (qPCR), the amplified DNA is detected as the reaction progresses in real time, compared to conventional PCR with its endpoint detection. Two common methods for the detection of products in qPCR are:

- Use of a non-specific fluorescent dye like SYBR green that intercalates between double-stranded DNA and emits fluorescence after exposure to a specific wavelength. A melting curve analysis is considered after the qPCR using SYBR green in order to determine the specificity of the product (Mackay et al. 2002).
- Use of sequence-specific DNA probes consisting of oligonucleotides labeled with a fluorescent reporter at the 5' end and a quencher at the 3' end which allow detection only after hybridization of the probe with its complementary sequence to quantify the target RNA or DNA. TaqMan probes are the most commonly-used fluorescently-labeled oligoprobes. They are hydrolysis probes and are designed to increase the specificity of quantitative PCR. When the probe is intact, the proximity of the reporter dye to the quencher dye suppresses the reporter fluorescence. Cleavage of the probe during the PCR reaction separates the reporter dye from the quencher dye, which allows detection of the fluorescence from reporter dye (Holland et al. 1991).

3.2.4. Droplet digital PCR. Droplet digital PCR (ddPCR) is a new approach for absolute quantification of nucleic acid with higher precision over qPCR. It involves the following steps: Droplet generation, PCR amplification, droplet reading and data analysis. During droplet generation, DNA or cDNA samples are partitioned into many individual droplets for parallel PCR reactions. This step is crucial as it uses a specific type of oil, ddPCR supermix and nanofluidic system to generate a sufficient number of droplets. The primers and probes used here are similar to that of qPCR reaction. Some droplets contain the target molecule (positive droplets) while others do not (negative droplets). In the PCR amplification step, an end point PCR is performed so that a single DNA or cDNA molecule is amplified to 2^{40} copies inside a positive droplet. In the droplet reading step, TaqMan chemistry with labeled probes is used to detect sequence-specific targets. During data analysis, positive droplets are fit into the poisson algorithm for absolute copy number quantification. Recently, Norovius has been detected in oyster (Polo et al. 2016) and contaminated water (Monteiro and Santos 2017) by this method.

3.3 Isothermal Amplification Methods

In contrast to PCR technology, isothermal amplification is carried out at a constant temperature, so it does not require a thermocycler.

3.3.1. Nucleic acid sequence-based amplification. Nucleic acid sequence-based amplification (NASBA) was developed by J. Compton in 1991 (Compton 1991). In this method, the reaction is performed at a single temperature, usually 41°C. This temperature is preferred for keeping the target genomic DNA in double-stranded form, so that it does not become a substrate for amplification. A NASBA reaction mixture contains: T7 RNA polymerase, reverse transcriptase from avian myeloblastosis virus, target specific primers flanking the sequence to be amplified and RNase H. The first primer (P1) has the sequence specific for the T7 RNA polymerase at its 5' end and is used to initiate the reverse-transcription reaction, catalyzed by a reverse-transcriptase. The RNA strand in the RNA–DNA hybrid, resulting from the RT reaction, is then degraded by RNase H. The remaining cDNA is accessible to the second primer (P2), which initiates the synthesis of the complementary strand. A third enzyme called T7 RNA Polymerase, docks the double stranded DNA on the sequence at the 5' end of P1 and produces many RNA copies of the sequence of interest. It includes repetition of the following cycle: First strand synthesis, hydrolysis of RNA, second strand synthesis and RNA transcription. RNA and double strand cDNA of appropriate size accumulate exponentially and can be separated and detected by agarose gel electrophoresis (Jean, D'Souza, and Jaykus 2004).

3.3.2. Molecular beacon in NASBA. Molecular beacons are DNA oligonucleotides with a hairpin loop sequence labeled with a fluorophore at the 5' end and a quencher at the 3' end. The sequence of the 3' end of the molecular beacon is complementary to that of the 5' fluorophore end and a hairpin stem is formed (Tyagi and Kramer 1996). The sequence of hairpin loop is complementary to the sequence of the target amplicon. If the beacon does not find a complimentary target sequence, it remains closed and does not produce fluorescence. When the loop sequence binds to the target, the hairpin stem opens up and the quencher becomes separated from the fluorophore.

The increase in light emitted can be detected by a fluorometer. Molecular beacons are specific to their targets. In a NASBA amplification reaction, molecular beacons form a stable hybrid with amplified target RNA. NASBA have been used successfully to detect HAV, rotavirus and NoV from contaminated lettuce, blueberry and sliced turkey (Jean et al. 2004, Moore et al. 2004, Jean et al. 2002).

3.3.3. Loop mediated isothermal amplification. Loop mediated isothermal amplification (LAMP) is a single tube isothermal amplification technique for the amplification of DNA at 59 to 65°C (Notomi et al. 2000). The amplicons contain many different sizes of stem loop DNAs, having numerous inverted repeats of the target sequence with several loops. LAMP products can be observed with the naked eye by employing SYBR green dye. Compared to PCR, LAMP is a low-cost molecular detection method which has the potential to be used as a simple point of care diagnostic test. Because LAMP is isothermal, no expensive thermocyclers are needed (Zhao et al. 2014).

4. Virus Inactivation

The majority of documented food borne viral outbreaks can be traced back to food that has been manually handled by an infected food handler, as demonstrated with ready-to-eat foods (Koopmans and Duizer 2004). Other outbreaks are due to environmental contamination of foods such as fresh produce and shellfish, which may become infected by virus-containing water or soil (Rzezutka and Cook 2004). Many traditional parameters for controlling bacterial levels in food, including pH, temperature and water activity are ineffective barriers against viral transmission to human hosts (Grove et al. 2006). Since enteric viruses must survive the enzymatic and extreme pH conditions of the gastrointestinal tract to infect a host, they tend to be resistant to a wide range of commonly-used food processing, preservation and storage treatments. Viruses can persist in foods and survive for days and weeks in most environments without loss of infectivity due to the presence of the capsid (Rzezutka and Cook 2004). The inactivation of food borne viruses processing should target inactivation of capsid proteins by conformational change and destruction of the nucleic acid inside the virus particle.

Some popular methods of viral inactivation include:

• Use of antimicrobial agents, such as chlorine, chlorine dioxide, etc.
• Ozone treatment
• Thermal processing
• Use of high pressure
• Ultra violet treatment
• Use of ionizing radiation
• Use of pulsed electric field

Maintaining a balance between inactivation of viruses and food properties is important. A single method alone cannot eliminate all enteric viruses from foods. The effect of virus inactivation technologies has been well studied in cell culture media, buffers and water, but still needs to be studied in various food matrices.

References

Aragones, L., A. Bosch and R.M. Pinto. 2008. Hepatitis A virus mutant spectra under the selective pressure of monoclonal antibodies: Codon usage constraints limit capsid variability. J. Virol. 82(4): 1688–700.

Atmar, R.L., A.R. Opekun, M.A. Gilger et al. 2008. Norwalk virus shedding after experimental human infection. Emerg. Infect. Dis. 14(10): 1553–7.

Ayodeji, M., M. Kulka, S.A. Jackson et al. 2009. A microarray based approach for the identification of common foodborne viruses. Open Virol. J. 3: 7–0.

Brundage, S.C. and A.N. Fitzpatrick. 2006. Hepatitis A. Am. Fam. Physician 73(12): 2162–8.

Butot, S., T. Putallaz and G. Sanchez. 2007. Procedure for rapid concentration and detection of enteric viruses from berries and vegetables. Appl. Environ. Microbiol. 73(1): 186–92.

Caul, E.O. 1996. Viral gastroenteritis: Small round structured viruses, caliciviruses and astroviruses. Part I. The clinical and diagnostic perspective. J. Clin. Pathol. 49(11): 874–80.

Chamberlain, J.S., R.A. Gibbs, J.E. Ranier, P.N. Nguyen and C.T. Caskey. 1988. Deletion screening of the Duchenne muscular dystrophy locus via multiplex DNA amplification. Nucleic Acids Res. 16(23): 11141–56.

Cheong, S., C. Lee, W.C. Choi, C.H. Lee and S.J. Kim. 2009. Concentration method for the detection of enteric viruses from large volumes of foods. J. Food Prot. 72(9): 2001–5.

Compton, J. 1991. Nucleic acid sequence-based amplification. Nature 350(6313): 91–2.

Cristina, J. and M. Costa-Mattioli. 2007. Genetic variability and molecular evolution of hepatitis A virus. Virus Res. 127(2): 151–7.

Dancho, B.A., H. Chen and D.H. Kingsley. 2012. Discrimination between infectious and non-infectious human norovirus using porcine gastric mucin. Int. J. Food Microbiol. 155(3): 222–6.

de Graaf, M., J. van Beek and M.P. Koopmans. 2016. Human norovirus transmission and evolution in a changing world. Nat. Rev. Microbiol. 14(7): 421–33.

Dubois, E., C. Agier, O. Traore et al. 2002. Modified concentration method for the detection of enteric viruses on fruits and vegetables by reverse transcriptase-polymerase chain reaction or cell culture. J. Food Prot. 65(12): 1962–9.

Escudero-Abarca, B.I., S.H. Suh, M.D. Moore, H.P. Dwivedi and L.A. Jaykus. 2014. Selection, characterization and application of nucleic acid aptamers for the capture and detection of human norovirus strains. PloS one 9(9): e106805.

Esseili, M.A., Q. Wang and L.J. Saif. 2012. Binding of human GII.4 norovirus virus-like particles to carbohydrates of romaine lettuce leaf cell wall materials. Appl. Environ. Microbiol. 78(3): 786–94.

Fino, V.R. and K.E. Kniel. 2008. Comparative recovery of foodborne viruses from fresh produce. Foodborne Pathog. Dis. 5(6): 819–25.

Grove, S.F., A. Lee, T. Lewis, C.M. Stewart, H. Chen and D.G. Hoover. 2006. Inactivation of foodborne viruses of significance by high pressure and other processes. J. Food Prot. 69(4): 957–68.

Haqqi, T.M., G. Sarkar, C.S. David and S.S. Sommer. 1988. Specific amplification with PCR of a refractory segment of genomic DNA. Nucleic Acids Res. 16(24): 11844.

Hirneisen, K.A., E.P. Black, J.L. Cascarino, V.R. Fino, D.G. Hoover and K.E. Kniel. 2009. Viral inactivation in foods: a review of traditional and novel food-processing technologies. Compr. Rev. Food Sci. Food Saf. 9(1): 3–20.

Holland, P.M., R.D. Abramson, R. Watson and D.H. Gelfand. 1991. Detection of specific polymerase chain reaction product by utilizing the 5'----3' exonuclease activity of *Thermus aquaticus* DNA polymerase. Proc. Natl. Acad. Sci. USA 88 16): 7276–80.

Hong, S.A., J. Kwon, D. Kim and S. Yang. 2015. A rapid, sensitive and selective electrochemical biosensor with concanavalin A for the preemptive detection of norovirus. Biosens. Bioelectron. 64: 338–44.

Irving, G.J., J. Holden, R. Yang and D. Pope. 2012. Hepatitis A immunisation in persons not previously exposed to hepatitis A. Cochrane Database Syst. Rev. (7): CD009051.

Jean, J., B. Blais, A. Darveau and I. Fliss. 2002. Simultaneous detection and identification of hepatitis A virus and rotavirus by multiplex nucleic acid sequence-based amplification (NASBA) and microtiter plate hybridization system. J. Virol. Methods 105(1): 123–32.

Jean, J., D. H. D'Souza and L. A. Jaykus. 2004. Multiplex nucleic acid sequence-based amplification for simultaneous detection of several enteric viruses in model ready-to-eat foods. Appl. Environ. Microbiol. 70(11): 6603–10.

Jones, M.K., K.R. Grau, V. Costantini et al. 2015. Human norovirus culture in B cells. Nat. Protoc. 10(12): 1939–47.

Jones, T.H., J. Brassard, M.W. Johns and M.J. Gagne. 2009. The effect of pre-treatment and sonication of centrifugal ultrafiltration devices on virus recovery. J. Virol. Methods 161(2): 199–204.

Kaikkonen, S., S. Rasanen, M. Ramet and T. Vesikari. 2010. Aichi virus infection in children with acute gastroenteritis in Finland. Epidemiol. Infect. 138(8): 1166–71.

Kapikian, A.Z., R.G. Wyatt, R. Dolin, T.S. Thornhill, A.R. Kalica and R.M. Chanock. 1972. Visualization by immune electron microscopy of a 27-nm particle associated with acute infectious nonbacterial gastroenteritis. J. Virol. 10(5): 1075–81.

Karst, S.M., C.E. Wobus, M. Lay, J. Davidson and H.W. 4th Virgin. 2003. STAT1-dependent innate immunity to a Norwalk-like virus. Science 299(5612): 1575–8.

Konduru, K. and G.G. Kaplan. 2006. Stable growth of wild-type hepatitis A virus in cell culture. J. Virol. 80(3): 1352–60.

Koopmans, M. and E. Duizer. 2004. Foodborne viruses: An emerging problem. Int. J. Food Microbiol. 90(1): 23–41.

Koopmans, M., C. H. von Bonsdorff, J. Vinje, D. de Medici and S. Monroe. 2002. Foodborne viruses. FEMS Microbiol. Rev. 26(2): 187–205.

Kostrzynska, M. and A. Bachand. 2006. Application of DNA microarray technology for detection, identification and characterization of food-borne pathogens. Can. J. Microbiol. 52(1): 1–8.

Le Guyader, F.S., C. Mittelholzer, L. Haugarreau et al. 2004. Detection of noroviruses in raspberries associated with a gastroenteritis outbreak. Int. J. Food Microbiol. 97(2): 179–86.

Lee, H.M., J. Kwon, J.S. Choi et al. 2013. Rapid detection of norovirus from fresh lettuce using immunomagnetic separation and a quantum dots assay. J. Food Prot. 76(4): 707–11.

Lees, D. 2000. Viruses and bivalve shellfish. Int. J. Food Microbiol. 59(1-2): 81–116.

Lemon, S.M., R.W. Jansen and E.A. Brown. 1992. Genetic, antigenic and biological differences between strains of hepatitis A virus. Vaccine 10 Suppl 1: S40-4.

Love, D.C., M.J. Casteel, J.S. Meschke and M.D. Sobsey. 2008. Methods for recovery of hepatitis A virus (HAV) and other viruses from processed foods and detection of HAV by nested RT-PCR and TaqMan RT-PCR. Int. J. Food Microbiol. 126(1-2): 221–6.

Mackay, I.M., K.E. Arden and A. Nitsche. 2002. Real-time PCR in virology. Nucleic Acids Res. 30(6): 1292–305.

Melgaco, F.G., M. Victoria, A.A. Correa et al. 2016. Virus recovering from strawberries: Evaluation of a skimmed milk organic flocculation method for assessment of microbiological contamination. Int. J. Food Microbiol. 217: 14–9.

Monteiro, S. and R. Santos. 2017. Nanofluidic digital PCR for the quantification of Norovirus for water quality assessment. PloS one 12(7): e0179985.

Moore, C., E.M. Clark, C.I. Gallimore, S.A. Corden, J.J. Gray and D. Westmoreland. 2004. Evaluation of a broadly reactive nucleic acid sequence based amplification assay for the detection of noroviruses in faecal material. J. Clin. Virol. 29(4): 290–6.

Moreno, L., R. Aznar and G. Sanchez. 2015. Application of viability PCR to discriminate the infectivity of hepatitis A virus in food samples. Int. J. Food Microbiol. 201: 1–6.

Notomi, T., H. Okayama, H. Masubuchi et al. 2000. Loop-mediated isothermal amplification of DNA. Nucleic Acids Res. 28(12): E63.

Oh, D.Y., P.A. Silva, B. Hauroeder, S. Diedrich, D.D. Cardoso and E. Schreier. 2006. Molecular characterization of the first Aichi viruses isolated in Europe and in South America. Arch. Virol. 151(6): 1199–206.

Pan, L., Q. Zhang, X. Li and P. Tian. 2012. Detection of human norovirus in cherry tomatoes, blueberries and vegetable salad by using a receptor-binding capture and magnetic sequestration (RBCMS) method. Food Microbiol. 30(2): 420–6.

Parrino, T.A., D.S. Schreiber, J.S. Trier, A.Z. Kapikian and N.R. Blacklow. 1977. Clinical immunity in acute gastroenteritis caused by Norwalk agent. N. Engl. J. Med. 297(2): 86–9.

Patterson, W., P. Haswell, P.T. Fryers and J. Green. 1997. Outbreak of small round structured virus gastroenteritis arose after kitchen assistant vomited. Commun. Dis. Rep. CDR Rev. 7(7): R101–3.

Polo, D., J. Schaeffer, N. Fournet et al. 2016. Digital PCR for quantifying norovirus in oysters implicated in outbreaks, France. Emerg. Infect. Dis. 22(12): 2189–2191.

Rzezutka, A. and N. Cook. 2004. Survival of human enteric viruses in the environment and food. FEMS Microbiol. Rev. 28(4): 441–53.

Salihah, N.T., M.M. Hossain, H. Lubis and M.U. Ahmed. 2016. Trends and advances in food analysis by real-time polymerase chain reaction. J. Food Sci. Technol. 53(5): 2196–209.

Sanchez, G., P. Elizaquivel and R. Aznar. 2012. A single method for recovery and concentration of enteric viruses and bacteria from fresh-cut vegetables. Int. J. Food Microbiol. 152(1-2): 9–13.

Shinohara, M., K. Uchida, S. Shimada et al. 2013. Application of a simple method using minute particles of amorphous calcium phosphate for recovery of norovirus from cabbage, lettuce, and ham. J. Virol. Methods 187(1): 153–8.

Siebenga, J.J., H. Vennema, D.P. Zheng, et al. 2009. Norovirus illness is a global problem: Emergence and spread of norovirus GII.4 variants, 2001–2007. J. Infect. Dis. 200(5): 802–12.

Siebenga, J., A. Kroneman, H. Vennema, E. Duizer and M. Koopmans. 2008. Food-borne viruses in Europe network report: The norovirus GII.4 2006b (for US named Minerva-like, for Japan Kobe034-like, for UK V6) variant now dominant in early seasonal surveillance. Euro surveillance: Bulletin Européen sur les maladies transmissibles = European Communicable Disease Bulletin 13(2).

Singh, A., S. Poshtiban and S. Evoy. 2013. Recent advances in bacteriophage based biosensors for food-borne pathogen detection. Sensors 13(2): 1763–86.

Stals, A., L. Baert, E. Van Coillie and M. Uyttendaele. 2012. Extraction of food-borne viruses from food samples: A review. Int. J. Food Microbiol. 153(1-2): 1–9.

Summa, M., C.H. von Bonsdorff and L. Maunula. 2012. Evaluation of four virus recovery methods for detecting noroviruses on fresh lettuce, sliced ham, and frozen raspberries. J. Virol. Methods 183(2): 154–60.

Tyagi, S. and F.R. Kramer. 1996. Molecular beacons: Probes that fluoresce upon hybridization. Nat. Biotechnol. 14 (3): 303–8.

Vinje, J. and M.P. Koopmans. 2000. Simultaneous detection and genotyping of "Norwalk-like viruses" by oligonucleotide array in a reverse line blot hybridization format. J. Clin. Microbiol. 38(7): 2595–601.

Ward, B.K., C.M. Chenoweth and L.G. Irving. 1982. Recovery of viruses from vegetable surfaces. Appl. Environ. Microbiol. 44(6): 1389–94.

Whetter, L.E., S.P. Day, O. Elroy-Stein, E.A. Brown and S.M. Lemon. 1994. Low efficiency of the 5' nontranslated region of hepatitis A virus RNA in directing cap-independent translation in permissive monkey kidney cells. J. Virol. 68(8): 5253–63.

Yamashita, T., S. Kobayashi, K. Sakae et al. 1991. Isolation of cytopathic small round viruses with BS-C-1 cells from patients with gastroenteritis. J. Infect. Dis. 164(5): 954–7.

Yamashita, T., K. Sakae, S. Kobayashi et al. 1995. Isolation of cytopathic small round virus (Aichi virus) from Pakistani children and Japanese travelers from Southeast Asia. Microbiol. Immunol. 39(6): 433–5.

Yamashita, T., M. Sugiyama, H. Tsuzuki, K. Sakae, Y. Suzuki and Y. Miyazaki. 2000. Application of a reverse transcription-PCR for identification and differentiation of Aichi virus, a new member of the Picornavirus family associated with gastroenteritis in humans. J. Clin. Microbiol. 38(8): 2955–61.

Yugo, D.M. and X.J. Meng. 2013. Hepatitis E virus: Foodborne, waterborne and zoonotic transmission. Int. J. Environ. Res. Public Health. 10(10): 4507–33.

Zhao, H.B., G.Y. Yin, G. P. Zhao et al. 2014. Development of Loop-Mediated Isothermal Amplification (LAMP) for Universal Detection of Enteroviruses. Indian J. Microbiol. 54(1): 80–6.

5

Bioinformatics in Food Microbiology

Marios Mataragas

1. Introduction

Nowadays, next-generation sequencing (NGS) has found several applications across the disciplines of life sciences and, therefore, has been widely integrated into food-based studies. The use of NGS in food microbiology has drastically increased, especially in the fields of food safety and food fermentations (van Hijum et al. 2013, Alkema et al. 2016), owing to the significant cost reduction of sequencing. The NGS applications encountered in food microbiology are usually referred to the investigation of either a single microorganism (genome, transcriptome, typing, comparative genomics, SNP identification, etc.), or a microbial community (taxonomic and functional profiles, comparative metagenomics and metatranscriptomics, etc.). For instance, whole genome sequencing (WGS) is a useful emerging tool for the control and study of microbiological hazards and foodborne illnesses. *"WGS is the biggest thing to happen to food microbiology since Pasteur showed us how to culture pathogens"* (Schlundt J., Executive Director and Founder of the Global Microbial Identifier – GMI). Study of the microbial ecology of food fermentation is another example of NGS application to gain insight and enhance the present knowledge regarding fermenting microbial ecosystems (Ferrocino and Cocolin 2017). To gain meaningful interpretations from the massive amount data produced by the different NGS platforms' specialized statistical tools is, however, required. The tools that are employed during the analysis of the NGS data fall within the discipline of bioinformatics.

A next-generation sequencing project is consisting of several steps (Fig. 1). The "Data analysis using bioinformatic tools" step, together with the first step of experimental design, constitute the most important phases. In the pre-NGS era, the generation of the sequencing data occupied a large part of the workload of an NGS

Hellenic Agricultural Organization "DEMETER", Institute of Technology of Agricultural Products, Department of Dairy Research, Ethnikis Antistaseos 3, 45221, Ioannina, Greece.
E-mail: mmatster@gmail.com

Sample collection and experimental design

Extraction and purification of genomic DNA (gDNA) and RNA. In the case of RNA, an additional step is included, the conversion to cDNA

Data analysis using bioinformatic tools

Taxonomic and functional profiles of a microbial community, genome and transcriptome assembly and annotation, comparative genomics, metagenomics and metatranscriptomics, etc.

Sequencing

Before sequencing of the DNA (DNA-seq) or RNA (RNA-seq) sample, library preparation is performed

Data management and Data reduction

The raw sequence reads are collected and preprocessed (quality control - QC)

Fig. 1. Graphical illustration of the basic workflow for the next-generation sequencing (NGS) procedure.

study: Sample collection and experimental design (5%), sequencing (70%), data reduction (5%), data management (5%) and downstream analysis (15%).Over the years, however, a shift has occurred since sequencing has gradually given way (from 70% to 5%) to sample collection,experimental design (from 5% to 30%), and data analysis (from 15% to 55%) (Sboner et al. 2011).

In this chapter, the available bioinformatics tools for the analysis of the genome and transcriptome of a single microorganism, as well as for the exploration of the taxonomic and functional profiles of microbial communities, are briefly introduced. To this end, this review does not present an exhaustive list of all analytical tools that exist today. Interested readers may find additional information on bioinformatics tools and pipelines in other studies (De Filippo et al. 2012, Scholz et al. 2012, Ladoukakis et al. 2014, Sharpton 2014, Dudhagara et al. 2015, Escobar-Zepeda et al. 2015, Oulas et al. 2015). Preceding the description of the bioinformatics software tools, a short report on and a comparison table between the NGS platforms are presented in the next section.

2. Platforms for Next Generation Sequencing

Currently, the platforms for NGS are available from different vendors, such as Illumina, Life Technologies/ThermoFisher Scientific, Qiagen, Pacific Biosciences and Oxford Nanopore Technologies. Roche was also one of the first companies to release an NGS platform (454 pyrosequencing platform), which belonged to the first-generation of the NGS instruments, but its production was discontinued in 2016.Other first-generation instruments,such as the Genome Analyzer and SOLiD, released by Illumina/Solexa and Applied Biosystems/ThermoFisher Scientific, respectively, appeared along with Roche's NGS platform. Today, Illumina and Life Technologies/ThermoFisher Scientific have already distributed their second-generation NGS platforms in an effort to overcome the limitations of the first-generation instruments. Based on the application needs, there are several different versions of the available instruments offered by each vendor (Table 1). Qiagen also made available to the market an NGS platform, known as GeneReader, which has the advantage of offering a complete NGS solution from the nucleic acid collection and extraction from samples, to DNA/RNA sequencing, data analysis and interpretation of the results.

The pipeline (wet-lab protocol) followed is similar among the different NGS platforms. The workflow begins with the DNA/RNA library preparation, continues with the amplification of the library, its sequencing and imaging, and ends with the NGS analysis (base calling and secondary analysis). Discrepancies in this general framework are found with the use of different chemistry solutions, PCR amplification techniques, detection methods and base-calling strategies. However, all NGS platforms present one common characteristic, which is the sequencing of small nucleic acid fragments (short-reads) (Table 1).With the advent of third/fourth-generation sequencing equipment (Pacific Biosciences and Oxford Nanopore Technologies), the way was paved for the production of long-reads fragments through the Single Molecule Real Time sequencing technology (SMRT sequencing) (Table 2). Tables 1 and 2 summarize and compare the technical specifications of the NGS platforms available today.

3. Bioinformatic Tools Applied to a Single Microorganism (Culture-Dependent Methods)

The application of multi-omics to a single microorganism requires its prior isolation from the food (culture-dependent). After isolation, there area variety of approaches for investigating different research questions: 16S gene amplicon sequencing, Whole Genome Sequencing (WGS) and transcriptome sequencing (Fig. 2). The 16S gene amplicon sequencing is a PCR-based (Polymerase Chain Reaction) molecular method which can be used for typing of the isolated strains like the other existing molecular typing methods, e.g., pulsed-field gel electrophoresis (PFGE), repetitive element sequence-based PCR (rep-PCR), restriction fragment length polymorphism PCR (RFLP-PCR), random amplified polymorphic DNA PCR (RAPD-PCR) and amplified fragment length polymorphism PCR (AFLP-PCR) (Olive and Bean 1999). The tools that can be used for the analysis of the 16S-based sequence data are the same asthose applied to a microbial community and are discussed in the next section. This is also the case for the analysis of the RNA-seq data. Two popular software tools for the analysis of the data obtained from RNA-seq studies are the Tuxedo (Trapnell et al. 2012) and Trinity (Haas et al. 2013) for genome-based and genome-free transcript reconstruction, respectively. In addition, these suites also include tools for differential gene expression (DGE) analysis.

One long-lasting issue in the food safety is the ability to rapidly identify the food source of the contamination. Different techniques, such as phage typing, serotyping, multilocus variable-number tandem repeat analysis (MLVA), multilocus sequence typing (MLST) and PFGE, are used to identify and characterize foodborne pathogens and to discriminate between different strains. So far, the technology of choice for this purpose was the PFGE; notwithstanding, PFGE frequently lacks the ability to precisely locate the source of an outbreak. Advances in molecular biology such as the advent

Table 1. Second-generation NGS platforms (short-reads) available today from different vendors, along with their technical specifications which were retrieved from the vendors' website.

Technical Specifications	Illumina							Life Technologies/ThermoFisher Scientific		
	iSeq 100	MiniSeq	MiSeq	NextSeq	HiSeq 2500	HiSeq 3000/4000	NovaSeq 6000	Ion PGM[a]	Ion S5/S5 XL	Ion Proton
Max output range	1.2 Gb	1.65–7.5 Gb	540 Mb–15 Gb	16.25–120 Gb	9 Gb–1 Tb	105–1500 Gb	167 Gb–6 Tb	30 Mb–1 Gb/60 Mb–2 Gb	300 Mb–15 Gb/600 Mb–8 Gb/500 Mb–4.5 Gb	15 Gb
Reads per run	4 M	7–25 M	1–25 M	130–400 M	300 M–4 B	2.1 M–5 B	1.6–20 B	400 K–5.5 M	2–80 M/2–20 M/3–12 M	60–80 M
Max read length[b]	2×150 bp	2×150 bp	2×300 bp	2×150 bp	2×250 bp	2×150 bp	2×150 bp	200 bp/400 bp	200 bp/400 bp/600 bp	200 bp
Total run time[c]	17.5 h	24 h	56 h	29 h	60 h	<1–3.5 days	44 h	2.3–4.4 h/3.7–7.3 h	2.5 h/4 h	2.5 h

[a] Ion Personal Genome Machine System

[b] The fragments can be read either from one end (single-end) (Life Technologies/ThermoFisher Scientific and Qiagen; the reads per run and the max read length for the Qiagen's GeneReader are 15 M and 1×150 bp, respectively) or from both ends (paired-end) (Illumina)

[c] It is referred to the max read length

Table 2. Third/Fourth-generation NGS platforms (long-reads) available today from different vendors, along with their technical specifications which were retrieved from the vendors' website.

Technical Specifications	Oxford Nanopore Technologies			Pacific Biosciences	
	MinION	GridION	PromethION	PacBioRS II	Sequel System
Max output range	< 10–20 Gb	< 50–100 Gb	< 233 Gb/< 11 Tb	500 Mb–1 Gb	5–8 Gb
Reads per run	≤ 4.4 M	≤ 4.4 M	≤ 26 M/≤ 1.25 B	55 K	365 K
Max read length	< 1 Mb	< 1 Mb	< 1 Mb	> 20 kb	> 20 kb
Total run time	1 min–48 h	1 min–48 h	1 min -48 h	30 min–6 h	30 min–10 h

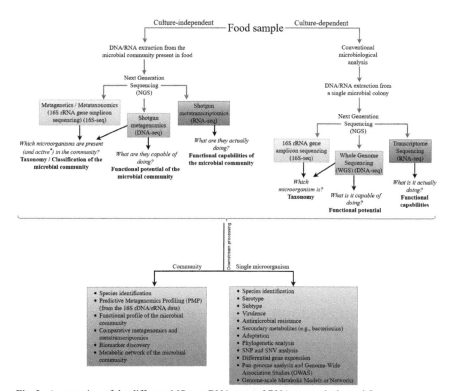

Fig. 2. An overview of the different 16S-seq, DNA-seq and RNA-seq analysis workflow.

of NGS platforms have paved the way for new exciting tools. WGS, i.e., the reading of the order of all the nucleotides one by one, is an emerging tool which holds a great promise in the way we will study, estimate and manage microbes, microbiological hazards and foodborne illnesses. The Executive Director and Founder of The Global Microbial Identifier (GMI), Schlundt J., said that *"WGS is the biggest thing to happen to food microbiology since Pasteur showed us how to culture pathogens"*. Even for WGS, the culturing of bacteria is a prerequisite, which may represent a major drawback because there are slow-growing, viable but nonculturable or low abundance bacteria. However, metagenomics (i.e., the study of the collective genomes of the members of a microbial community) could provide a solution to this matter as a culture-independent technique (Fig. 2) as it is discussed in the next section. Instead of sequencing the bacterial pathogens isolated from a food product, the DNA for all the microorganisms present in food is sequenced.

WGS has several applications,from detection and identification to antimicrobial testing and epidemiological typing (Köser et al. 2012). All these applications also require specialized software tools for the analysis of the sequences obtained. The analysis of the WGS data obtained from the various NGS machines available today includes several steps, such as the quality check of the raw sequence data, genome assembly (reference-based or *de novo*), ordering of the assembled contigs and genome annotation. Afterwards, the annotated genomes can be used for further analysis (Fig. 2). Each analysis step involves dedicated software tools. The review paper from Ekblom and Wolf (2014) is an excellent guide to WGS, assembly and annotation of raw sequences, providing information on tips and software tools associated with the preprocessing (quality check), assembly and annotation steps. For the assembly step in particular, there are several bioinformatics tools available which differ in their performance, so it is advisable to pre-consider the assembly approach that will be followed based on the WGS project needs (Earl et al. 2011, Narzisi and Mishra 2011, Bradnam et al. 2013, Magoc et al. 2013, Liao et al. 2015). Finally, for downstream analysis such as comparative genomics, a recent paper reviews the bioinformatics platforms and their recent developments (Yu et al. 2017).

4. Bioinformatic Tools Applied to a Microbial Community (Culture-Independent Methods)

NGS can be used in food microbiology for the determination of the presence of microorganisms in a microbial community. It is a culture-independent method since the food is directly sampled for nucleic acid extraction and sequencing. There are two main strategies currently employed in studying the microbial community diversity of a foodstuff by means of NGS (Fig. 2): The 16S rRNA gene amplicon sequencing (metagenetics – metataxonomics) and the shotgun metagenomic sequencing (Jovel et al. 2016). The first approach is the most widely and frequently used method when the assessment of the taxonomic diversity constitutes the primary interest of the study. Therefore, 16S sequencing mainly responds to the question *"Which microorganisms are present in the food?"*. In addition, the objective of the study determines the nucleic acid thatwill be used as a template, i.e., the genomic DNA (gDNA), if the goal is to explore which microbes are (alive) or were (non-alive) present in the food sample, or the RNA (cDNA), if the interest is focused only on the active (alive) microorganisms. Although amplicon sequencing can give more detail in the taxonomic composition of a microbial community, the biological functions associated with the microorganisms identified cannot be reliably inferred. However, there are some statistical tools which can be applied to 16S sequencing data in order to predict potential biological functions related to the identified taxa, as long as the abundance of new or unknown species in the community is relatively low (Hao et al. 2017), responding to the question *"what they are presumably doing?"*. PICRUSt (Langille et al. 2013), Tax4Fun (Aßhauer et al. 2015), Paprica (Bowman and Ducklow 2015), Piphillin (Iwai et al. 2016) and MMinte (Mendes-Soares et al. 2016) are bioinformatics tools which can predict the

potential function of the microbial community that inhabits a specific food product and drive future experiments for the elucidation of the actual functional diversity of the community. This approach is known as Predictive Metagenomics Profiling (PMP) (Wood 2016).

Although 16S rRNA sequencing has become a low-cost, robustand widely-applied tool for taxonomic classification of the metagenome samples, it presents some drawbacks, such as its low discrimination power at strain level, or even at species level if there are several closely related species within the microbial community. There are species in the *Enterobacteriaceae* family, for example, which have up to 99% sequence similarity of their entirely sequenced 16S rRNA gene (Větrovský and Baldrian 2013, Jovel et al. 2016). In this way, the total taxonomic diversity of the community in a food sample is underestimated because closely related species, even they are different, are grouped into one taxonomic unit (Větrovský and Baldrian 2013). The computational method known as oligotyping overcomes this limited taxonomic resolution (Eren et al. 2013, 2014). Another issue is the presence of a different number of 16S gene copies within the genome of a microorganism. This number can be highly variable between species (Coenye and Vandamme 2003). As a result, the abundance of species determined by the 16S rRNA gene amplicon sequencing is related to the different copies of the 16S rRNA gene and not to the different bacterial cells present in the food. To computationally accommodate for this problem, acorrection for the 16S gene copy number which will improve the accuracy of the microbial abundance within a community is required (Kembel et al. 2012, Angly et al. 2014). Lastly, the presence of Polymerase Chain Reaction (PCR) gene amplification during the 16S rRNA sequencing may introduce a bias which is mainly hingedon the primers matching to the different species present in the sample (Tremblay et al. 2015, Fouhy et al. 2016, Laursen et al. 2017). Nevertheless, careful selection of the primer pairs for PCR amplification and optimization of the PCR conditions during library preparation may reduce the bias to a minimum (Klindworth et al. 2013, Laursen et al. 2017). Recently, a new PCR-independent technique has been introducedin order to evaluate the microbial diversity in a metagenomic sample through the direct 16S rRNA sequencing without the inclusion of a primer-dependent step (Rosselli et al. 2016). Despite the bioinformatics tools which may overcome these difficulties in 16S rRNA sequencing, the latter remains a method with a limited resolution regarding taxonomic and functional profiling of a microbial community. On the other hand, metagenomic shotgun sequencing or shotgun metagenomics provide a higher resolution of the taxonomic composition, even at the strain level, and more specific functional profiles of the microbial communities (Jovel et al. 2016). However, if someone wants to focus on the actively-transcribing microorganisms of a community, then RNA sequencing should be used instead (Fig. 2).

4.1 Metagenetics/Metataxonomics—16S rRNA Gene Amplicon Sequencing

Metagenetics or Metataxonomics include the analysis of a single marker gene for microbial identification; usually the bacterial 16S ribosomal RNA gene. The gene is 1542 base pairs (bp) long and consists of nine (V1–V9) hypervariable regions with variable conservation degree. Highly conserved sequences between the hypervariable regions are used to design universal primers in order to amplify a part of the 16S rRNA gene from all microorganisms present in the food sample. While the information provided by the sequencing of these hypervariable regions is informative enough for bacterial classification, the resolution of the taxonomic classification is not the same across the nine regions. The regions V1–V3, V3–V4 or V4 alone have been widely used for differentiating microorganisms in foods (Mayo et al. 2014, Połka et al. 2015, Calasso et al. 2016, Dalmasso et al. 2016, Lee et al. 2016, Levante et al. 2017, Ramezani et al. 2017). Chakravorty et al. (2007) found that V2, V3 and V6 regions were the most useful in identifying the genus of all pathogenic bacterial species examined (110 different bacterial species), except for closely related *Enterobacteriaceae*. However, by combining these three hypervariable regions, the discrimination at species level of 97 out of the 110 pathogenic bacterial species was possible.

There are several tools available for the analysis of the 16S rRNA sequencing data, such as QIIME (Quantitative Insights Into Microbial Ecology), Mothur, SINA and RDP Classifier (Schloss et al. 2009, Wang et al. 2007, Caporaso et al. 2010, Pruesse et al. 2012). These tools preprocess the raw data (e.g., quality control) and classify the 16S sequences in specific Operational Taxonomic Units (OTUs) using databases which contain 16S rRNA sequences: NCBI, SILVA, GreenGenes, Ribosomal Database Project and EzTaxon (Maidak et al. 1996, Cole et al. 2005, 2007, 2014, DeSantis et al. 2006, Chun et al. 2007, Pruesse et al. 2007, NCBI 2017).

4.2 Metagenomics and Metatranscriptomics—Shotgun sequencing

Shotgun metagenomics is the alternative strategy of investigating microbial communities. While 16S rRNA sequencing targets a specific gene marker, shotgun metagenomics targets all DNA which is sheared in fragments and sequenced. Since all DNA is sequenced, shotgun metagenomics enables the investigation of both the taxonomy of the microorganisms present in the community and their biological function (Fig. 2) (Sharpton 2014). However, this strategy produces far more complex biological data, making its computational analysis very challenging (e.g., alignment of a read to the genome from which the sequence was initially originated, large quantity of data is required to sufficiently represent the diversity of the community, get rid of host DNA and contaminants genomes). Also, shotgun metagenomics is more expensive than amplicon sequencing. However, library preparation methodology and template quantity may again impact on metagenomic reconstruction of a microbial community (Bowers et al. 2015, Jones et al. 2015).

The analysis of the data originated from shotgun metagenomics includes different steps, such as preprocessing, assembly of sequences into contigs, contigs binning (grouping), classification and functional annotation. There are several bioinformatic tools available, which serve the purpose of each step of the analysis. In addition, there are analysis pipelines integrating the above steps in a single bioinformatic workflow. In the review papers of Mande et al. (2012), Scholz et al. (2012), Miller et al. (2013), Sharpton (2014), Dudhagara et al. (2015), Oulas et al. (2015), Aguiar-Pulido et al. (2016), Olson et al. (2017), Sedlar et al. 2017 and Vincent et al. (2017), bioinformatic tools and pipelines available for shotgun metagenomics and comparative metagenomics are discussed.

Metatranscriptomics shows which genes are expressed by the whole microbial community, providing information on the active functional profile of the community (Fig. 2). Therefore, the target is the total RNA (cDNA) and the produced fragments are sequenced. Different variation sources in RNA-seq studies may have an impact on the detection of differentially expressed genes (Li et al. 2014). The cost per sample for metatranscriptomics is as high as it is for metagenomics because of the highamount of datarequired in order to geta representative overview of the microbial community, including low abundance genomes and low expressing genes (Haas et al. 2012, Ni et al. 2013). Data analysis of the metatranscriptomics includes quality control of the raw reads, read alignment against a reference or *de novo*, and quantification of transcripts expression level. Conesa et al. (2016) and Aguiar-Pulido et al. (2016) provide useful information on the best practices for RNA-seq data analysis and software tools, respectively, which can be used for the preprocessing, processing and downstream processing of the RNA-seq data. Finally, Seyednasrollah et al. (2015) compared various software packages associated with differential expression of genes, which are frequently used during the downstream analysis of the RNA-seq data.

Future Perspectives

Today there are several analytical tools for investigating the diversity of microbial communities in food products or microorganisms' biology. These tools allow researchers to detect genes or proteins that are regulated and detect patterns that are associated with various phenotypic traits, such as antimicrobial resistance. However, such alterations do not necessarily mean that a specific phenotype is also expressed. Therefore, phenotypic experiments should complement molecular techniques, like genomic, transcriptomic or proteomic analysis. Phenomics is the discipline which quantitatively assesses the physiological responses of a microorganism to different environmental conditions. OmniLog® Phenotype Microarray System (Biolog Inc., Hayward, CA, USA) achieve this by measuring several cellular phenotypes at the same time. Such technology allows for the integration of phenotypic results with data obtained from NGS sequencing (DNA-seq and RNA-seq) (Fig. 3). DuctApe and PhenoLink are computational suites for the analysisand association of genomic/transcriptomic data with phenotype microarray data (Bayjanov et al. 2012, Galardini et al. 2014).

Another rapidly growing field is the utilization of biological data for the (re)construction of genome-scale metabolic models or networks of a microorganism or a microbial community, which is more challenging (Bordbar et al. 2014, Hanemaaijer et al. 2017). During the (re)construction of such models, different datasets, obtained from the various omics technologies, are combined in order to provide a quantitative tool, through which the metabolism of microbial cells can be investigated in more detail and even manipulated. Genomics, transcriptomics and proteomics, which constitute a measure of the metabolic capacity of the network, are integrated with fluxomics (it is the metabolic flux of each reaction in the network, providing a measure of the metabolic phenotype) and metabolomics (the distribution of the metabolite profiles) to obtain an overview of the microbial metabolic system (Patil and Nielsen 2005, Krömer et al. 2009, Henson and Hanly 2014). Genome-scale models represent an *in silico* illustration of the biochemical reactions occurring in microbial cells and the genes involved in these reactions, as obtained from the annotation of the genome sequence of a microorganism (Hamilton and Reed 2014). Currently, there are several software tools available to facilitate the process of the genome-scale (re)constructions of the

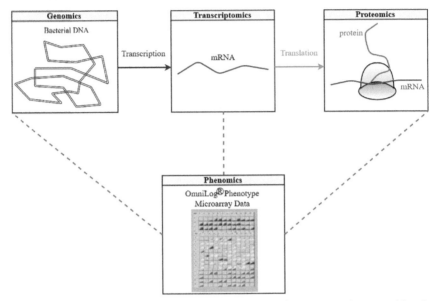

Fig. 3. Integration of phenomics with genomics, transcriptomics or proteomics to combine the genotype with thephenotype of a microorganism.

microorganisms' metabolic networks,as well as their analysis,such as the COBRA Toolbox (Schellenberger et al. 2011, Vlassis et al. 2014), Model SEED (Henry et al. 2010), RAVEN Toolbox (Argen et al. 2013), Pathway Tools (Karp et al. 2002, 2010), FAME (Boele et al. 2012), VANTED (Junker et al. 2006), FBA-SimVis (Grafahrend-Belau et al. 2009), CellNetAnalyzer (Klamt et al. 2007), OptFlux (Rocha et al. 2010) and GEMSiRV (Liao et al. 2012).

References

Aßhauer, K.P., B. Wemheuer, R. Daniel and P. Meinicke. 2015. Tax4Fun: Predicting functional profiles from metagenomic 16S rRNA data. Bioinformatics 31: 2882–2884.

Aguiar-Pulido, V., W. Huang, V. Suarez-Ulloa, T. Cickovski, K. Mathee and G. Narasimhan. 2016. Metagenomics, Metatranscriptomics, and Metabolomics approaches for microbiome analysis. Evol. Bioinform 12: 5–16.

Alkema, W., J. Boekhorst, M. Wels and S.A.F.T. van Hijum. 2016. Microbial bioinformatics for food safety and production. Brief. Bioinform. 17: 283–292.

Andrews, S. 2010. FastQC: A quality control tool for high throughput sequence data. Available online at: http://www.bioinformatics.babraham.ac.uk/projects/fastqc.

Angly, F.E., P.G. Dennis, A. Skarshewski, I. Vanwonterghem, P. Hugenholtz and G.W. Tyson. 2014. CopyRighter: A rapid tool for improving the accuracy of microbial community profiles through lineage-specific gene copy number correction. Microbiome 2: 11.

Argen, R., L. Liu, S. Shoaie, W. Vongsangnak, I. Nookaew and J. Nielsen. 2013. The RAVEN Toolbox and its use for generating a genome-scale metabolic model for *Penicillium chrysogenum*. PLoS Comput. Biol. 9: e1002980.

Bayjanov, J.R., D. Molenaar, V. Tzeneva, R.J. Siezen and S.A.F.T. van Hijum. 2012. PhenoLink – a web-tool for linking phenotype to ~omics data for bacteria: Application to gene-trait matching for *Lactobacillus plantarum* strains. BMC Genomics 13: 170.

Boele, J., B.G. Olivier and B. Teusink. 2012. FAME, the Flux Analysis and Modeling Environment. BMC Syst. Biol. 6: 8.

Bordbar, A., J.M. Monk, Z.A. King and B.Ø. Palsson. 2014. Constraint-based models predict metabolic and associated cellular functions. Nat. Rev. Genet. 15: 107–120.

Bowers, R.M., A. Clum, H. Tice, J. Lim, K. Singh, D. Ciobanu, C.Y. Ngan, J.F. Cheng, S.G. Tringe and T. Woyke. 2015. Impact of library preparation protocols and template quantity on the metagenomic reconstruction of a mock microbial community. BMC Genomics 16: 856.

Bowman, J.S. and H.W. Ducklow. 2015. Microbial communities can be described by metabolic structure: A general framework and application to a seasonally variable, depth-stratified microbial community from the coastal west antarctic peninsula. PLoS One 10: e0135868.

Bradnam, K.R., J.N. Fass, A. Alexandrov, P. Baranay, M. Bechner, I. Birol, S. Boisvert, J.A. Chapman, G. Chapuis, R. Chikhi, H. Chitsaz, W.C. Chou, J. Corbeil, C. Del Fabbro, T.R. Docking, R. Durbin, D. Earl, S. Emrich, P. Fedotov, N.A. Fonseca, G. Ganapathy, R.A. Gibbs, S. Gnerre, E. Godzaridis, S. Goldstein, M. Haimel, G. Hall, D. Haussler, J.B. Hiatt, I.Y. Ho, J. Howard, M. Hunt, S.D. Jackman, D.B. Jaffe, E.D. Jarvis, H. Jiang, S. Kazakov, P.J. Kersey, J.O. Kitzman, J.R. Knight, S. Koren, T.W. Lam, D. Lavenier, F. Laviolette, Y. Li, Z. Li, B. Liu, Y. Liu, R. Luo, I. Maccallum, M.D. Macmanes, N. Maillet, S. Melnikov, D. Naquin, Z. Ning, T.D. Otto, B. Paten, O.S. Paulo, A.M. Phillippy, F. Pina-Martins, M. Place, D. Przybylski, X. Qin, C. Qu, F.J. Ribeiro, S. Richards, D.S. Rokhsar, J.G. Ruby, S. Scalabrin, M.C. Schatz, D.C. Schwartz, A. Sergushichev, T. Sharpe, T.I. Shaw, J. Shendure, Y. Shi, J.T. Simpson, H. Song, F. Tsarev, F. Vezzi, R. Vicedomini, B.M. Vieira, J. Wang, K.C. Worley, S. Yin, S.M. Yiu, J. Yuan, G. Zhang, H. Zhang, S. Zhou and I.F. Korf. 2013. Assemblathon 2: Evaluating *de novo* methods of genome assembly in three vertebrate species. Gigascience 2: 10.

Calasso, M., D. Ercolini, L. Mancini, G. Stellato, F. Minervini, R. Di Cagno, M. De Angelis and M. Gobbetti. 2016. Relationships among house, rind and core microbiotas during manufacture of traditional Italian cheeses at the same dairy plant. Food Microbiol. 54: 115–126.

Caporaso, J.G., J. Kuczynski, J. Stombaugh, K. Bittinger, F.D. Bushman, E.K. Costello, N. Fierer, A.G. Pena, J.K. Goodrich, J.I. Gordon, G.A. Huttley, S.T. Kelley, D. Knights, J.E. Koenig, R.E. Ley, C.A. Lozupone, D. McDonald, B.D. Muegge, M. Pirrung, J. Reeder, J.R. Sevinsky, P.J. Turnbaugh, W.A. Walters, J. Widmann, T. Yatsunenko, J. Zaneveld and R. Knight. 2010. QIIME allows analysis of high-throughput community sequencing data. Nat. Methods 7: 335-336.

Chakravorty, S., D. Helb, M. Burday, N. Connell and D. Alland. 2007. A detailed analysis of 16S ribosomal RNA gene segments for the diagnosis of pathogenic bacteria. J. Microbiol. Meth. 69: 330–339.

Chun, J., J.H. Lee, Y. Jung, M. Kim, S. Kim, B.K. Kim and Y.W. Lim. 2007. EzTaxon: A web-based tool for the identification of prokaryotes based on 16S ribosomal RNA gene sequences. Int. J. Syst. Evol. Microbiol. 57: 2259–2261.

Coenye, T. and P. Vandamme. 2003. Intragenomic heterogeneity between multiple 16S ribosomal RNA operons in sequenced bacterial genomes. FEMS Microbiol. Lett. 228: 45–49.

Cole, J.R., B. Chai, R.J. Farris, Q. Wang, S.A. Kulam, D.M. McGarrell, G.M. Garrity and J.M. Tiedje. 2005. The Ribosomal Database Project (RDP-II): sequences and tools for high-throughput rRNA analysis. Nucleic Acids Res. 33: D294–D296.

Cole, J.R., B. Chai, R.J. Farris, Q. Wang, S.A. Kulam-Syed-Mohideen, D.M. McGarrell, A.M. Bandela, E. Cardenas, G.M. Garrity and J.M. Tiedje. 2007. The ribosomal database project (RDP-II): Introducing *myRDP* space and quality controlled public data. Nucleic Acids Res. 35: D169–D172.

Cole, J.R., Q. Wang, J.A. Fish, B. Chai, D.M. McGarrell, Y. Sun, C.T. Brown, A. Porras-Alfaro, C.R. Kuske and J.M. Tiedje. 2014. Ribosomal Database Project: Data and tools for high throughput rRNA analysis. Nucleic Acids Res. 42: D633–D642.

Conesa, A., P. Madrigal, S. Tarazona, D. Gomez-Cabrero, A. Cervera, A. McPherson, M.W. Szcześniak, D.J. Gaffney, L.L. Elo, X. Zhang and A. Mortazavi. 2016. A survey of best practices for RNA-seq data analysis. Genome Biol. 17: 13.

Dalmasso, A., M. de los Dolores Soto del Rio, T. Civera, D. Pattono, B. Cardazzo and M.T. Bottero. 2016. Characterization of microbiota in Plaisentif cheese by high-throughput sequencing. LWT – Food Sci. Technol. 69: 490–496.

De Filippo, C., M. Ramazzotti, P. Fontana and D. Cavalieri. 2012. Bioinformatic approaches and pathway inference in metagenomics data. Brief. Bioinform. 13: 696–710.

DeSantis, T.Z., P. Hugenholtz, N. Larsen, M. Rojas, E.L. Brodie, K. Keller, T. Huber, D. Dalevi, P. Hu and G.L. Andersen. 2006. Greengenes, a chimera-checked 16S rRNA gene database and workbench compatible with ARB. Appl. Environ. Microbiol. 72: 5069–5072.

Dudhagara, P., S. Bhavsar, C. Bhagat, A. Ghelani, S. Bhatt and R. Patel. 2015. Web resources for metagenomics studies. Genomics Proteomics Bioinformatics 13: 296–303.

Earl, D., K. Bradnam, J. St John, A. Darling, D. Lin, J. Fass, H.O. Yu, V. Buffalo, D.R. Zerbino, M. Diekhans, N. Nguyen, P.N. Ariyaratne, W.K. Sung, Z. Ning, M. Haimel, J.T. Simpson, N.A. Fonseca, Í. Birol, T.R. Docking, I.Y. Ho, D.S. Rokhsar, R. Chikhi, D. Lavenier, G. Chapuis, D. Naquin, N. Maillet, M.C. Schatz, D.R. Kelley, A.M. Phillippy, S. Koren, S.P. Yang, W. Wu, W.C. Chou, A. Srivastava, T.I. Shaw, J.G. Ruby, P. Skewes-Cox, M. Betegon, M.T. Dimon, V. Solovyev, I. Seledtsov, P. Kosarev, D. Vorobyev, R. Ramirez-Gonzalez, R. Leggett, D. MacLean, F. Xia, R. Luo, Z. Li, Y. Xie, B. Liu, S. Gnerre, I. MacCallum, D. Przybylski, F.J. Ribeiro, S. Yin, T. Sharpe, G. Hall, P.J. Kersey, R. Durbin, S.D. Jackman, J.A. Chapman, X. Huang, J.L. DeRisi, M. Caccamo, Y. Li, D.B. Jaffe, R.E. Green, D. Haussler, I. Korf and B. Paten. 2011. Assemblathon 1: A competitive assessment of *de novo* short read assembly methods. Genome Res. 21: 2224–2241.

Ekblom, R. and J.B. Wolf. 2014. A field guide to whole-genome sequencing, assembly and annotation. Evol. Appl. 7: 1026–1042.

Eren, A.M., L. Maignien, W.J. Sul, L.G. Murphy, S.L. Grim, H.G. Morrison and M.L. Sogin. 2013. Oligotyping: Differentiating between closely related microbial taxa using 16S rRNA gene data. Methods Ecol. Evol. 4: 1111–1119.

Eren, A.M., G.G. Borisy, S.M. Huse and J.L.M. Welch. 2014. Oligotyping analysis of the human oral microbiome. PNAS 111: E2875–E2884.

Escobar-Zepeda, A., A. Vera-Ponce de Leon and A. Sanchez-Flores. 2015. The road to Metagenomics: From microbiology to DNA sequencing technologies and bioinformatics. Front. Genet. 6: 348.

Ferrocino, I. and L. Cocolin. 2017. Current perspectives in food-based studies exploiting multi-omics approaches. Curr. Opin. Food Sci. 13: 10–15.

Fouhy, F., A.G. Clooney, C. Stanton, M.J. Claesson and P.D. Cotter. 2016. 16S rRNA gene sequencing of mock microbial populations – impact of DNA extraction method, primer choice and sequencing platform. BMC Microbiol. 16: 123.

Galardini, M., A. Mengoni, E.G. Biondi, R. Semeraro, A. Florio, M. Bazzicalupo, A. Benedetti and S. Mocali. 2014. DuctApe: A suite for the analysis and correlation of genomic and OmniLog™ Phenotype Microarray data. Genomics 103: 1–10.

Grafahrend-Belau, E., C. Klukas, B.H. Junker and F. Schreiber. 2009. FBA-SimVis: Interactive visualization of constraint-based metabolic models. Bioinformatics 25: 2755–2757.

Haas, B.J., M. Chin, C. Nusbaum, B.W. Birren and J. Livny. 2012. How deep is deep enough for RNA-Seq profiling of bacterial transcriptomes? BMC Genomics 13: 734.

Haas, B.J., A. Papanicolaou, M. Yassour, M. Grabherr, P.D. Blood, J. Bowden, M.B. Couger, D. Eccles, B. Li, M. Lieber, M.D. Macmanes, M. Ott, J. Orvis, N. Pochet, F. Strozzi, N. Weeks, R. Westerman, T. William, C.N. Dewey, R. Henschel, R.D. Leduc, N. Friedman and A. Regev. 2013. *De novo* transcript sequence reconstruction from RNA-seq using the Trinity platform for reference generation and analysis. Nat. Protoc. 8: 1494–1512.

Hamilton, J.J. and J.L. Reed. 2014. Software platforms to facilitate reconstructing genome-scale metabolic networks. Environ. Microbiol. 16: 49–59.

Hanemaaijer, M., B.G. Olivier, W.F.M. Röling, F.J. Bruggeman and B. Teusink. 2017. Model-based quantification of metabolic interactions from dynamic microbial-community data. PLoS One 12: e0173183.

Hao, Y., Z. Pei and S.M. Brown. 2017. Bioinformatics in Microbiome Analysis. pp. 1–18. *In*: C. Harwood (ed.). Methods in Microbiology of The Human Microbiome, volume 44, Elsevier Ltd.

Henry, C.S., M. DeJongh, A.A. Best, P.M. Frybarger, B. Linsay and R.L. Stevens. 2010. High-throughput generation, optimization and analysis of genome-scale metabolic models. Nat. Biotechnol. 28: 977–982.

Henson, M.A. and T.J. Hanly. 2014. Dynamic flux balance analysis for synthetic microbial communities. IET Syst. Biol. 8: 214–229.

Iwai, S., T. Weinmaier, B.L. Schmidt, D.G. Albertson, N.J. Poloso, K. Dabbagh and T.Z. DeSantis. 2016. Piphillin: improved prediction of metagenomic content by direct inference from human microbiomes. PLoS One 11: e0166104.

Jones, M.B., S.K. Highlander, E.L. Anderson, W. Li, M. Dayrit, N. Klitgord, M.M. Fabani, V. Seguritan, J. Green, D.T. Pride, S. Yooseph, W. Biggs, K.E. Nelson and J.C. Venter. 2015. Library preparation methodology can influence genomic and functional predictions in human microbiome research. PNAS 112: 14024–14029.

Jovel, J., J. Patterson, W. Wang, N. Hotte, S. O'Keefe, T. Mitchel, T. Perry, D. Kao, A.L. Mason, K.L. Madsen and G.K.-S. Wong. 2016. Characterization of the gut microbiome using 16S or shotgun metagenomics. Front. Microbiol. 7: 459.

Junker, B.H., C. Klukas and F. Schreiber. 2006. VANTED: A system for advanced data analysis and visualization in the context of biological networks. BMC Bioinformatics 7: 109.

Karp, P.D., S.M. Paley and P. Romero. 2002. The Pathway Tools software. Bioinformatics 18: S225–S232.

Karp, P.D., S.M. Paley, M. Krummenacker, M. Latendresse, J.M. Dale, T.J. Lee, P. Kaipa, F. Gilham, A. Spaulding, L. Popescu, T. Altman, I. Paulsen, I.M. Keseler and R. Caspi. 2010. Pathway Tools version 13.0: Integrated software for pathway/genome informatics and systems biology. Brief. Bioinform. 11: 40–79.

Kembel, S.W., M. Wu, J.A. Elsen and J.L. Green. 2012. Incorporating 16S gene copy number information improves estimates of microbial diversity and abundance. PLoS Comput. Biol. 8: e1002743.

Klamt, S., J. Saez-Rodriguez and E.D. Gilles. 2007. Structural and functional analysis of cellular networks with CellNetAnalyzer BMC Syst. Biol. 1: 2.

Klindworth, A., E. Pruesse, T. Schweer, J. Peplies, C. Quast, M. Horn and F.O. Glöckner. 2013. Evaluation of general 16S ribosomal RNA gene PCR primers for classical and next-generation sequencing-based diversity studies. Nucleic Acid Res. 41: e1.

Köser, C.U., M.J. Ellington, E.J. Cartwright, S.H. Gillespie, N.M. Brown, M. Farrington, M.T. Holden, G. Dougan, S.D. Bentley, J. Parkhill and S.J. Peacock. 2012. Routine use of microbial whole genome sequencing in diagnostic and public health microbiology. PLoS Pathog. 8: e1002824.

Krömer, J., L.-E. Quek and L. Nielsen. 2009. ^{13}C-Fluxomics: A tool for measuring metabolic phenotypes. Aust. Biochem. 40: 17–20.

Ladoukakis, E., F.N. Kolisis and A.A. Chatziioannou. 2014. Integrative workflows for metagenomic analysis. Front. Cell. Dev. Biol. 2: 70.

Langille, M.G., J. Zaneveld, J.G. Caporaso, D. McDonald, D. Knights, J.A. Reyes, J.C. Clemente, D.E. Burkepile, R.L. Vega Thurber, R. Knight, R.G. Beiko and C. Huttenhower. 2013. Predictive functional profiling of microbial communities using 16S rRNA marker gene sequences. Nat. Biotechnol. 31: 814–821.

Laursen, M.F., M.D. Dalgaard and M.I. Bahl. 2017. Genomic GC-content affects the accuracy of 16S rRNA gene sequencing based microbial profiling due to PCR bias. Front. Microbiol. 8: 1934.

Lee, J.Y., J.Y. Joung, Y.-S. Choi, Y. Kim and N.S. Oh. 2016. Characterization of microbial diversity and chemical properties of Cheddar cheese prepared from heat-treated milk. Int. Dairy J. 63: 92–98.

Levante, A., F. De Filippis, A. La Storia, M. Gatti, E. Neviani, D. Ercolini and C. Lazzi. 2017. Metabolic gene-targeted monitoring of non-starter lactic acid bacteria during cheese ripening. Int. J. Food Microbiol. 257: 276–284.

Li, S., P.P. Łabaj, P. Zumbo, P. Sykacek, W. Shi, L. Shi, J. Phan, P.Y. Wu, M. Wang, C. Wang, D. Thierry-Mieg, J. Thierry-Mieg, D.P. Kreil and C.E. Mason. 2014. Detecting and correcting systematic variation in large-scale RNA sequencing data. Nat. Biotechnol. 32: 888–895.

Liao, Y.-C., M.-H. Tsai, F.-C. Chen and C.A. Hsiung. 2012. GEMSiRV: A software for Genome-scale metabolic model simulation, reconstruction and visualization. Bioinformatics 28: 1752–1758.

Liao, Y.-C., S.-H. Lin and H.-H. Lin. 2015. Completing bacterial genome assemblies: Strategy and performance comparisons. Sci. Rep. 5: 8747.

Magoc, T., S. Pabinger, S. Canzar, X. Liu, Q. Su, D. Puiu, L.J. Tallon and S.L. Salzberg. 2013. GAGE-B: An evaluation of genome assemblers for bacterial organisms. Bioinformatics 29: 1718–1725.

Maidak, B.L., G.J. Olsen, N. Larsen, R. Overbeek, M.J. McCaughey and C.R. Woese. 1996. The Ribosomal Database Project (RDP). Nucleic Acids Res. 24: 82–85.

Mande, S.S., M.H. Mohammed and T.S. Ghosh. 2012. Classification of metagenomic sequences: Methods and challenges. Brief. Bioinform. 13: 669–681.

Mayo, B., C.T.C.C. Rachid, A. Alegría, A.M.O. Leite, R.S. Peixoto and S. Delgado. 2014. Impact of next generation sequencing techniques in food microbiology. Curr. Genomics 15: 293–309.

Mendes-Soares, H., M. Mundy, L. Mendes-Soares and N. Chia. 2016. MMinte: An application for predicting metabolic interactions among the microbial species in a community. BMC Bioinformatics 17: 343.

Miller, R.R., V. Montoya, J.L. Gardy, D.M. Patrick and P. Tang. 2013. Metagenomics for pathogen detection in public health. Genome Med. 5: 81.

Narzisi, G. and B. Mishra. 2011. Comparing *de novo* genome assembly: The long and short of it. PLoS One 6: e19175.

NCBI Resource Coordinators. 2017. Database Resources of the National Center for Biotechnology Information. Nucleic Acids Res. 45: D12–D17.

Ni, J., Q. Yan and Y. Yu. 2013. How much metagenomic sequencing is enough to achieve a given goal? Sci. Rep. 3: 1968.

Olive, D.M. and P. Bean. 1999. Principles and applications of methods for DNA-based typing of microbial organisms. J. Clin. Microbiol. 37: 1661–1669.

Olson, N.D., T.J. Treangen, C.M. Hill, V. Cepeda-Espinoza, J. Ghurye, S. Koren and M. Pop. 2017. Metagenomic assembly through the lens of validation: Recent advances in assessing and improving the quality of genomes assembled from metagenomes. Brief. Bioinform., doi:10.1093/bib/bbx098.

Oulas, A., C. Pavloudi, P. Polymenakou, G.A. Pavlopoulos, N. Papanikolaou, G. Kotoulas, C. Arvanitidis and I. Iliopoulos. 2015. Metagenomics: Tools and insights for analyzing next-generation sequencing data derived from biodiversity studies. Bioinform. Biol. Insights 9: 75–88.

Patil, K.R. and J. Nielsen. 2005. Uncovering transcriptional regulation of metabolism by using metabolic network topology. PNAS 102: 2685–2689.

Połka, J., A. Rebecchi, V. Pisacane, L. Morelli and E. Puglisi. 2015. Bacterial diversity in typical Italian salami at different ripening stages as revealed by high-throughput sequencing of 16S rRNA amplicons. Food Microbiol. 46: 342–356.

Pruesse, E., C. Quast, K. Knittel, B.M. Fuchs, W. Ludwig, J. Peplies and F.O. Glöckner. 2007. SILVA: A comprehensive online resource for quality checked and aligned ribosomal RNA sequence data compatible with ARB. Nucleic Acids Res. 35: 7188–7196.

Pruesse, E., J. Peplies and F.O. Glöckner. 2012. SINA: Accurate high-throughput multiple sequence alignment of ribosomal RNA genes. Bioinformatics 28: 1823–1829.

Ramezani, M., S.M. Hosseini, I. Ferrocino, M.A. Amoozegar and L. Cocolin. 2017. Molecular investigation of bacterial communities during the manufacturing and ripening of semi-hard Iranian Liqvan cheese. Food Microbiol. 66: 64–71.

Rocha, I., P. Maia, P. Evangelista, P. Vilaça, S. Soares, J.P. Pinto, J. Nielsen, K.R. Patil, E.C. Ferreira and M. Rocha. 2010. OptFlux: An open-source software platform for in silico metabolic engineering. BMC Syst. Biol. 4: 45.

Rosselli, R., O. Romoli, N. Vitulo, A. Vezzi, S. Campanaro, F. de Pascale, R. Schiavon, M. Tiarca, F. Poletto, G. Concheri, G. Valle and A. Squartini. 2016. Direct 16S rRNA-seq from bacterial communities: A PCR-independent approach to simultaneously assess microbial diversity and functional activity potential of each taxon. Sci. Rep. 6: 32165.

Sboner, A., X.J. Mu, D. Greenbaum, R.K. Auerbach and M.B. Garstein. 2011. The real cost of sequencing: higher than you think! Genome Biol. 12: 125.

Schellenberger, J., R. Que, R.M. Fleming, I. Thiele, J.D. Orth, A.M. Feist, D.C. Zielinski, A. Bordbar, N.E. Lewis, S. Rahmanian, J. Kang, D.R. Hyduke and B.Ø. Palsson. 2011. Quantitative prediction of cellular metabolism with constraint-based models: The COBRA Toolbox v2.0. Nat. Prot. 6: 1290–1307.

Schloss, P.D., S.L. Westcott, T. Ryabin, J.R. Hall, M. Hartmann, E.B. Hollister, R.A. Lesniewski, B.B. Oakley, D.H. Parks, C.J. Robinson, J.W. Sahl, B. Stres, G.G. Thallinger, D.J. Van Horn and C.F. Weber. 2009. Introducing mothur: Open-source, platform-independent, community-supported software for describing and comparing microbial communities. Appl. Environ. Microbiol. 75: 7537–7541.

Scholz, M.B., C.-C. Lo and P.S.G. Chain. 2012. Next generation sequencing and bioinformatic bottlenecks: The current state of metagenomic data analysis. Curr. Opin. Biotechnol. 23: 9–15.

Sedlar, K., K. Kupkova and I. Provaznik. 2017. Bioinformatics strategies for taxonomy independent binning and visualization of sequences in shotgun metagenomics. Comput. Struct. Biotechnol. J. 15: 48–55.

Seyednasrollah, F., A. Laiho and L.L. Elo. 2015. Comparison of software packages for detecting differential expression in RNA-seq studies. Brief. Bioinform. 16: 59–70.

Sharpton, T.J. 2014. An introduction to the analysis of shotgun metagenomic data. Front. Plant Sci. 5: 209.

Trapnell, C., A. Roberts, L. Goff, G. Pertea, D. Kim, D.R. Kelley, H. Pimentel, S.L. Salzberg, J.L. Rinn and L. Pachter. 2012. Differential gene and transcript expression analysis of RNA-seq experiments with TopHat and Cufflinks. Nat. Prot. 7: 562–578.

Tremblay, J., K. Singh, A. Fern, E.S. Kirton, S. He, T. Woyke, J. Lee, F. Chen, J.L. Dangl and S.G. Tringe. 2015. Primer and platform effects on 16S rRNA tag sequencing. Front. Microbiol. 6: 771.

van Hijum, S.A.F.T., E.E. Vaughan and R.F. Vogel. 2013. Application of state-of-art sequencing technologies to indigenous food fermentations. Curr. Opin. Biotechnol. 24: 178–186.

Větrovský, T. and P. Baldrian. 2013. The variability of the 16S rRNA gene in bacterial genomes and its consequences for bacterial community analyses. PLoS One: 8: e57923.

Vincent, A.T., N. Derome, B. Boyle, A.I. Culley and S.J. Charette. 2017. Next-generation sequencing (NGS) in the microbiological world: How to make the most of your money. J. Microbiol. Methods 138: 60–71.

Vlassis, N., M.P. Pacheco and T. Sauter. 2014. Fast reconstruction of compact context-specific metabolic network models. PLoS Comput. Biol. 10: e1003424.

Wang, Q., G.M. Garrity, J.M. Tiedje and J.R. Cole. 2007. Naïve bayesian classifier for rapid assignment of rRNA sequences into the new bacterial taxonomy. Appl. Environ. Microbiol. 73: 5261–5267.

Wood, J. 2016. Predictive metagenomics profiling: Why, what and how? Bioinformatics Rev. 2: 1–4.

Yu, J., J. Blom, S.P. Glaeser, S. Jaenicke, T. Juhre, O. Rupp, O. Schwengers, S. Spänig and A. Goesmann. 2017. A review of bioinformatics platforms for comparative genomics. Recent developments of the EDGAR 2.0 platform and its utility for taxonomic and phylogenetic studies. J. Biotechnol. 261: 2–9.

6

Advanced 'Omics Approaches Applied to Microbial Food Safety & Quality

From Ecosystems to the Emerging Foodborne Pathogen *Campylobacter*

Alizée Guérin,[1,*] *Amélie Rouger,*[1,*] *Raouf Tareb,*[1]
Jenni Hultman,[2] *Johanna Björkroth,*[2] *Marie-France Pilet,*[1]
Monique Zagorec[1] and *Odile Tresse*[1,†]

1. Introduction

Food safety and quality concerns the whole food supply chain from the primary production to human consumption. Each step between farm and fork is critical, specifically for microbial food safety and quality. Microbial safety refers to acceptable standards for microbiological hazards while microbial quality refers to acceptable sensory characteristics. Basically, microorganisms classified as hazards are called

[1] SECALIM, INRA, Oniris, Université Bretagne Loire, 44307, Nantes, France.
[2] Department of Food Hygiene and Environmental Health, Faculty of Veterinary Medicine, University of Helsinki, 00014 Helsinki, Finland.
* Two first co-authors
† Corresponding author: odile.tresse@inra.fr

pathogens while microorganisms responsible for food alteration are called spoilers. The reality is more complex as certain bacterial species described as harmless can become pathogenic in specific microbiota or host contexts and others described as pathogenic can act as etiological factors in specific diseases.

Since food products are considered to be open matrices, foodstuffs cannot be disconnected from microorganism colonization. Even though safety practices are implemented in order to reduce the environmental impact on microorganism colonization and development, food products always have an opportunity to be colonized by microorganisms. The interaction between microorganisms and food products depends on the biology of the microorganisms, the food product composition and the environment. The colonization of food matrices by microorganisms can be sought when they are part of the food processing (e.g., fermented food products) or fought when they could alter food quality or affect consumer health. To ensure the high quality and innocuousness of foodstuffs, specific regulations resulting from food safety authorities and food safety risk analyses about the quantitative levels of microorganism acceptance are applicable to food producers and food distributors all over the world. These regulations concern both the food processing plants and the foodstuffs. To keep the microbial risk hazards below the recommended thresholds, safety practices are continuously implemented and maintained throughout the food chain. With the evolution of the international food trade, primary production, agrofourniture, demand and habits of consumers associated with ecological awareness, and the emergence of new bacteria, microbial food safety is a constantly developing field.

As viable organisms, bacteria can adapt to various environments including food matrices. The fitness of a bacterial population results from the regulation/deregulation of cellular processes in response to biotic and abiotic environmental cues, which will subsequently affect the survival and growth of cells. In fact, foodstuffs are colonized by multiple bacterial populations forming a sustainable ecosystem, the so-called food microbiota. The dynamics of these ecosystems are based on multifactorial, synergetic/antagonistic and complex biological mechanisms that affect microbial food safety and quality. With the advent of 'omics cutting-edge technologies, scientific researchers are addressing questions concerning food microbiota dynamics and foodborne pathogen adaptation that could not be raised a decade ago.

In this chapter, the first section is dedicated to the appraisal of microbial community dynamics on food products using 'omics analyses. This section includes conventional and new methods, that are applied in order to decipher the microbiota composition, and gives examples of their use in food products. The second section deals with the 'omics approaches to address questions about biological functions within microbial communities on food products. This section presents metatranscriptomics analyses, bacterial interactions and crosstalk analyses. The third section gives an overall description of cutting-edge 'omics approaches to explore bacterial fitness at the level of a single population. It focuses on the analyses of the foodborne pathogen *Campylobacter*, which is currently recognized as the leading cause of foodborne infections worldwide (EFSA 2016, Epps et al. 2013). In the last section, the opportunities and weaknesses of these advanced 'omics methodologies are discussed.

2. Appraisal of Microbial Community Dynamics on Food Products

2.1 From Conventional Methods to Advanced Techniques to Decipher Food Microbiota

Determining which bacterial species are present on food products and their dynamics during storage is important in guaranteeing food safety. This includes the detection/enumeration of foodborne pathogens that may be present at very low levels or bacteria that may cause spoilage when present in high concentrations. The determination of the total bacterial population is also an important criterion, particularly in establishing the shelf life of highly perishable foods, such as raw meat or seafood products. Until recently, food microbiology was mainly investigated by cultural methods, consisting of bacterial enumeration on various selective media. For some pathogens present at low levels, as is the case for *Campylobacter* in poultry cuts, an enrichment step in liquid medium is required prior to plating in order to increase the detection threshold (Katsav et al. 2008). However, although widely used, cultural methods give a reductive vision of the complex microbial communities hosted in foodstuffs (Ercolini 2004). For instance, in fermented food it has been estimated that, besides the well-known and cultivable bacterial species, 25 to 50% of the bacteria could not grow in the media and laboratory conditions commonly used (Juste et al. 2008). Several hypotheses can explain these limitations, such as the selectivity of the media or the incubation conditions, particularly temperature or atmosphere (Doulgeraki et al. 2012). Furthermore, some bacterial species cannot yet be cultivated as no selective media targeting them have been developed (Doulgeraki et al. 2012). As an example, one of the major bacterial populations of *Fusobacteriaceae* encountered on spoiled cod fillet has been identified only by metabarcoding sequencing methods (Chaillou et al. 2015). Therefore, various molecular methods, mostly based on DNA rather than on Petri dishes, have been developed for a better appraisal of food microbiology.

Real-time quantitative PCR (RT-q-PCR) can be used to quantify various species from bacterial DNA extracted from food samples, as reported for *Salmonella enterica* in poultry meat (Agrimonti et al. 2013), or *Salmonella* Typhimurium and *Escherichia coli* O157:H7 in ground beef (Chaillou et al. 2014). Various spoilage bacteria in both meat and seafood products have also been quantified through RT-q-PCR (Fougy et al. 2016, Jones et al. 2009, Mamlouk et al. 2012). However, the DNA from dead bacterial cells can also be amplified and may introduce a bias in the detection or quantification. On the other hand, such methods can detect or quantify non-cultivated bacteria by using appropriate primers. Hybridization to DNA microarrays and Fluorescence In Situ Hybridization (FISH) have also been reported to identify bacteria present in food ecosystems (Juste et al. 2008). These methods require primers specific to the bacteria to be identified (Diaz-Sanchez et al. 2013). Hybridization to DNA microarrays of RNA extracted during sausage fermentation has also been used to investigate the functions transcribed by *Staphylococcus xylosus* in meat (Vermassen et al. 2014). Because of their specificity even when targeting several species, the methods mentioned above, based on an *a priori* approach, are not suitable for describing the microbial communities composing complex ecosystems as a whole. Other methods, based on a first step of DNA extraction followed by PCR amplification and subsequent analysis, have recently emerged, aiming at an overall description of microbial (essentially bacterial) species of various ecosystems, including food products.

PCR-DGGE (denaturing gradient gel electrophoresis) performed after amplification on whole DNA extracted from food can be used to compare different samples or to follow the dynamics of the bacterial communities during storage. The sequencing of fragments obtained after PCR targeting the bacterial 16S rRNA gene or the yeast 26S rRNA gene followed by PCR-DGGE migration led to the identification of the major bacterial and yeast species in Italian dry sausage (Villani et al. 2007). The data obtained after PCR-DGGE and 16S rRNA gene pyrosequencing on the same subset of DNA extracted from seafood products have been compared (Chaillou et al. 2015, Roh et al. 2010). A quantitative comparison was not possible, but PCR-DGGE profiles partially confirmed pyrosequencing observations. Hence, the most exhaustive methods for describing the microbial ecology of complex ecosystems are based on high throughput sequencing. This has led to new possibilities for detecting, identifying and quantifying bacteria from food products and the food processing environment without the prerequisite of a culture step (for reviews, see Doulgeraki et al. 2012, Juste et al. 2008, Cao et al. 2017). It has also greatly modified the experimental design compared to cultural methods, as well as the techniques and skills to be applied. In Fig. 1, the main steps for the experimental design from food sample collection to the identification and relative abundance of the different members composing food microbial communities using next generation sequencing are presented.

2.2 DNA Extraction for Next Generation Sequencing

Sampling can vary depending on the research objective and the nature of the food. The chosen food samples should be representative of the environment under study. Bacteria can be on the surface or in the food matrix. They may also be organized in biofilms that can protect them. Swabbing, rinsing, grinding or stomaching methods, followed by filtering and centrifugation, are routinely used for separating bacteria from the matrix. Choosing the separation procedure depends on the physical nature and the texture of the food and the accessibility of the bacteria. Food matrix residues (particularly fats) can inhibit subsequent PCR amplification and should therefore be eliminated (Abu Al-Soud and Radstrom 2000, Lübeck et al. 2003, Rossen et al. 1992). As high throughput sequencing may generate a large number of reads depending on the method used, the bacterial contamination level and the sequencing methodology to be used subsequently must be considered when choosing the required size of the sample. For instance, if the expected number of reads is about 10^4, a 1 g sample contaminated with 10^8 bacteria/g will enable access to only the dominant bacteria (i.e., only the bacteria present at a level higher than 10^4/g will be represented in the reads obtained).

After bacteria collection, the lysis step can be carried out by mechanical and/or chemical procedures. Mechanical lysis with glass, zirconium, or silica beads may help lyse bacteria but can also fragment or degrade nucleic acids, leading to misassembly during the annealing step of PCR amplification, and constitute chimera sequences. In addition, the type of beads may influence the sequencing data obtained (Costea et al. 2017).

In contrast to human microbiota studies, which led to a standardized protocol for human fecal sample processing (Costea et al. 2017), several procedures have been reported for bacterial DNA extraction from food matrices (for examples, see Chaillou

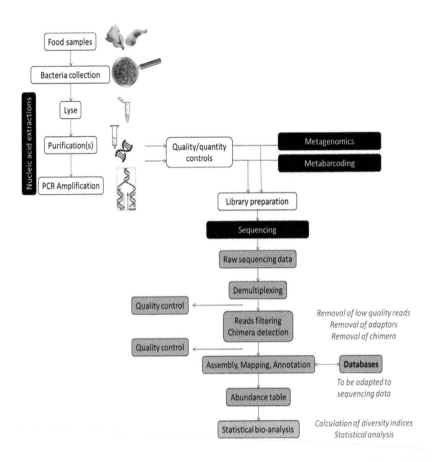

Fig. 1. Schematic representation of the main steps in deciphering microbial communities of food products, based on high throughput sequencing methods. White boxes represent "wet-laboratory" experiments. Gray boxes represent *in silico* experiments. Black boxes represent the main steps.

et al. 2015, Diaz-Sanchez et al. 2013, Rouger et al. 2017b). As reported by Costea et al. (2017) for fecal samples, the DNA extraction procedure and the various kits available may have an impact on the quantity and quality of DNA extracted from food, as also observed for poultry meat samples (Rouger et al. 2017b). Several commercial purification kits are available, some of which are specifically dedicated to food samples. In general, nucleic acids are captured after lysis by fixation on affinity columns, and are then washed and eluted in adapted buffers that are provided in the kits. As food matrices may also contain eukaryotic nucleic acids from food tissues, columns may be overloaded with a large quantity of nucleic acids. Therefore, the routine procedures recommended by kit manufacturers often need to be improved in order to collect nucleic acids from food (Pinto et al. 2007) and to limit the presence of PCR inhibitors (Faber et al. 2013). The quantity and quality of nucleic acids obtained at the end of the process depend on the initial bacterial concentration as well as the recovery after each step.

Once the nucleic acids have been extracted, various methods can be used. In metagenomic analyses, the whole DNA is sequenced without any PCR amplification step. PCR amplification can be performed simply to detect the presence of various bacteria using primers specifically designed for targeting a single species, genus, or family. However, Taq polymerase inhibitors and contamination by eukaryotic DNA or RNA may impact amplification efficiency (Glassing et al. 2016). In fact, difficulties in PCR amplification for *Salmonella* detection in meat and the description of bacterial communities of chicken cuts have been reported (Bülte and Jakob 1995, Rouger et al. 2017a). In metabarcoding approaches, universal primers are used most often for PCR amplification, targeting the conserved regions of the bacterial 16S rRNA gene, and then the fragments are sequenced. Variable regions of 16S rDNA are currently used to sequence and identify bacterial populations with V1, V3, V4, and V9 being the most frequently reported. Many factors can impact the final results. Biases when using molecular techniques or during PCR amplification and sequencing were noticed 20 years ago. As the rRNA gene copy number varies between bacterial species, quantifying the relative abundance of different populations may be biased (Klappenbach et al. 2000). In addition, the requirement for tagged primers, such as for pyrosequencing, leads to quite long (30 bp) primers being used, which may limit the amplification yield. The fidelity and efficacy of the polymerases used to perform the PCR are also important (Abu Al-Soud and Rådström 1998), particularly for long fragment amplification (Cline et al. 1996, Keohavong and Thilly 1989). Facilitators such as T4 gene 32 have been proposed as a way to improve PCR efficiency (Abu Al-Soud and Radstrom 2000, Wilson 1997). During this step, chimeric DNA fragments can be synthesized, being at the origin of chimeric sequence reads. If misassembly occurs during the first PCR cycles (Haas et al. 2011), the detection of chimeric sequences is required during the filtering of reads after sequencing.

2.3 Quantification and Quality of Extracted Nucleic Acids

Quality and quantity controls are required before sequencing. The fluorimeters or spectrophotometers mainly used for measuring nucleic acid concentration are NanoDrop and Qubit apparatus. The quality of both DNA and RNA is usually checked by automatic capillary electrophoresis, such as the Bioanalyzer from Agilent or Experion from Biorad.

2.4 Nucleic Acid Sequencing

Several sequencing methods are now used for food microbiome studies (for a recent review, see Cao et al. 2017). The sequence quality depends on the DNA purity and the library preparation method (Tyler et al. 2016). The optimization of library preparation has been studied (Oyola et al. 2012). For example, the Illumina TruSeq kit generated data with a better uniformity of sequence than the Nextera XT method. In metagenomics approaches, the bias most often reported is at the fragmentation step or during the random amplification (van Dijk et al. 2014).

Since 2005, following the pyrosequencing development that revolutionized the access to bacterial genome sequences (Margulies et al. 2005), many techniques have emerged and are still in constant evolution. A large (or even huge) number of sequencing reads can be obtained in a short time from only a small quantity of DNA, without cloning steps and at an affordable price. There are two main approaches: the most used is based on the sequencing of a short fragment, obtained by PCR amplification of a region that is common to the microbial communities, but with variable regions that enable the different populations to be distinguished (metabarcoding) (Taberlet et al. 2012). A second approach is based on cDNA sequencing after retrotranscription of 16S rRNA obtained from total RNA.

With metabarcoding, the microbial species present in an ecosystem are determined by comparison with sequence databases, and their relative quantification is possible. This approach has been mainly used in environmental microbiology and to describe the microbiota of the digestive tract of many animals. It has only recently emerged in food science, with reports still being mostly restricted to bacterial 16S rRNA gene pyrosequencing. Pipelines and software for data analyses (e.g., Mothur, Qiime, Frogs) are in constant development and are based on the same principles. Nevertheless, the nature of the database and the pipeline or software used on the same subset of pyrosequencing data may generate different identification results, as shown from poultry cuts (Rouger et al. 2018). With metabarcoding, depending on the number of reads obtained and the diversity of the samples, the depth can reach 10^4–10^5 reads (i.e., within a bacterial community of 10^x, only a bacterial population present at up to 10^{x-4} or 10^{x-5} will be detected). Identification of bacteria through the partial 16S rRNA gene sequence can reach not only the genus level but also the species level. Identification accuracy depends on the quality of the sequence database used to assign sequence reads to operational taxonomic units (OTU) and on the 16S rRNA gene variable regions amplified prior to sequencing. This method can also generate errors, resulting from wrong PCR amplifications or contamination by the food matrix DNA (mitochondrial DNA of the animals from which the food is produced or chloroplast DNA from spices). In fact, the number of reads finally assigned to chloroplasts reached more than half of the total reads obtained from poultry sausage (Chaillou et al. 2015). These were attributed to the spices added to the sausage formula. Nevertheless, this method is useful for a more accurate assessment of the diversity of food ecosystems. It should be noted that the presence of other taxa, such as archaea, has also been found in fermented seafood via pyrosequencing (Roh et al. 2010). With metagenomics, the whole DNA sequence is determined in order to assess what is there and which functions are potentially present. Such a method does not only focus on bacteria and may reveal the presence of other microorganisms, such as yeasts, archaea, or viruses. As an example, the virome of chicken skin has been assessed by metagenomics (Denesvre et al. 2015, Nieminen et al. 2012). It should also be noted that with metagenomics, depending on the samples, more than half of the sequences may come from food tissues; for example, 50 to 80% of the reads obtained from poultry meat studies could be aligned to the *Gallus gallus* genome (Denesvre et al. 2015, Nieminen et al. 2012).

3. Deciphering Functions of Microbial Communities on Food Products

3.1 Metatranscriptomics Analyses

3.1.1. Scientific context in food products. Microbial ecology has traditionally been an area associated with environmental microbiology. More recently, food processing chains have been studied as man-made ecological niches in which microbes contaminate our foods during primary production, harvesting and processing, followed by formation of different food-associated communities. In the case of meat, products are often packaged, resulting in a temporarily closed environment in which the members of a spoilage microbial community grow and succession of species happens. Even though shelf life of meat products is usually short and the circumstances are not as diversifying as in the nature, there are many reasons to study food spoilage also in the context of microbial ecology.

To prevent food from spoiling, the modern food industry applies two major obstacles, namely refrigerated temperatures and carbon dioxide. They are the main selective pressures that have an effect on food spoilage organisms growing in perishable foods packaged under modified atmospheres and cold stored throughout the distribution chains from manufacturers to consumers. Increasingly, perishable food items are manufactured "case ready" allowing retail distributors to place ready-made packages on shelves. This omits handling of the food in shops and has been an increasing trend in many countries, especially what comes to meat products. From the microbial ecology perspective, microbiota in these packages develops through a time-dependent succession, which is also associated with development of food spoilage. Studying the functions of the bacterial community in this type of foods provides insights on behavioral patterns of different spoilage organisms. This knowledge helps us to understand interactions between different spoilage organisms and may provide us with new tools for interfering with the growth of some bacteria and, thus, reduce spoilage.

3.1.2. Metatranscriptomics of high-oxygen packaged beef. Metatranscriptomics approach is currently under application to understand the activities of a developing microbial meat spoilage community. Time-dependent metatranscriptome of beef packaged under high oxygen modified atmosphere (MA) was analyzed during shelf life and after spoilage. Our earlier metagenomics study (Nieminen et al. 2012) prompted the transcriptomics work in order to understand the community functions, as the DNA-level data was too limited to answer all questions. Using metagenomics, we are able to show the diversity of the microbiome and its possible interacting pathways, but it was not clear which of the pathways had active roles during meat spoilage.

For metatranscriptomics studies, methods of extracting the RNA from the foodborne bacterial communities were first developed (Fig. 2) working with beef, pork, chicken and fish. To begin with all of these samples, the bacterial cells and RNA were separated from the host species. Since the bacterial cells can be both on the meat surface or deeper under the meat surface, homogenization was required before RNA extraction. When RNAs are extracted, protection from RNAses that can be present in food tissues is necessary in avoiding their degradation (McCarthy et al. 2015). Soaking the sample to RNAlater® (Sigma) and homogenizing with stomaching is very useful. Following this, two rounds of centrifugation were performed in order to ensure removal of host cells from the bacterial cells (Hultman et al. 2015). However, meats from different animal species behave differently. There are texture-dependent properties, at least in pork and salmon, which cause problems regarding the removal of host cells. In addition, fat from pork and chicken causes technological locks.

The diversity of the active community was analyzed by mapping the 16S rRNA transcripts to the Silva database (Quast et al. 2013). This 16S rRNA database was used as it contains the largest diversity of the bacterial and archaeal sequences. With this step, we have a more comprehensive understanding on the active species in a

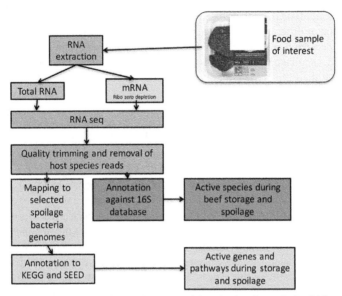

Fig. 2. Schematic representation of the main steps to define the functions of microbial communities of food products.

food sample. Therefore, ribosomal RNA depletion from the samples was not applied so as to keep the knowledge on the active species. Quite often in the case of meat, succession leads first to dominance of *Leuconostoc gelidum* subsp. *gasicomitatum*, followed by the presence and activity of *Lc. piscium*. Following species level identification, the metabolic pathways active in the samples were analyzed. This was conducted by annotating the genes against COG (clusters of orthologous genes). For example, in communities dominated by *Leuconostoc* spp. and *Lactococcus* spp., the following main activities are usually detected: (i) activity of pentose phosphate pathway throughout growth and spoilage phase (subsp. *gasicomitatum*), (ii) more active respiration metabolism close to the end of shelf life (subsp. *gasicomitatum*), (iii) active metabolism associated with diacetyl/acetoin formation at the end of shelf life (subsp. *gasicomitatum*) and (iv) active pyruvate utilization pathways at the end of shelf life (*Lc. piscium*). The communities under high-oxygen atmosphere are also facing oxidative stress since adaptive tolerance response (ATR) to oxidative stress associated to genes highly active at the end of shelf life was detected.

According to our results regarding high oxygen atmosphere for pork and beef conversation, species able to resist to oxygen stress are the best growing microorganisms and are responsible for food spoilage. The heme-dependent respiration associated with *Ln. gelidum* subspecies *gasicomitatum* (Jaaskelainen et al. 2015), together with other stress responses, is likely to explain why this species is able to grow well and spoil high-oxygen MA packaged meat products. The co-existence with *Lc. piscium* is probably beneficial what comes to the pyruvate dissipating pathways and acid tolerance, but more studies are needed in order to understand the interplay of these species. Without the knowledge from the active species and pathways, it is not possible to develop targeted approaches for suppressing the growth of the most important spoilers.

3.2 Bacterial Interaction/Crosstalk Analyses

3.2.1. Scientific context in food products. As stated in part 2, bacterial ecosystems of food and beverages are extremely diverse and include spoilage bacteria, foodborne pathogens in some cases and bacteria showing positive effects on the product quality. For many years, the majority of studies on the ecology of food-associated microbial communities have focused on analyzing the population composition or understanding the behavior of a single species, including its interactions with the food matrix. Many of these studies are mainly descriptive and relatively little is known about the mechanisms governing population dynamics and the molecular interactions between the consortium members. Generally, microbial community interactions occur via different strategies, divided into five main classes (Sieuwerts et al. 2008) (Fig. 3). Commensalism and mutualism correspond to positive interactions with benefit for one or both parts. Amensalism and competition describe adverse interactions affecting one or both organisms. Parasitism has not been described in food bacteria and mainly concerns their interactions with phages,which will not be discussed further in this section.

These interactions between microbial ecosystems may have important consequences for the final quality of food. Amensalism and competition between bacteria usually lead to improving microbial safety and/or quality of food by limiting the growth of spoilage or pathogenic bacteria. Mutualism or commensalism may enhance food quality with the production of flavor metabolites in the case of fermented products, or, in contrast, may contribute to damaging food when spoilage microorganisms are concerned.

These exchanges between bacterial populations imply the transfer molecular and genetic information and many mechanisms can be involved, including primary

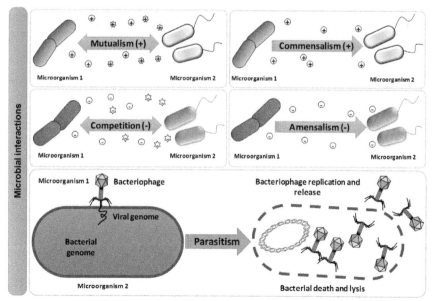

Fig. 3. Summary of ecological interactions between microorganisms. Five distinct types of social interactions between two microorganisms are described: Mutualism, commensalism, amensalism, competition, and parasitism. The chemicals synthesized by one microorganism may be deleterious (-) or beneficial (+) to the growth of the other.

or secondary metabolites, siderophores and signaling molecules (Braga et al. 2016). In most studies concerning food microbiology, bacterial interactions between two or a limited number of species are considered. When molecular tools are implemented to characterize the interactions, model media or media simulating the food environment are generally used. However,with the development of Next Generation Sequencing approaches for characterizing bacterial communities and their activities, new methodologies have been considered in recent years,these methodologies have produced new insights on the bacterial interactions between complex ecosystems.

In the following section, the implementation of 'omics methods is considered for positive (i.e., mutualism and commensalism) and negative (i.e., amensalism and competition) interactions. Then complementary approaches investigating bacterial communication and microbial networks in food are presented.

3.2.2. Positive interactions. Bacterial cooperation in food has mostly been studied in fermented products. Yogurt is probably one of the simplest food models for studying positive bacterial interactions as milk fermentation involves only two bacterial species growing and acting together to give the final product characteristics. In fact, in this fermented milk, metabolic interactions occur between *Streptococcus thermophilus* and *Lb. delbrueckii* subsp. *bulgaricus*, enabling higher acidification and better organoleptic characteristics when both species are together (Elabbassy and Sitohy 1993). The respective effects of the proteolytic activity of *Lb. delbrueckii* subsp. *bulgaricus* and formic acid production by *S. thermophilus* on the metabolic activity of the other species were unknown until 'omics approaches were applied to this co-culture. Analyses combining 2-DE proteomic and gene expression using microarrayswere performed on the co-culture of both yogurt lactic acid bacteria (LAB) after growth on skimmed micro-filtered milk in order to compare the protein and gene expression of *S. thermophilus* alone or with *Lactobacillus* (Herve-Jimenez et al. 2008). This study confirmed the enhancement of *S. thermophilus* growth in co-culture but revealed that the main changes concerning sugar metabolism, Fe-S biosynthesis, amino acid transport and amino acid metabolism were equivalent in mono- and co-culture (Herve-Jimenez et al. 2008). Using the same approaches, these authors demonstrated that *S. thermophilus* growth was stimulated in milk co-culture with *Lb. delbrueckii* and noticed that genes and proteins involved in aminoacid metabolism were up-regulated. These modifications notably concerned the branched-chain aminoacids and arginine. The 'omics approach also enabled the authors to suggest that modifications occurring in iron metabolism regulation are part of the adaptive response of *S. thermophilus* to the production of H_2O_2 by *Lactobacillus* (Herve-Jimenez et al. 2009). Complementary results were obtained concerning the response of *Lb. delbrueckii* in the study of Sieuwerts et al. (2010) in which the transcriptome evaluated using microarrays for both species in milk co-culture was performed. They confirmed that, in a mixed culture, *Streptococcus* provides purine and folic acid for *Lactobacillus*,which down-regulates their biosynthesis pathways compared to its growth in mono-culture.

Positive interactions have also been investigated within more complex microbial food ecosystems using global methods. Kefir grains represent a small microbial community in which the succession of two main species, *Lb. kefiranofasciens* and *Leuconostoc mesenteroides*, has been evidenced using 16S rDNA barcoding. Metagenomics analysis has revealed a possible cooperation between both species since they implement different metabolic pathways and have a complementary participation in the volatilome composition when they grow together (Walsh et al. 2016). In a more complex ecosystem, like cheese, interactions between some Gram-negative bacteria known to play a role in flavor compound production and the bacterial community of the product have been investigated (Irlinger et al. 2012). These authors used model cheese inoculated with a consortium of 11 microorganisms involved in cheese ripening in the presence or absence of the Gram-negative strains. The volatilome analysis performed in the different co-culture combinations indicated an increase in flavor compound production in the presence of one of the Gram-negative strains, *Psychrobacter celer* (Irlinger et al. 2012).

When positive interactions occur between specific spoilage organisms, they are generally followed by negative consequences in terms of the quality of affected food. The production of metabolic compounds or the use of specific substrates providing favorable conditions for the production of spoilage metabolites by another species is usually called metabiosis (Gram et al. 2002). It has been described mainly in seafood products by co-inoculation of specific spoilage organisms and evaluation of the resulting effect by sensory analysis and spoilage indicator measurements (Macé et al. 2013, 2014). Following the same approach, Laursen et al. (2006) performed a volatilome analysis on cooked peeled shrimps stored under modified atmosphere after co-inoculation by *Brochothrix thermosphacta* in association with *Carnobacterium maltaromaticum, C. divergens* or *C. mobile*. They showed that specific off-odors, occurring only when two species are associated, are probably linked to the interaction between metabolites produced by both species. The increase in genome sequences of specific spoilage microorganisms (Remenant et al. 2015) will probably promote transcriptomics approaches to investigate the mechanisms involved in these interactions in the future.

3.2.3. Negative/adverse interactions. In complex environments, the behavior of foodborne pathogens or spoilage microorganisms is impacted by endogenous microbiota or specific microorganisms. Interactions between foodborne pathogens or spoilage bacteria and inhibiting bacteria have been extensively studied in culture media or food products by demonstrating inhibitory effects on bacterial growth or survival. Many studies have focused on a case of amensalism widely observed between LAB involving antimicrobial peptides called bacteriocins (Chikindas et al. 2017). During the last ten years, the specific responses of the target bacteria in antagonist interactions have been investigated using transcriptome, proteome or volatilome analyses.

In some cases, the antimicrobial components have already been identified and the associated methods have determined their molecular targets on the inhibited bacteria. For example, the RNA-seq transcriptome analysis of *Listeria monocytogenes* submitted to pediocin action revealed the over-expression of genes involved in the stress response, suggesting a global defense mechanism (Laursen et al. 2015). Similarly, the same method applied to co-cultures of *Lc. garviae* and *Staphylococcus aureus* in laboratory medium afforded a better understanding of the mechanisms involved in H_2O_2 inhibition (Delpech et al. 2017).

In other cases, these methods have produced new insights regarding the action of inhibiting bacteria on the specific activities of target bacteria. Concerning food pathogens, the expression profiles of virulence-related genes, such as *S. aureus*, have been investigated in the presence of inhibiting LAB in laboratory medium, milk or cheese (Cretenet et al. 2011, Delpech et al. 2015, Even et al. 2009, Nouaille et al.

2014, Zdenkova et al. 2016). Cretenet et al. (2011) explored the dynamics of *S. aureus in situ* by coupling a microbiological and, for the first time, a transcriptomics approach in a cheese matrix. This study highlighted the intimate link between environment, metabolism and virulence, as illustrated by the influence of the cheese matrix context, including the presence of *Lc. lactis*, on two major virulence regulators, environment-sensing systems like Agr and the two-component system SaeR/S. Concerning spoilage bacteria inhibition, volatilome analysis performed on a mono- or co-culture of *B. thermosphacta* and/or *Lc. piscium* on shrimp indicated that specific volatile compounds were not, or less, produced by the spoiler when co-cultured with the LAB with positive consequences on the sensory evaluation of the product (Fall et al. 2012).

Lastly, 'omics approaches may trigger new hypotheses about the inhibition mechanism, when not attributed to organic acids, bacteriocins or hydrogen peroxide. In this way, the antifungal effect of *Lb. plantarum* strains against *Aspergillus* species investigated by 2-D proteomics and microarray transcriptomics evidenced modifications in the expression of proteins involved in the respiratory chain, and under-expression of genes involved in ergosterol biosynthesis (Crowley et al. 2013, Strom et al. 2005). Also using microarray experiments, Nilsson et al. (2005) demonstrated that the inhibition of *L. monocytogenes* by a non-bacteriocinogenic *C. maltaromaticum* strain was due to glucose competition.

3.2.4. Recent studies of bacterial interactions in complex food ecosystems: Bacterial communication and network inference. Besides the characterization of metabolites produced by bacteria and involved in their interactions, the communication between bacteria and their hosts or environments could determine their behavior and should be considered. It occurs via signaling molecules, such as the autoinducer-2 family, which is found in all bacteria; peptides, which are used by Gram-positive bacteria; acyl-homoserine lactones and autoinducer 3, which are used by Gram-negative bacteria (Rul and Monnet 2015). It is generally dependent on population density and is commonly known as quorum sensing (QS). There are two kinds of quorum sensing: Species-specific and interspecies. A variety of phenotypes are controlled by quorum sensing: The synthesis of bacteriocins and the conjugal transfer of plasmids, genetic competence and the expression of virulence factors, biofilm formation and stress responses (Braga et al. 2016, Di Cagno et al. 2011, Skandamis and Nychas 2012). Proteomics and transcriptomics are the principal approaches used to study the multiple mechanisms of QS (Rul and Monnet 2015, Tian 2014).

Communication via signaling molecules is only beginning to be explored in food ecosystems. Microbial communication can potentially influence food quality by affecting both sensory quality, through the development of beneficial bacteria, and safety, through limiting the growth of pathogens and spoilers. For example, the involvement of QS in the regulation of exopolymeric substances (EPS) production and biofilm formation has been reported in spoilers and pathogenic microorganisms in foods or food processing devices, which makes them resistant to antimicrobial agents (Rul and Monnet 2015). This leads to food processing problems and consumer health risks. Microbial food spoilage is characterized by undesired changes in food components (e.g., off-odor, textural defects) that primarily result from the action of various microbial enzymes. In general, it is known that QS regulates the secretion of some extracellular enzymes (De Angelis et al. 2016, Van Houdt et al. 2007). Many bacteria isolated from foods produce signaling molecules such as acyl-homoserine lactones (Bruhn et al. 2004). Nevertheless, in meat products, in which *Enterobacteriaceae*, *Pseudomonadaceae*, and some LAB are the main producers of spoiling enzymes such as lipases and proteases, no link with QS controls *in situ* has yet been observed (Bruhn et al. 2004, Liu et al. 2006). Another example of intra- and interspecies communication via QS between food bacteria is bacteriocin-producing-dependent. Bacteriocin production is often regulated by peptide-dependent QS. The broad activity spectrum of bacteriocins has been exploited for inhibiting the outgrowth of spoilage microbes and foodborne pathogens such as *L. monocytogenes* (Mills et al. 2017).

The analytical methods of met-genomics or metabarcoding data for food microbial ecosystem characterization will also highlight specific concerns linked to bacterial interactions. For example, the screening of genes involved in the production of aromatic compounds in cheese or those linked to bacteriocin synthesis may help to identify the bacterial groups exhibiting potentially positive or negative interactions with other members of the community (Escobar-Zepeda et al. 2016). More generally, the application of the microbial network inference approach to food microbial ecosystems is emerging. These statistical models have been developed and applied to evidence co-occurrences and co-exclusions between species in environmental ecosystems and, more recently, in marine soil and gut microbial communities (Faust and Raes 2012). Alessandria et al. (2016) have recently applied these statistical analyses to metabarcoding 16S data obtained from Italian cheese microbial ecosystems. They observed strong exclusion between species of *Lactobacillus* and spoilage flora, such as *Pseudomonas*.

Most of the studies described above are related to the interactions between microbes in simplified model systems or defined systems with a limited microbial complexity. Only a few studies are available in which transcriptomics and proteomics approaches are used to study the interactions in complex environments. This can be explained by the microbial complexity of food ecosystems and the physicochemical composition of food matrices, which lead to technical challenges. The evolution of 'omics approaches and the recent emergence of genomics may boost research into the different strategies used by food microbes to interact, and their communication methods. Microbial interactions are key components in understanding the dynamics of microorganisms in any ecosystem and they require similar scientific approaches for environmental, gut and food microbial communities. Cooperation between microbiologists working in these areas will also open up new avenues leading to the development of knowledge about interactions in food microbial ecosystems and their consequences on quality and safety.

4. 'Omics Approaches to Study Foodborne Pathogens: The Example of *Campylobacter*

4.1 Scientific Context

This section deliberately focuses on *Campylobacter* as this microorganism is reported to be the main foodborne pathogen responsible for bacterial gastroenteritis in humans worldwide (EFSA 2016, Epps et al. 2013, Silva et al. 2011, Moore et al. 2005). The disease it causes is called campylobacteriosis and it contributes 58% of the global foodborne disease burden (Hald et al. 2016). The number of campylobacteriosis cases has increased significantly since 2005: In Europe, 236,851 cases were confirmed in 2014, 63% more than cases of salmonellosis and 99% more than those of listeriosis (EFSA 2016). For the first time, the European Commission amended Regulation (EC) No 2073/2005 on the hygiene of foodstuffs in 2017 with regard to *Campylobacter* on broiler carcasses, stating a limit of 1000 cfu/g. Human symptoms after ingestion of food contaminated by *Campylobacter* are fever and watery diarrhea,which can be bloody, with abdominal pain and possiblynausea and vomiting (Allos 2001, Altekruse et al. 1999, Epps et al. 2013, Hofreuter 2014, Moore et al. 2005, Silva et al. 2011). *Campylobacter jejuni* in particular can cause serious or disruptive sequelae with late on-set complications, such as the immune-mediated polyneuropathy Guillian Barré Syndrome (GBS) (Alshekhlee et al. 2008, Mortensen et al. 2009, Nyati and Nyati 2013, Sivadon et al. 2005, Shamshiev et al. 2000), Miller Fisher Syndrome (MFS) (Ang et al. 2001) and the reactive arthropathy Reiter Syndrome (ReS) (Altekruse et al. 1999).

The *Campylobacter* pathogen resides in a commensal way in poultry, pork and cattle and, to a lesser extent, in sheep, horses and pets (Corry and Atabay 2001, Denis et al. 2009, Guevremont et al. 2004, Janssen et al. 2006, 2008, Messaoudi et al. 2013, Stanley and Jones 2003, Hue et al. 2011). This Gram-negative spiral-shaped bacterium requires fastidious growth conditions, including a low level of oxygen, a high level of carbon dioxide, a temperature between 30°C and 45°C and a pH between 6.5 and 7.5 (Garenaux et al. 2008, Kaakoush et al. 2007, Silva et al. 2011, Mace et al. 2015). It remains a puzzle to scientists as to how an obligate microaerobic bacterium can survive from farms to retail outlets. The cutting edge of 'omics approaches applied to single bacterial strain populations (Fig. 4) are contributing to understandingthe diversity of genotypes and phenotypes according to the questions addressed in Table 1. The key results obtained on *Campylobacter* are developed below.

Table 1. From genotype to phenotypes. The targeted molecules include DNA, RNA, proteins, lipids, carbohydrates and secondary metabolites. Various phenotypes could result from one genotype according to environmental cues.

	What can happen?	What appear to happen?	What makes it happen?	What has happened?	What triggers host infections?	
G E N O T Y P E	Genomics Epigenomics	Transcriptomics	Proteomics Subproteomics	Metabolomics	Immunoproteomics	**P H E N O T Y P E S**
			Proteomics on PTMs			
			Proteocomplexomics Lipidomics			
			Glycomics			

4.2 What Can Happen?

The molecule responsible for what can happen in cells is deoxyribonucleic acid (DNA), the bio-information support that is divided into units named genes. This genetic code, formed by different genes, is bacterial strain-specific; although, similarities between bacterial strains can be found. This genetic code now constitutes the basis of strain identification.

To date, various methods of DNA sequencing to study bacterial genetic codes haveco-existed. Sequencing methods have evolved quickly during the last years

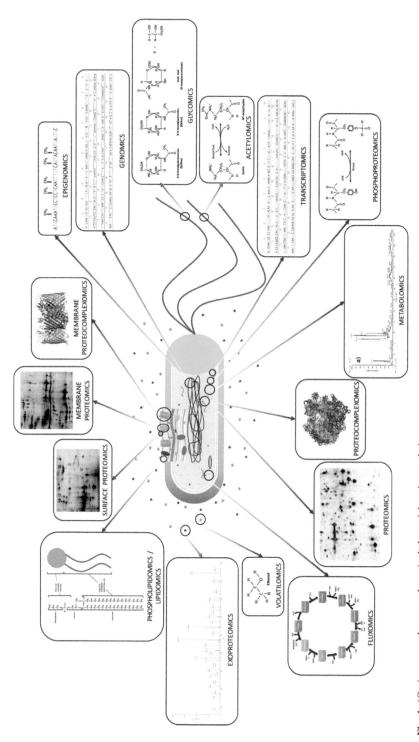

Fig. 4. 'Omics approaches to analyze single bacterial strain populations.

(Table 2). The first automated methods were implemented in 1990-2000. In 2001, it was possible to sequence 1500 nucleotides in a few days. Since 2016, around 1500 billion nucleotides can be sequenced in a few hours with the Illumina sequencer (Shendure et al. 2017). Concerning single-cell sequencing (SCS), there are various methods that can reveal functional differences such as pathogenicity, antibiotic resistance or transmissibility. The first-generation sequencing methodology was based on Sanger's technique (Shendure et al. 2017) and used didesoxyribonucleotides (ddNTP) with an atom of hydrogen instead of the OH grouping in carbon 3' of ribose. This method required a primer marked radioactively and four sequencing reactions were carried out in parallel with only one ddNTP (ddTTP, ddATP, ddCTP or ddGTP). The extension of the DNA molecule was stopped when a ddNTP was incorporated and several DNA fragments with different lengths were formed and separated by electrophoresis on polyacrylamide gel (Sanger et al. 1977, Shendure et al. 2017). For instance, this method was used to analyze the sequence and expression of the tetracycline-resistance gene from *C. jejuni* (Manavathu et al. 1988).

Later, the second generation of sequencing methodologies, also named NGS, was developed. These new technologies provide genomics and metagenomics data with an information production that is faster than with Sanger's method, less dangerous and more cost-effective (Escalona et al. 2016, Shendure et al. 2017, Diaz-Sanchez et al. 2013). NGS can be used for known or unknown genomes and genome annotation. Moreover, these sequencing methods enable the study of genetic variability and the polymorphism of single nucleotides. Various types of NGS are available for the study of DNA sequences, with Illumina currently being the most popular one on the market. Several complete genomes of *Campylobacter* have been described with this sequencing technology, such as a variant of *C. jejuni* NCTC 11168 (Revez et al. 2012, 2014). DNA sequencing can also be achieved by the Oligonucleotide Ligation and Detection technology (SOLiD), the Roche 464 or the IonTorrent technologies (Escalona et al. 2016, Liu et al. 2012, Shendure and Ji 2008, Diaz-Sanchez et al. 2013). These NGS technologies may differ according to the read length, the time of sequencing or the capacity of sequencing (Mb) (Diaz-Sanchez et al. 2013, Escalona et al. 2016, Liu et al. 2012, Shendure and Ji 2008).

The third generation of sequencing, with Pacific Biosciences (PacBio) technology (Quail et al. 2012) and Oxford Nanopore technologies such as MinIon (Laver et al. 2015, Quick et al. 2014), has recently been implemented. These technologies provide read lengths between 3,000 bp and 10,0000 bp although error rates are more frequent when compared to NGS because they do not use a cyclic method (Laver et al. 2015, Quail et al. 2012, Quick et al. 2014). Complete genome sequences of *Campylobacter*

Table 2. Characteristics of NGS technologies.

Technology	Illumina	SOLiD	Ion Torrent	Roche 454
Sequencing time/run	26 hr–14 d	8 d–12 d	2 hr	10 hr–20 hr
Sequencing capacity (Mb)/run	1,500–200,000	70,000–80,000	1,000	500–900
Reads lenght (bp)	300	75	400	700
Cost/run (€)	750–20,000	6,000–10,000	500 - 950	6,200
Error rates	0.0034%–1%	0.01%–1%	1.78%	1.07%– 1.7 %

d, days ; hr, hours

have been obtained with PacBio technology, such as for the *C. jejuni* RM1285 genome (Gunther et al. 2015). Later, the methylome of bacteria was investigated using these sequencing methods (NGS, PacBio, Nanopore) combined with single molecule real-time (SMRT) analysis (Murray et al. 2012). Methylation of the bacterial genome plays an important role in cells as it is the main system involved in restriction modification (RM) (O'Loughlin et al. 2015). Several studies of the methylome have been performed in various strains of *Campylobacter*, using mainly PacBio technology (Anjum et al. 2016, Mou et al. 2015, 2017, Murray et al. 2012, O'Loughlin et al. 2015, Zautner et al. 2015).

Whole-genome sequencing (WGS) provides the entire bacterium genome sequence in order to compare it to known reference sequences. Sequencing bacterial genomes can be useful for food in public health tracking, infectious disease surveillance, molecular epidemiology studies and environmental metagenomics. The conventional typing methods (serotyping, phenotyping, PFGE and AFLP) are gradually being replaced by WGS. Furthermore, the low cost of NGS facilitates the application of WGS for routine surveillance and outbreak investigations of bacterial infectious diseases, such as *Campylobacter* infections, by public health authorities (Koser et al. 2012, Llarena et al. 2017, Pendleton et al. 2013, Lefebure et al. 2010). WGS can be used to compare several *Campylobacter* genomes with target genes, such as for Multilocus Sequence Typing (MLST) (Dingle et al. 2005, Sheppard et al. 2012, van Rensburg et al. 2016). Metagenome sequencing applications encompass the discovery of unknown microorganisms, the characterization of uncultivable microorganisms and the discovery of unknown gene products that operate only in the presence of other microbial populations. WGS provides opportunities to detect genes and proteins involved in the pathogenesis of *Campylobacter* by pinpointing specific genes already described in other foodborne pathogens (Diaz-Sanchez et al. 2013, Parkhill et al. 2000, Stahl and Stintzi 2011). Furthermore, WGS could be used to predict the antimicrobial resistance phenotype in *Campylobacter* spp., as explored amongst 114 *Campylobacter* species isolates by Zhao et al. (2016).

4.3 What Appears to Happen?

Metagenomics studies are crucial in understanding the bacterial genome organization. However, genes are not constantly transcribed. Housekeeping genes are constitutively expressed, while the expression of other genes can be enhanced or silenced. Gene expression transition is mainly operated by environmental cues. To obtain more information on what can happen in the bacterial cell, transcriptomics approaches constitute the studies of choice.

The transcriptome corresponds to all the RNA stemming from the genome transcription. It subsequently varies according to the transient conditions undergone by bacterial cells. Two technologies are currently used: Microarrays and RNA-sequencing (RNA-seq). Microarray technology has provided results that have improved our understanding of foodborne disease progression and the development of therapeutic methods. This approach enables the expression of thousands of genes to be simultaneously analyzed with probes targeting unique sequences of mRNA transcripts. Only the transcripts of known genes can be detected and analyzed, which is why the transcriptome analysis is incomplete. Several studies concerning the ATR to acid of *C. jejuni* have been performed using transcriptomics approaches. These were aimed at monitoring the expression of all genes or characterizing the impact of one molecule, such as the ferric uptake regulator Fur in *C. jejuni* under acid stress (Askoura et al. 2016, Birk et al. 2012, Reid et al. 2008a,b, Varsaki et al. 2015). Other transcriptomics analyses on *Campylobacter* explored the effect of oxidative stress exposure (Flint et al. 2014, Koolman et al. 2016, Palyada et al. 2009) by carrying out a step of RNA extraction by real-time reverse transcription quantification PCR (RT-qPCR) before the microarray analysis (Askoura et al. 2016, Birk et al. 2012, Reid et al. 2008a,b, Varsaki et al. 2015).

The RNA-seq technology leads to rapid profiling of the transcriptome, for any bacterial species. This approach is ten-hundred times more specific and sensitive in detecting differential expression than microarray technology. Moreover, RNA-seq does not require transcript-specific probes and can detect novel transcripts, small RNAs, single nucleotide variants or gene fusion in contrast to microarrays. It can also detect and quantify RNA expressed at very low levels, unlike microarray analysis. However, few studies related to food have yet been performed on *Campylobacter* using RNA-seq technology, contrary to microarrays. Chaudhuri et al. (2011) compared the ability of RNA-seq and microarray technologies in order to identify differentially expressed genes in *C. jejuni* NCTC 11168 with a focus on the transcriptomics analysis of an *rpoN* mutant (Chaudhuri et al. 2011).

4.4 What Makes it Happen?

4.4.1. General proteomics analyses.
The molecules responsible for what happens in cells are proteins. In bacteria, proteins are present in all compartments. They can mediate cellular function or be part of the cell structure. As enzymes, they are responsible for metabolic reactions. When encased in the membrane, they participate in the bacterial structure. When associated with other proteins, they form protein machineries involved in molecule trafficking, the respiration process, translation or DNA replication and repair. As regulators, they can control gene expression and protein translation.

Proteins can be continuously expressed for housekeeping reasons or specifically expressed in response to any stress, such as cold/heat shock proteins, and so contribute to cell adaptation or cell survival. To explore protein expression at the cellular level, proteomics studies in bacteria have developed over the last three decades. These investigations have resulted in a snapshot view of a set of proteins.

In *Campylobacter*, the main proteomics studies have analyzed the adaptive tolerance response (ATR) to stressful environmental conditions or the characterization of the mutant phenotype. The ATR to acidic exposure was investigated using proteomics in order to understand better the survival of *C. jejuni* in gastric conditions or in food containing organic acids for food preservation (Birk et al. 2012). Since ambient oxygen concentrations are detrimental to *C. jejuni* survival, the proteomics approach was able to pinpoint specific enzymes, membrane proteins or regulators involved in the ATR to oxidative stress (Garenaux et al. 2007, 2008, Kang et al. 2016, Sulaeman et al. 2012, Rodrigues et al. 2016). Furthermore, the effect of temperature was compared between *C. jejuni* cells cultivated at 37°C (human body temperature) and 42°C (chicken body temperature) using proteomics (Rathbun et al. 2009, Zhang et al. 2009). These studies revealed that the immunogenic membrane protein Peb4 was more abundant at 37°C than at 42°C. The passage of *C. jejuni* into the human intestine is also affected by bile, mainly due to the presence of salts. By comparing the 2-D electrophoretic profiles of *C. jejuni* after exposure to 2.5% ox-bile, bile was considered to stimulate the expression of virulence host factors, such as Cia and flagellum proteins (Clark et al. 2014, Fox et al. 2007, Malik-Kale et al. 2008). Since iron homeostasis is critical for *C. jejuni* survival, as it is directly connected to reactive oxygen species accumulation during oxidative stress via the Fenton reaction, the proteomics approach was also chosen by Holmes et al. (2005) in order to define proteins differently expressed under iron starvation in *C. jejuni*. Proteins involved in iron uptake and iron pumping systems were detected, including ChuA, CfrA, P19 and CfbpA (Holmes et al. 2005). However, not all partners of these iron trafficking systems could be detected in this study. This could be explained by the limitations of the first applications of proteomics, which detected mainly cytosoluble proteins as the gel-based technologies using isoelectrofocalization for the first dimension required non-ionic detergents and chaotropic reagents, which were not compatible with the solubilization of proteins harboring hydrophobic regions. Mutant characterization of *C. jejuni* was used in order to determine in which phenotype or molecular process specific proteins were involved. This was the case with the proteomics analyses of *C. jejuni* NCTC 11168 and 81–176 mutants defective in the *peb4* gene, and consequently in the expression of Peb4, which impairs epithelial cell adhesion, culturability after oxidative stress, biofilm formation and mouse model colonization (Asakura et al. 2007, Rathbun et al. 2009). With the elucidation of the crystal structure of Peb4 (Kale et al. 2011) and, more recently, its PPIase activity (Taylor et al. 2017), this outer membrane (OM) protein was defined as a chaperone involved in OM biogenesis and assembly. In another example, the proteomics analysis of a mutant defective in the putative polynucleotide phosphorylase (PNPase) resulted in an alteration of protein expression in virulence, motility, stress response and translation (Haddad et al. 2012). Subsequently, the exoribonucleotic activity of potential ribonucleases in *C. jejuni* was demonstrated (Haddad et al. 2014).

With the advent of high throughput spectrometry mass analyses, gel-free methods have emerged in the last ten years as an alternative to 2-D electrophoresis, in which the relative quantification of proteins is based on chemical or metabolic labeling. Chemical labeling was applied, using iTRAQ to quantify protein abundance changes after the passage of *C. jejuni* in chicken, phage insertion, iron starvation or bile salt stress (Clark et al. 2014, Asakura et al. 2016). Notably, Clark et al. (2014) detected more proteins involved in pumping and iron uptake systems under iron deprivation. In fact, gel-free and gel-based methods are complementary. For instance, in *Campylobacter concisus*, proteins extracted using LTQ-FT/MS were more enriched in protein complexes while the gel-based technique detected a lower protein abundance (Kaakoush et al. 2010).

4.4.2. Compartmentalization analyses. Specific detergents compatible with isoelectrofocalization, such as tributylphosphine (TBP), amidosulfobetaine-14 (ASB-14) and lauryl-sarkonisate, have been proposed and tested in order to enhance the solubilization of hydrophobic proteins for proteomics studies. In *C. jejuni,* these detergents were applied in order to extract and identify whole membrane proteins (Asakura et al. 2007, Cordwell et al. 2008) or to separate outer from inner membrane proteins (Sulaeman et al. 2012, Watson et al. 2014). These new tools have opened up new perspectives in proteomics for exploring different compartments of bacterial cells. For instance, Sulaeman et al. (2012) compared the membrane proteome of

C. jejuni under microaerobic and oxygen-enriched conditions and found that the expression of the virulent protein CadF is mediated by oxygen and involved in adhesion to inert surfaces. Other compartments have been explored in *C. jejuni*, such as the exoproteome and the outer membrane vesicle (OMV) proteome (Elmi et al. 2012, 2016, Kaakoush et al. 2010). OMVs are small exosomes, around 50 nm in diameter, released by bacterial cells. A proteomics approach to *C. jejuni* OMVs led to the identification of three proteases able to cleave epithelial cell E-cadherin and occludin, indicating a potential function of these OMVs in epithelial cell invasion (Elmi et al. 2012, 2016).

Protein expression is the result of gene expression and post-transcriptional regulations. Various isoform species of proteins can be detected. Switching from one isoform to another contributes to the tuning of bacterial adaptability by the activation of proteins.

4.4.3. Post-translational modifications (PTM). Multiple post-translational modifications (PTMs) occur in bacteria, which could compensate for the simpler organization of prokaryotes compared to eukaryotes (Cain et al. 2014). In order to detect proteins transformed by PTMs, proteomics approaches can be applied when a pretreatment is available to isolate these specifically transformed proteins and/or when mass spectrometry can identify the modification. The diversity of glycosylation is especially studied in *Campylobacter* as both *O*-linked and *N*-linked glycosylation systems are present (Nothaft and Szymanski 2013). In this microorganism, evenness is higher for *O*-linked proteins while richness is higher for *N*-linked proteins. To detect *N*-glycosylated proteins at the cell level, the *N*-linked glycoproteome was first analyzed by Young et al. (2002) using specific LC-MS hydrophilic interaction liquid chromatography (HILIC) strategies. In this study, the authors described 38 glycoproteins localized in the periplasmic space. Then, by separating *N*-glycoproteins on lectin affinity columns, followed by 2-DE gels before identification by LC-MS HILIC, 81 individual *N*-glycosylation sites on 53 proteins were described (Scott et al. 2011), and later 30 additional sites (Scott et al. 2014). These studies contributed to the elucidation of the biosynthesis pathway of *N*-glycoconjugates in *C. jejuni* (Nothaft and Szymanski 2013). The advent of cutting-edge technologies for MS has also provided new opportunities for analyzing the phosphoproteome in order to detect which proteins could be activated by the phosphorylation process. In *C. jejuni*, Voisin et al. (2007) have identified 36 phosphoproteins using LC-MS/MS of tryptic-digested phosphoproteins previously separated on SDS-PAGE.

PTMs also include subtle modifications of proteins, such as acetylation and/ or methylation, in response to environmental cues, which can be investigated using proteomics approaches: Protein acetylomics and methylomics, respectively. Acetylated and methylated proteins can be enriched by immunoprecipitation and identified using LC-MS/MS (Ouidir et al. 2016). In bacteria, *N*-acetylation of proteins was detected at the N-terminus or on lysine residues and *O*-acetylation on serine and threonine residues (Ouidir et al. 2016). Lysine acetylation has been investigated using proteomics tools in several bacterial species, including foodborne pathogens such as *Salmonella* (Wang et al. 2010) and *Vibrio parahaemolyticus* (Pan et al. 2014), and is currently being studied in *C. jejuni* (personal communication). Besides histone methylation in eukaryotes, research has shown that protein methylation occurs in bacteria with the discovery of *N*-methyl lysine in the flagella of *Salmonella* Typhimurium in 1959 (Ambler and Rees 1959). Mono-, di- or tri-methylation of proteins are among the processing mechanisms involved in protein activity or turnover and protein-protein interactions (Murn and Shi 2017). Lysine and arginine are the amino acids most frequently exposed to methylation via methyltransferases. Although the quantification of methylation sites from the proteome became accessible with the development of the heavy methyl SILAC method (Ong et al. 2004), studies of the decoration of proteins by methyl residues in bacteria using proteomics approaches are still in their infancy.

4.4.4. Cellular process functionalization. Beyond protein expression and protein processing for its activation, many proteins are not able to trigger cellular processes by themselves. These are mainly carried out by sophisticated multi-subunit protein machineries, i.e., different protein complexes maintained by stable protein interactions. These functional entities can be defined as protein complexes composed of a minimal biological structure of assembled protein subunits necessary for a specific cellular process (Reisinger and Eichacker 2008a,b). Identifying functional multi-subunit entities at the proteome level—apart from salient exceptions such as ribosomal proteins or mRNA transcription—is a new challenge. To achieve this, two-dimensional (2-D) blue native (BN)/SDS-PAGE (Dresler et al. 2011) or high resolution clear native electrophoresis (hrCNE) (Dieguez-Casal et al. 2014) of extracted intact protein complexes using mild non-ionic and non-denaturing detergents have been developed. They have been successfully applied to monitor the oligomeric state, stoichiometry and protein subunit composition of protein complexes at the cell level (Bernarde et al. 2010) and at the membrane level (Wohlbrand et al. 2016). The study of the membranar proteocomplexome of *Campylobacter* using 'omics approaches is currently under investigation (personal communication).

4.4.5. Lipidomics/phopholipidomics. Lipids are molecules involved in the membrane structuring of bacterial cells and are part of the energy storage. They are overall soluble in organic solvents. Lipids are classified according to their sequence and branched forms, consequently most studies are lipid family-dependent. For instance, the plasmalogen family of phospholipids is used for glycerophospholipid studies (Kolek et al. 2015, Rezanka et al. 2012). Global multi-dimensional analyses on all lipids are called lipidomics, while sublipidomics analyses are named according to the studied family, e.g., phospholipidomics for phospholipids. Technologies for the identification and quantification of lipids are essentially based on gas or liquid chromatography (GC, LC) coupled to MS or nuclear magnetic resonance (NMR) (Tam 2013). Few lipidomics studies have been conducted onbacteria and none of them has yet concerned *Campylobacter*. For instance, a phospholipidomics analysis was performed on *Clostridium pasteurianum* in order to decipher the membrane phospholipid composition using high resolution electrospray mass spectrometry (ESI-MS) (Kolek et al. 2015). Another example involved the pathogen *Pseudomonas aeruginosa* with a comparison of the distribution of lipid classes during biofilm and planktonic growth using GC-MS analyses (Benamara et al. 2014).

4.5 What has Happened?

Secondary metabolite production represents the signature of the biological activity of bacterial cells. The global approach of making an inventory of the small molecules produced by bacterial populations is metabolome analysis or metabolomics. Secondary metabolites are separated and identified using mainly mass spectrometry (MS) and nuclear magnetic resonance (NMR) spectroscopy, which have been developed according to analytical descriptors (Covington et al. 2017). Recently, combined analyses of genomics data and the small molecule prediction approach to metabolomics data have enhanced research into the nature and function of intracellular secondary metabolites. In *C. jejuni,* metabolomics was applied in order to define the metabolic fingerprint of mutants defective in the main carbon-source amino acid supply (Howlett et al. 2014) and to profile the metabolic changes in antibiotic-resistant strains (Li et al. 2015).

4.6 What Triggers Host Infections?

Infection of humans by *Campylobacter* is triggered during intestine cell adhesion and invasion. The immune response of the human body is mediated by specific bacterial antigenic proteins. Immunoproteomics analyses enable large sets of immunoreactive proteins that could trigger the immune response to be studied. Technologies to carry out immunoproteomics analyses include two-dimensional electrophoresis (2-DE) and two-dimensional Western blotting (2D-WB) for protein separation and immunogenic detection, quantitative real time RT-PCR for gene expression analysis and matrix-assisted laser desorption ionization-time of flight mass spectrometry (MALDI-TOF MS) for protein identification (Hu et al. 2013). Immunogenic proteins of *C. jejuni* were screened also using subproteomics analyses of surface proteins (Prokhorova et al. 2006, Mehla and Ramana 2017).

5. Weaknesses and Opportunities of Global Approaches in the Context of Food Safety and Quality

5.1 Weaknesses

5.1.1. Data generation and handling. Although global approaches have already revolutionized our view of food microbiology, some weaknesses of such approaches should be mentioned. Most studies dealing with metagenomics or metabarcoding have been dedicated to the description of bacterial communities present on various foodstuffs. Only a few were targeted or revealed other organisms as viruses, fungi or *Archaea*. This results mainly from the databases that are available and from the ease of identification through DNA sequence targets. Identification of bacterial species through 16S rDNA sequence now has a long history of use and vast curated 16S rDNA sequence databases are publicly available. However, fungi identification still requires attention and the development of accurate tools. Also, identification of viruses cannot yet be assessed only through metagenomics and RNA viruses still escape most studies which are based on DNA sequencing. The use of other targets, such as housekeeping genes instead of rDNA, has been reported for a better description of bacterial communities in complex environments. For instance, *gyr*B has proven to be an interesting alternative in assessing bacterial species or genera present on seeds (Barret et al. 2015). Although not yet reported for food analysis, this alternative is worth considering as the variable regions of the 16S rDNA sequence sometimes fail to discriminate some species, like *Pseudomonas* sp. or LAB, present in foodstuffs. However, alternative genes require the generation and curation of dedicated databases. Global approaches have also sometimes revealed the presence of unknown species (Chaillou et al. 2015) that could not be identified precisely, simply because no sequence was available for them yet. So, whatever the targeted gene used for identifying microbiota in food, the availability of databases to be used may impact the robustness of the results.

For metagenomics and metatranscriptomics analyses, the existence of complete draft genome sequences is a prerequisite. Nowadays, plenty of bacterial genome sequences are available. However, draft genomes with putative sequencing errors and poor or automatic annotations represent the majority of biases and may lead to erroneous conclusions. Conversely, the absence of genome sequence of a species impairs its transcriptome analysis.

Another weakness of those big data analyses is their requirement of strong bioinformatics, bio-computing and bio-statistics skills. Indeed, several pipelines are still being developed for 'omics and meta-omics approaches which require expertise for raw data handling and for generation and interpretation of the results. Such investigations require heavy laboratory experiments and computing. The storage of a huge amount of data must be also mentioned. In addition, the large variety of procedures (sampling, nucleic acid extractions, sequencing procedures, databases and pipelines) that are applied make it difficult to compare the results reported in the literature.

As was the case for genome sequencing a decade ago, 'omics and meta-' omics are still expensive, although they become more and more affordable as time progresses. Nevertheless, their cost and the burden of the experiments they require often lead to the performance of a small number of replicates. As a consequence, the robustness of the data must be considered critically.

In addition, when compared to other complex ecosystems, such as environmental ones of animal digestive tract microbiota, food matrices present specific features: (i) they represent systems with a short shelf life (usually from few days to few weeks); (ii) the microbial communities they host are usually composed of a relatively small number of species, most of them being known and cultivated; (iii) a fast dynamics of microbiota, shaped by storage or manufacturing conditions is observed. As an example, poultry cuts are often poorly contaminated after slaughtering, i.e., 2–3 log CFU/g (Rouger et al. 2017a). During shelf life, usually 1–2 weeks long, the bacterial load can reach up to 8 log CFU/g and diversity changes depending on the storage conditions (Rouger et al. 2017a, 2018). Thus, global analyses at the beginning of the process may be impaired because of the low amount of DNA and data must be considered depending on the time period of sampling.

5.1.2. Technical limitations. The first limitations are directly linked to the extraction and quality of the biological material, nucleic acids, proteins or metabolites that are required to apply the 'omics' approaches.

Food matrices are characterized by their diversity in composition and structure and the recovery of bacterial DNA requires appropriate extraction methods. In most cases, bacterial DNA kit extraction is applied after different preparation steps, including filtration on columns retaining eukaryotic DNA, mechanical lysis (Chaillou et al. 2014) and proteolysis (Escobar-Zepeda et al. 2016). The optimization of extraction protocols may be optimized for each type of food according to diversity indices calculation, or using challenge-tests with known bacterial species or bacterial communities already characterized (Rouger et al. 2017b). The protocols of DNA extraction also have to be reconsidered for some bacteria occurring as spores that are particularly resistant to lysis.

The question of RNA extraction for transcriptomics or metatranscriptomics analysis remains a critical step in food. RNA does not have the stability of DNA and is rapidly degraded in the food matrix by RNAses coming from autolysis and apoptosis of somatic cells. The main solutions are to use RNA protecting agents during extraction but, for DNA, the characteristics of each matrix have to be considered with a specific optimization.

For both metagenomics and metatranscriptomics methods, the extraction of extra-bacterial nucleic acids, coming from the animal or vegetal tissues, will disturb the analysis. The efficiency of bioinformatics data analysis is then crucial in removing sequences coming from eukaryotic DNA. With the integration of an amplification step in 16S metabarcoding analysis, this eukaryotic DNA contamination is reduced but may, therefore, occur with chloroplast DNA for food from vegetal origin. On the other hand, PCR amplification will also add some limitations with the presence of Taq polymerase inhibitors, primer selection and variability of target gene copy number, as discussed previously in § 2.2.

Concerning bacterial protein extraction that is usually performed on growth media, the difficulties are related to the ability to separate proteins from different compartments of the cells and to avoid protein cross contamination when studies are dedicated to a specific compartment. The main steps that required adjustments, depending of the bacterial strains, are protein extraction, purification and separation. To overcome fastidious requirements of gel-based method which are too limited to detect hydrophobic proteins, gel-free methods (shotgun) were developed. However, quantification is still critical using isotope or isobaric labelling, as they might disturb the bacterial response or protein identification. In addition, detection of isoforms suggesting PTMs could not be detected in protein lysates when shotgun approaches are applied.

The second set of limitations concerns the proportion of different species inside the communities and their physiological state, especially for DNA based methods. The detection of subdominant communities, like pathogenic bacteria, using metagenomics or metabarcoding methods remains a challenge in food, considering their low abundance compared to the other bacteria. The main actual solutions are to complete NGS analysis by specific detection of target pathogens with specific q-PCR methods, including an enrichment step if necessary. However, the focus on specific genes involved in virulence or antibiotic resistance may be investigated in order to consider sub-populations related to safety risks in food (Escobar-Zepeda et al. 2016).

One of the main criticisms addressed at bacterial DNA-based methods in food is usually that they do not give information about the viable or non-viable status of bacteria. Some technical solutions have been suggested as a way to circumvent this problem, such as using intercalants like propidium monoazide (PMA) that are able to enter dead cells and block further DNA amplification (Mamlouk et al. 2012).

The other approaches are to pass through cDNA for metagenomics or to combine metagenomics or metabarcoding data with RNA-based methods (transcriptomics or metatranscriptomics).

The DNA and RNA-based 'omics approaches will undoubtedly bring a better understanding of food microbiota dynamics and interactions, however their combination with classical microbiology, molecular analysis targeted on specific species or function, metabolic activities measurements, is recommended for overcoming the technical limitations linked to these new technologies.

5.2 Opportunities

Holistic approaches are not paradigm-driven and create conditions for serendipity. These global 'omics' methods move towards an exhaustive response of a biological system at one point. They require a quantity of raw material that could not be obtained from a unique bacterial cell. This is the reason why a single bacterial population constitutes the base of the definition of these 'omics' approaches. When the question is addressed at a microbial community, this supra-approach is called meta-omics analysis, while when it concerns a specific compartment of the bacteria, this infra-approach is called sub-omics analysis. In the former case, the studied molecules define the analysis (e.g., metatranscriptomics, metaproteomics) and in the latter, the biological compartment is mentioned (e.g., membrane proteomics, surface proteomics). Future analyses might combine both approaches, for instance, the analysis of membrane proteins of a bacterial community (e.g., membrane metaproteomics). When the physical interaction or assemblage between subunits is investigated, the methodology is called complexome (ex. for proteins: proteocomplexomics). If the question is addressed to only membrane proteins, the methodology will be called membrane proteocomplexomics. Further analyses might explore metaexoproteocomplexomes (the extracellular protein complexes of the bacterial community).

The priority of these methods is to describe, at the molecular level, the biological state of a viable microorganism or microbial community. The cutting edge technologies developed for this purpose are complex, expensive and require scientific expertise. As a consequence, to date, these methodologies are limited to dynamics experiments and industrial applications. They currently produce results that are used as starters to decipher biological mechanisms or biologically active molecules. In the future, these technologies would be valuable to be applied to the food safety and quality field in order to quantify all the potential benefit and harmful bacteria present on or in food products in an all-in-one analysis. As mentioned above, major pitfalls, biases and locks remain to be solved. Then, tracking microbial dynamics of food microbiota throughout the life of the food product would be feasible. Meanwhile, description of food bacterial consortia, their function and interaction might help to identify main traits, molecular markers or sentinel bacteria that would be more easily incorporated into food safety risk assessment models and meet food safety and quality objectives.

Acknowledgments

This work was financially supported by Pays de la Loire Project CompCamp RFI Food for Tomorrow.

References

Abu Al-Soud, W. and P. Rådström. 1998. Capacity of nine thermostable DNA polymerases to mediate DNA amplification in the presence of PCR-inhibiting samples. Appl. Environ. Microbiol. 64(10): 3748–3753.

Abu Al-Soud, W. and P. Radstrom. 2000. Effects of amplification facilitators on diagnostic PCR in the presence of blood, feces, and meat. J. Clin. Microbiol. 38(12): 4463–4470.

Agrimonti, C., L. Bortolazzi, E. Maestri, A.M. Sanangelantoni and N. Marmiroli. 2013. A real-time PCR/SYBR green I method for the rapid quantification of Salmonella enterica in poultry meat. Food Analytical Methods 6(4): 1004–1015.

Alessandria, V., I. Ferrocino, F. De Filippis, M. Fontana, K. Rantsiou, D. Ercolini and L. Cocolin. 2016. Microbiota of an Italian Grana-like cheese during manufacture and ripening, unraveled by 16S rRNA-based approaches. Appl. Environ. Microbiol. 82(13): 3988–3995.

Allos, B.M. 2001. *Campylobacter jejuni* infections: Update on emerging issues and trends. Clin. Infect. Dis. 32(8): 1201–1206.

Alshekhlee, A., Z. Hussain, B. Sultan and B. Katirji. 2008. Guillain-Barre syndrome - incidence and mortality rates in US hospitals. Neurology 70(18): 1608–1613.

Altekruse, S.F., N.J. Stern, P.I. Fields and D.L. Swerdlow. 1999. *Campylobacter jejuni* - An emerging foodborne pathogen. Emerg. Infect. Dis. 5(1): 28–35.

Ambler, R.P. and M.W. Rees. 1959. Epsilon-N-methyl-lysine in bacterial flagellar protein. Nature 184: 56–57

Ang, C.W., M.A. de Klerk, H.P. Endtz, B.C. Jacobs, J.D. Laman, F.G. van der Meché and P.A. van Doorn. 2001. Guillain-Barre syndrome- and Miller Fisher syndrome-associated *Campylobacter jejuni* lipopolysaccharides induce anti-GM(1) and anti-GQ(1b) antibodies in rabbits. Infect. Immun. 69(4): 2462–2469.

Anjum, A., K.J. Brathwaite, J. Aidley, P.L. Connerton, N.J. Cummings, J. Parkhill, I. Connerton and C.D. Bayliss. 2016. Phase variation of a Type IIG restriction-modification enzyme alters site-specific methylation patterns and gene expression in *Campylobacter jejuni* strain NCTC11168. Nucleic Acids Res. 44 (10): 4581–4594.

Asakura, H., M. Yamasaki, S. Yamamoto and S. Igimi. 2007. Deletion of peb4 gene impairs cell adhesion and biofilm formation in *Campylobacter jejuni*. FEMS Microbiol. Lett. 275(2): 278–285.

Asakura, H., K. Kawamoto, S. Murakami, M. Tachibana, H. Kurazono, S. Makino, S. Yamamoto and S. Igimi. 2016. *Ex vivo* proteomics of *Campylobacter jejuni* 81–176 reveal that FabG affects fatty acid composition to alter bacterial growth fitness in the chicken gut. Res. Microbiol. 167(2): 63–71.

Askoura, M., S. Sarvan, J.F. Couture and A. Stintzi. 2016. The *Campylobacter jejuni*ferric uptake regulator promotes acid survival and cross-protection against oxidative stress. Infect. Immun. 84(5): 1287–1300.

Barret, M., M. Briand, S. Bonneau, A. Préveaux, S. Valièred, O. Bouchez, G. Hunault, P. Simoneau and M.-A. Jacques. 2015. Emergence shapes the structure of the seed microbiota. Appl. Environ. Microbiol. 81(4): 1257–1266.

Benamara, H., C. Rihouey, I. Abbes, M.A. Ben Mlouka, J. Hardouin, T. Jouenne and S. Alexandre. 2014. Characterization of membrane lipidome changes in *Pseudomonas aeruginosa* during biofilm growth on glass wool. Plos One 9(9):e108478.

Bernarde, C., P. Lehours, J.P. Lasserre, M. Castroviejo, M. Bonneu, F. Mégraud and A. Ménard. 2010. Complexomics study of two *Helicobacter pylori* strains of two pathological origins: Potential targets for vaccine development and new insight in bacteria metabolism. Mol. Cell. Proteomics 9(12): 2796–2826.

Birk, T., M.T. Wik, R. Lametsch and S. Knochel. 2012. Acid stress response and protein induction in *Campylobacter jejuni* isolates with different acid tolerance. BMC Microbiol. 12: 174.

Braga, R.M., M.N. Dourado and W.L. Araujo. 2016. Microbial interactions: Ecology in a molecular perspective. Braz. J. Microbiol. 47 Suppl 1: 86–98.

Bruhn, J.B., A.B. Christensen, L.R. Flodgaard, K.F. Nielsen, T.O. Larsen, M. Givskov and L. Gram. 2004. Presence of acylated homoserine lactones (AHLs) and AHL-producing bacteria in meat and potential role of AHL in spoilage of meat. Appl. Environ. Microbiol. 70 (7):4293–302.

Bülte, M. and P. Jakob. 1995. The use of a PCR-generated invA probe for the detection of *Salmonella* spp. in artificially and naturally contaminated foods. Int. J. Food Microbiol. 26(3): 335–344.

Cain, J.A., N. Solis and S.J. Cordwell. 2014. Beyond gene expression: The impact of protein post-translational modifications in bacteria. J. Proteomics 97: 265–286.

Cao, Y., S. Fanning, S. Proos, K. Jordan and S. Srikumar. 2017. A review on the applications of next generation sequencing technologies as applied to food-related microbiome studies. Front. Microbiol. 8: 1829.

Chaillou, S., S. Christieans, M. Rivollier, I. Lucquin, M.C. Champomier-Vergès and M. Zagorec. 2014. Quantification and efficiency of *Lactobacillus sakei* strain mixtures used as protective cultures in ground beef. Meat Sci. 97(3): 332–338.

Chaillou, S., A. Chaulot-Talmon, H. Caekebeke, M. Cardinal, S. Christieans, C. Denis, M.H. Desmonts, X. Dousset, C. Feurer, E. Hamon, J.J. Joffraud, S. La Carbona, F. Leroi, S. Leroy, S. Lorre, S. Macé, M.F. Pilet, H. Prévost, M. Rivollier, D. Roux, R. Talon, M. Zagorec and M.C. Champomier-Vergès. 2015. Origin and ecological selection of core and food-specific bacterial communities associated with meat and seafood spoilage. ISME J 9(5): 1105–1118.

Chaudhuri, R.R., L. Yu, A. Kanji, T.T. Perkins, P.P. Gardner, J. Choudhary, D.J. Maskell and A.J. Grant. 2011. Quantitative RNA-seq analysis of the *Campylobacter jejuni* transcriptome. Microbiol. SGM 157: 2922–2932.

Chikindas, M.L., R. Weeks, D. Drider, V.A. Chistyakov and L.M. Dicks. 2017. Functions and emerging applications of bacteriocins. Curr. Opin. Biotechnol. 49: 23–28.

Clark, C.G., P.M. Chong, S.J. McCorrister, P. Simon, M. Walker, D.M. Lee, K. Nguy, K. Cheng, M.W. Gilmour and G.R. Westmacott. 2014. The CJIE1 prophage of *Campylobacter jejuni* affects protein expression in growth media with and without bile salts. BMC Microbiol. 14: 70.

Cline, J., J.C. Braman and H.H. Hogrefe. 1996. PCR fidelity of pfu DNA polymerase and other thermostable DNA polymerases. Nucleic Acids Res. 24(18): 3546–3551.

Cordwell, S.J., C.L.L. Alice, R.G. Touma, N.E. Scott, L. Falconer, D. Jones, A. Connolly, B. Crossett and S.P. Djordjevic. 2008. Identification of membrane-associated proteins from *Campylobacter jejuni* strains using complementary proteomics technologies. Proteomics 8(1): 122–139.

Corry, J.E.L. and H.I. Atabay. 2001. Poultry as a source of *Campylobacter* and related organisms. J. Appl. Microbiol. 90: 96S–114S.

Costea, P.I., G. Zeller, S. Sunagawa et al. 2017. Towards standards for human fecal sample processing in metagenomic studies. Nat. Biotechnol. 35(11): 1069–1076.

Covington, B.C., J.A. McLean and B.O. Bachmann. 2017. Comparative mass spectrometry-based metabolomics strategies for the investigation of microbial secondary metabolites. Nat. Prod. Rep. 34(1): 6–24.

Cretenet, M., S. Nouaille, J. Thouin, L. Rault, L. Stenz, P. François, J.A. Hennekinne, M. Piot, M.B. Maillard, J. Fauquant, P. Loubière, Y. Le Loir and S. Even. 2011. *Staphylococcus aureus* virulence and metabolism are dramatically affected by *Lactococcus lactis* in cheese matrix. Environ. Microbiol. Rep. 3(3): 340–351.

Crowley, S., J. Mahony, J.P. Morrissey and D. van Sinderen. 2013. Transcriptomic and morphological profiling of *Aspergillus fumigatus* Af293 in response to antifungal activity produced by *Lactobacillus plantarum* 16. Microbiol. 159: 2014–2024.

De Angelis, M., M. Calasso, N. Cavallo, R. Di Cagno and M. Gobbetti. 2016. Functional proteomics within the genus *Lactobacillus*. Proteomics 16(6): 946–962.

Delpech, P., S. Bornes, E. Alaterre, M. Bonnet, G. Gagne, M.C. Montel and C. Delbès. 2015. *Staphylococcus aureus* transcriptomic response to inhibition by H_2O_2-producing *Lactococcus garvieae*. Food Microbiol. 51: 163–170.

Delpech, P., E. Rifa, G. Ball, S. Nidelet, E. Dubois, G. Gagne, M.C. Montel, C. Delbès and S. Bornes. 2017. New insights into the anti-pathogenic potential of *Lactococcus garvieae* against *Staphylococcus aureus* based on RNA sequencing profiling. Front. Microbiol. 8: 359.

Denesvre, C., M. Dumarest, S. Remy, D. Gourichon and M. Eloit. 2015. Chicken skin virome analyzed by high-throughput sequencing shows a composition highly different from human skin. Virus Genes 51(2): 209–216.

Denis, M., B. Chidaine, M.J. Laisney, I. Kempf, K. Rivoal, F. Mégraud and P. Fravalo. 2009. Comparison of genetic profiles of *Campylobacter* strains isolated from poultry, pig and *Campylobacter* human infections in Brittany, France. Pathol. Biol. (Paris) 57 (1):23–29.

Di Cagno, R., M. De Angelis, M. Calasso and M. Gobbetti. 2011. Proteomics of the bacterial cross-talk by quorum sensing. J. Proteomics 74(1): 19–34.

Diaz-Sanchez, S., I. Hanning, S. Pendleton and D. D'Souza. 2013. Next-generation sequencing: The future of molecular genetics in poultry production and food safety. Poult. Sci. 92(2): 562–572.

Dieguez-Casal, E., P. Freixeiro, L. Costoya, M.T. Criado, C. Ferreiros and S. Sanchez. 2014. High resolution clear native electrophoresis is a good alternative to blue native electrophoresis for the characterization of the *Escherichia coli* membrane complexes. J. Microbiol. Methods 102:45–54.

Dingle, K.E., F.M. Colles, D. Falush and M.C.J. Maiden. 2005. Sequence typing and comparison of population biology of *Campylobacter coli* and *Campylobacter jejuni*. J. Clin. Microbiol. 43 (1):340-347.

Doulgeraki, A., D. Ercolini, F. Villani and G.-J.E. Nychas. 2012. Spoilage microbiota associated to the storage of raw meat in different conditions. Int. J. Food Microbiol. 157(2): 130–141.

Dresler, J., J. Klimentova and J. Stulik. 2011. *Francisella tularensis* membrane complexome by blue native/SDS-PAGE. J. Proteomics 75(1): 257–269.

EFSA. 2016. Scientific report of EFSA and ECDC - The European Union summary report on trends and sources of zoonoses, zoonotic agents and food-borne outbreaks in 2014. EFSA J. 13(12): 4329.

Elabbassy, M.Z. and M. Sitohy. 1993. Metabolic interaction between *Streptococcus-thermophilus* and *Lactobacillus-bulgaricus* in single and mixed starter yogurts. Die Nahrung 37(1): 53–58.

Elmi, A., E. Watson, P. Sandu, O. Gundogdu, D.C. Mills, N.F. Inglis, E. Manson, L. Imrie, M. Bajaj-Elliott, B.W. Wren, D.G. Smith and N. Dorrell. 2012. *Campylobacter jejuni* outer membrane vesicles play an important role in bacterial interactions with human intestinal epithelial cells. Infect. Immun. 80(12): 4089–4098.

Elmi, A., F. Nasher, H. Jagatia, O. Gundogdu, M. Bajaj-Elliott, B. Wren and N. Dorrell. 2016. *Campylobacter jejuni* outer membrane vesicle-associated proteolytic activity promotes bacterial invasion by mediating cleavage of intestinal epithelial cell E-cadherin and occludin. Cell. Microbiol. 18(4): 561–572.

Epps, S.V.R., R.B. Harvey, M.E. Hume, T.D. Phillips, R.C. Anderson and D.J. Nisbet. 2013. Foodborne *Campylobacter*: Infections, metabolism, pathogenesis and reservoirs. Int. J. Environ. Res. Public Health 10(12): 6292–6304.

Ercolini, D. 2004. PCR-DGGE fingerprinting: Novel strategies for detection of microbes in food. J. Microbiol. Methods 56(3): 297–314.

Escalona, M., S. Rocha and D. Posada. 2016. A comparison of tools for the simulation of genomic next-generation sequencing data. Nat. Rev. Gen. 17(8): 459–469.

Escobar-Zepeda, A., A. Sanchez-Flores and M.Q. Baruch. 2016. Metagenomic analysis of a Mexican ripened cheese reveals a unique complex microbiota. Food Microbiol. 57: 116–127.

Even, S., C. Charlier, S. Nouaille, N.L. Ben Zakour, M. Cretenet, F.J. Cousin, M. Gautier, M. Cocaign-Bousquet, P. Loubière and Y. Le Loir. 2009. *Staphylococcus aureus* virulence expression is impaired by *Lactococcus lactis* in mixed cultures. Appl. Environ. Microbiol. 75(13): 4459–4472.

Faber, K.L., E.C. Person and W.R. Hudlow. 2013. PCR inhibitor removal using the NucleoSpin® DNA Clean-Up XS kit. Forensic Sci. Int. 7(1): 209–213.

Fall, P.A., M.F. Pilet, F. Leduc, M. Cardinal, G. Duflos, C. Guérin, J.J. Joffraud and F. Leroi. 2012. Sensory and physicochemical evolution of tropical cooked peeled shrimp inoculated by *Brochothrix thermosphacta* and *Lactococcus piscium* CNCM I-4031 during storage at 8°C. Int. J. Food Microbiol. 152(3): 82–90.

Faust, K. and J. Raes. 2012. Microbial interactions: From networks to models. Nat. Rev. Microbiol. 10 (8): 538–550.

Flint, A., Y.Q. Sun, J. Butcher, M. Stahl, H.S. Huang and A. Stintzi. 2014. Phenotypic screening of a targeted mutant library reveals *Campylobacter jejuni* defenses against oxidative stress. Infect. Immun. 82(6): 2266–2275.

Fougy, L., M.H. Desmonts, G. Coeuret, C. Fassel, E. Hamon, B. Hézard, M.-C. Champomier-Vergès and S. Chaillou. 2016. Reducing salt in raw pork sausages increases spoilage and correlates with reduced bacterial diversity. Appl. Environ. Microbiol. 82(13): 3928–3939.

Fox, E.M., M. Raftery, A. Goodchild and G.L. Mendz. 2007. *Campylobacter jejuni* response to ox-bile stress. FEMS Immunol. Med. Microbiol. 49(1): 165–172.

Garenaux, A.G., G. Ermel, S. Guillou, M. Federighi and M. Ritz. 2007. Proteomic analysis of *Campylobacter jejuni* response to oxidative stress reveals a possible influence on virulence. Zoonoses Public Health 54: 60–60.

Garenaux, A., S. Guillou, G. Ermel, B. Wren, M. Federighi and M. Ritz. 2008. Role of the Cj1371 periplasmic protein and the Cj0355c two-component regulator in the *Campylobacter jejuni* NCTC 11168 response to oxidative stress caused by paraquat. Res. Microbiol. 159(9-10): 718–726.

Glassing, A., S.E. Dowd, S. Galandiuk, B. Davis and R.J. Chiodini. 2016. Inherent bacterial DNA contamination of extraction and sequencing reagents may affect interpretation of microbiota in low bacterial biomass samples. Gut Pathog. 8:24.

Gram, L., L. Ravn, M. Rasch, J.B. Bruhn, A.B. Christensen and M. Givskov. 2002. Food spoilage—interactions between food spoilage bacteria. Int. J. Food Microbiol. 78(1-2): 79–97.

Guevremont, E., R. Higgins and S. Quessy. 2004. Characterization of *Campylobacter* isolates recovered from clinically healthy pigs and from sporadic cases of campylobacteriosis in humans. J. Food Prot. 67(2): 228–234.

Gunther, N.W., J.L. Bono and D.S. Needleman. 2015. Complete genome sequence of *Campylobacter jejuni* RM1285, a rod-shaped morphological variant. Genome Announc. 3(6): e01361–15.

Haas, B.J., D. Gevers, A.M. Earl, M. Feldgarden, D.V. Ward, G. Giannoukos, D. Ciulla, D. Tabbaa, S.K. Highlander, E. Sodergren, B. Methé, T.Z. DeSantis, Human Microbiome Consortium, J.F. Petrosino, R. Knight and B.W. Birren. 2011. Chimeric 16S rRNA sequence formation and detection in Sanger and 454-pyrosequenced PCR amplicons. Genome Res. 21(3): 494–504.

Haddad, N., O. Tresse, K. Rivoal, D. Chevret, Q. Nonglaton, C.M. Burns, H. Prévost and J.M. Cappelier. 2012. Polynucleotide phosphorylase has an impact on cell biology of *Campylobacter jejuni*. Front. Cell. Infect. Microbiol. 2: 30.

Haddad, N., R. G. Matos, T. Pinto, P. Rannou, J.M. Cappelier, H. Prévost and C.M. Arraiano. 2014. The RNase R from *Campylobacter jejuni* has unique features and is involved in the first steps of infection. J. Biol. Chem. 289(40): 27814–27824.

Hald, T., W. Aspinall, B. Devleesschauwer, R. Cooke, T. Corrigan, A.H. Havelaar, H.J. Gibb, P.R. Torgerson, M.D. Kirk, F.J. Angulo, R.J. Lake, N. Speybroeck and S. Hoffmann. 2016. World Health Organization estimates of the relative contributions of food to the burden of disease due to selected foodborne hazards: A structured expert elicitation. PLoS One 11(1): e0145839.

Herve-Jimenez, L., I. Guillouard, E. Guedon, S. Boudebbouze, P. Hols, V. Monnet, E. Maguin and F. Rul. 2009. Postgenomic analysis of *Streptococcus thermophilus* cocultivated in milk with *Lactobacillus delbrueckii* subsp *bulgaricus*: Involvement of nitrogen, purine, and iron metabolism. Appl. Environ. Microbiol. 75(7): 2062–2073.

Herve-Jimenez, L., I. Guillouard, E. Guedon, C. Gautier, S. Boudebbouze, P. Hols, V. Monnet, F. Rul and E. Maguin. 2008. Physiology of *Streptococcus thermophilus* during the late stage of milk fermentation with special regard to sulfur amino-acid metabolism. Proteomics 8(20): 4273–4286.

Hofreuter, D. 2014. Defining the metabolic requirements for the growth and colonization capacity of *Campylobacter jejuni*. Front. Cell. Infect. Microbiol. 4: 137.

Holmes, K., F. Mulholland, B.M. Pearson, C. Pin, J. McNicholl-Kennedy, J.M. Ketley and J.M. Wells. 2005. *Campylobacter jejuni* gene expression in response to iron limitation and the role of Fur. Microbiol SGM 151: 243–257.

Howlett, R.M., M.P. Davey, W. Paul Quick and D.J. Kelly. 2014. Metabolomic analysis of the foodborne pathogen *Campylobacter jejuni*: Application of direct injection mass spectrometry for mutant characterisation. Metabolomics 10(5): 887–896.

Hu, Y.Q., Y.W. Shang, J.L. Huang, Y. Wang, F. Ren, Y. Jiao, Z. Pan and X.A. Jiao. 2013. A novel immunoproteomics method for identifying *in vivo*-induced *Campylobacter jejuni* antigens using pre-adsorbed sera from infected patients. Biochim. Biophys. Acta 1830(11): 5229–5235.

Hue, O., V. Allain, M.J. Laisney, S. Le Bouquin, F. Lalande, I. Petetin, S. Rouxel, S. Quesne, P.Y. Gloaguen, M. Picherot, J. Santolini, S. Bougeard, G. Salvat and M. Chemaly. 2011. *Campylobacter* contamination of broiler caeca and carcasses at the slaughterhouse and correlation with *Salmonella* contamination. Food Microbiol. 28(5): 862–868.

Hultman, J., R. Rahkila, J. Ali, J. Rousu and K.J. Bjorkroth. 2015. Meat processing plant microbiome and contamination patterns of cold-tolerant bacteria causing food safety and spoilage risks in the manufacture of vacuum-packaged cooked sausages. Appl. Environ. Microbiol. 81(20): 7088–7097.

Irlinger, F., S.A. Yung, A.S. Sarthou, C. Delbès-Paus, M.C. Montel, E. Coton, M. Coton and S. Helinck. 2012. Ecological and aromatic impact of two Gram-negative bacteria (*Psychrobacter celer* and *Hafnia alvei*) inoculated as part of the whole microbial community of an experimental smear soft cheese. Int. J. Food Microbiol. 153(3): 332–338.

Jaaskelainen, E., S. Vesterinen, J. Parshintsev, P. Johansson, M.L. Riekkola and J. Bjorkroth. 2015. Production of buttery-odor compounds and transcriptome response in *Leuconostoc gelidum* subsp. *gasicomitatum* LMG18811[T] during growth on various carbon sources. Appl. Environ. Microbiol. 81(6): 1902–1908.

Janssen, R., K.A. Krogfelt, S.A. Cawthraw, W. van Pelt, J.A. Wagenaar and R.J. Owen. 2008. Host-pathogen interactions in *Campylobacter* infections: The host perspective. Clin. Microbiol. Rev. 21(3): 505–518.

Jensen, A.N., A. Dalsgaard, D.L. Baggesen and E.M. Nielsen. 2006. The occurrence and characterization of *Campylobacter jejuni* and *C. coli* in organic pigs and their outdoor environment. Vet. Microbiol. 116(1-3): 96–105.

Jones, R.J., M. Zagorec, G. Brightwell and J.R. Tagg. 2009. Inhibition by *Lactobacillus sakei* of other species in the flora of vacuum packaged raw meats during prolonged storage. Food Microbiol. 26(8): 876–881.

Juste, A., B.P. Thomma and B. Lievens. 2008. Recent advances in molecular techniques to study microbial communities in food-associated matrices and processes. Food Microbiol. 25(6): 745–761.

Kaakoush, N.O., S.M. Man, S. Lamb, M.J. Raftery, M.R. Wilkins, Z. Kovach and H. Mitchell. 2010. The secretome of *Campylobacter concisus*. FEBS J. 277(7): 1606–1617.

Kaakoush, N.O., W.G. Miller, H. De Reuse and G.L. Mendz. 2007. Oxygen requirement and tolerance of *Campylobacter jejuni*. Res. Microbiol. 158(8-9): 644–650.

Kale, A., C. Phansopa, C. Suwannachart, C.J. Craven, J.B. Rafferty and D.J. Kelly. 2011. The virulence factor PEB4 (Cj0596) and the periplasmic protein Cj1289 are two structurally related SurA-like chaperones in the human pathogen *Campylobacter jejuni*. J. Biol. Chem. 286(24): 21254–21265.

Kang, S., J.L.M. Rodrigues, J.P. Ng and T.J. Gentry. 2016. Hill number as a bacterial diversity measure framework with high-throughput sequence data. Sci. Rep. 6: 38263.

Katzav, M., P. Isohanni, M. Lund, M. Hakkinen and U. Lyhs. 2008. PCR assay for the detection of Campylobacter in marinated and non-marinated poultry products. Food Microbiol. 25(7): 908–914.

Keohavong, P. and W.G. Thilly. 1989. Fidelity of DNA polymerases in DNA amplification. Proc. Natl. Acad. Sci. USA 86(23): 9253–9257.

Klappenbach, J., J.M. Dunbar and T.M. Schmidt. 2000. rRNA gene copy number predicts ecological strategies in bacteria. Appl. Environ. Microbiol. 66: 1328–1333.

Kolek, J., P. Patakova, K. Melzoch, K. Sigler and T. Rezanka. 2015. Changes in membrane plasmalogens of *Clostridium pasteurianum* during butanol fermentation as determined by lipidomic analysis. Plos One 10(3): e0122058.

Koolman, L., P. Whyte, C. Burgess and D. Bolton. 2016. Virulence gene expression, adhesion and invasion of *Campylobacter jejuni* exposed to oxidative stress (H_2O_2). Int. J. Food Microbiol. 220: 33–38.

Koser, C.U., M.J. Ellington, E.J.P. Cartwright, S.H. Gillespie, N.M. Brown, M. Farrington, M.T.G. Holden, G. Dougan, S.D. Bentley, J. Parkhill and S.J. Peacock. 2012. Routine use of microbial whole genome sequencing in diagnostic and public health microbiology. Plos Pathog. 8(8): e1002824.

Laursen, B.G., J.J. Leisner and P. Dalgaard. 2006. *Carnobacterium* species: effect of metabolic activity and interaction with *Brochothrix thermosphacta* on sensory characteristics of modified atmosphere packed shrimp. J. Agric. Food Chem. 54(10): 3604–3611.

Laursen, M.F., M.I. Bahl, T.R. Licht, L. Gram and G.M. Knudsen. 2015. A single exposure to a sublethal pediocin concentration initiates a resistance-associated temporal cell envelope and general stress response in *Listeria monocytogenes*. Environ. Microbiol. 17(4): 1134–1151.

Laver, T., J. Harrison, P.A. O'Neill, K. Moore, A. Farbos, K. Paszkiewicz and D.J. Studholme. 2015. Assessing the performance of the Oxford Nanopore Technologies MinION. Biomol. Detect. Quantif. 3(Supplement C): 1–8.

Lefebure, T., P.D.P. Bitar, H. Suzuki and M.J. Stanhope. 2010. Evolutionary dynamics of complete *Campylobacter* pan-genomes and the bacterial species concept. Genome Biol. Evol. 2: 646–655.

Li, H., X. Xia, X. Li, G. Naren, Q. Fu, Y. Wang, C. Wu, S. Ding, S. Zhang, H. Jiang, J. Li and J. Shen. 2015. Untargeted metabolomic profiling of amphenicol-resistant *Campylobacter jejuni* by ultra-high-performance liquid chromatography-mass spectrometry. J. Proteome Res. 14(2): 1060–1068.

Liu, M., J.M. Gray and M.W. Griffiths. 2006. Occurrence of proteolytic activity and N-acyl-homoserine lactone signals in the spoilage of aerobically chill-stored proteinaceous raw foods. J. Food Prot. 69(11): 2729–37.

Liu, X.Y., B. Gao, V. Novik and J.E. Galan. 2012. Quantitative proteomics of intracellular *Campylobacter jejuni* reveals metabolic reprogramming. Plos Pathog. 8(3): e1002562.

Llarena, A.K., E. Taboada and M. Rossi. 2017. Whole-Genome sequencing in epidemiology of *Campylobacter jejuni* infections. J. Clin. Microbiol. 55(5): 1269–1275.

Lübeck, P.S., P. Wolffs, S.L.W. On, P. Ahrens, P. Rådström and J. Hoorfar. 2003. Toward an international standard for PCR-based detection of food-borne thermotolerant *Campylobacters*: assay development and analytical validation. Appl. Environ. Microbiol. 69(9): 5664–5669.

Mace, S., N. Haddad, M. Zagorec and O. Tresse. 2015. Influence of measurement and control of microaerobic gaseous atmospheres in methods for *Campylobacter* growth studies. Food Microbiol. 52: 169–176.

Macé, S., J.-J. Joffraud, M. Cardinal, M. Malcheva, J. Cornet, V. Lalanne, F. Chevalier, T. Sérot, M.-F. Pilet and X. Dousset. 2013. Evaluation of the spoilage potential of bacteria isolated from spoiled raw salmon (*Salmo salar*) fillets stored under modified atmosphere packaging. Int. J. Food Microbiol. 160: 227–238.

Macé, S., E. Jaffrès, M. Cardinal, J. Cornet, V. Lalanne, F. Chevalier, T. Sérot, M.F. Pilet, X. Dousset and J.J. Joffraud. 2014. Evaluation of the spoilage potential of bacteria isolated from spoiledcooked whole tropical shrimp (*Penaeus vannamei*) stored under modified atmosphere packaging. Food Microbiol. 40: 9–17.

Malik-Kale, P., C.T. Parker and M.E. Konkel. 2008. Culture of *Campylobacter jejuni* with sodium deoxycholate induces virulence gene expression. J. Bacteriol. 190(7): 2286–2297.

Mamlouk, K., S. Macé, M. Guilbaud, E. Jaffrès, M. Ferchichi, H. Prévost, M.F. Pilet and X. Dousset. 2012. Quantification of viable *Brochothrix thermosphacta* in cooked shrimp and salmon by real-time PCR. Food Microbiol. 30(1): 173–179.

Manavathu, E.K., K. Hiratsuka and D.E. Taylor. 1988. Nucleotide sequence analysis and expression of a tetracycline-resistance gene from *Campylobacter jejuni*. Gene 62(1): 17–26.

Margulies, M., M. Egholm, W.E. Altman et al. 2005. Genome sequencing in microfabricated high-density picolitre reactors. Nature 437(7057): 376–380.

McCarthy, A., E. Chiang, M.L. Schmidt and V.J. Denef. 2015. RNA preservation agents and nucleic acid extraction method bias perceived bacterial community composition. PLoS ONE 10(3): e0121659.

Mehla, K. and J. Ramana. 2017. Surface proteome mining for identification of potential vaccine candidates against *Campylobacter jejuni*: An *in silico* approach. Funct. Integr. Genomics 17(1): 27–37.

Messaoudi, S., M. Manai, M. Federighi and X. Dousset. 2013. *Campylobacter*: Control in poultry breedings. Rev. Med. Vet. 164(2): 90–99.

Mills, S., R.P. Ross and C. Hill. 2017. Bacteriocins and bacteriophage:A narrow-minded approach to food and gut microbiology. FEMS Microbiol. Rev. 41(Supp_1): S129–S153.

Moore, J.E., D. Corcoran, J.S.G. Dooley, S. Fanning, B. Lucey, M. Matsuda, D.A. McDowell, F. Mégraud, B.C. Millar, R. O'Mahony, L. O'Riordan, M. O'Rourke, J.R. Rao, P.J. Rooney, A. Sails and P. Whyte. 2005. *Campylobacter*. Vet. Res. 36(3): 351–382.

Mortensen, N.P., M.L. Kuijf, C.W. Ang, P. Schiellerup, K.A. Krogfelt, B.C. Jacobs, A. van Belkum, H.P. Endtz and M.P. Bergman. 2009. Sialylation of *Campylobacter jejuni* lipo-oligosaccharides is associated with severe gastro-enteritis and reactive arthritis. Microbes Infect. 11(12): 988–994.

Mou, K.T., U.K. Muppirala, A.J. Severin, T.A. Clark, M. Boitano and P.J. Plummer. 2015. A comparative analysis of methylome profiles of *Campylobacter jejuni* sheep abortion isolate and gastroenteric strains using PacBio data. Front. Microbiol. 5: 15.

Mou, K.T., T.A. Clark, U.K. Muppirala, A.J. Severin and P.J. Plummer. 2017. Methods for genome-wide methylome profiling of *Campylobacter jejuni*. In: *Campylobacter Jejuni: Methods and Protocols*, edited by J. Butcher, and A. Stintzi. Totowa: Humana Press Inc.

Murn, J. and Y. Shi. 2017. The winding path of protein methylation research: milestones and new frontiers. Nat. Rev. Mol. Cell Biol. 18(8): 517–527.

Murray, I.A., T.A. Clark, R.D. Morgan, M. Boitano, B.P. Anton, K. Luong, A. Fomenkov, S.W. Turner, J. Korlach and R.J. Roberts. 2012. The methylomes of six bacteria. Nucleic Acids Res. 40 (22): 11450–11462.

Nieminen, T.T., K. Koskinen, P. Laine, J. Hultman, E. Säde, L. Paulin, A. Paloranta, P. Johansson, J. Björkroth and P. Auvinen. 2012. Comparison of microbial communities in marinated and unmarinated broiler meat by metagenomics. Int. J. Food Microbiol. 157(2): 142–149.

Nilsson, L., T.B. Hansen, P. Garrido, C. Buchrieser, P. Glaser, S. Knøchel, L. Gram and A Gravesen. 2005. Growth inhibition of *Listeria monocytogenes* by a non bacteriocinogenic *Carnobacterium piscicola*. J. Appl. Microbiol. 98: 172–183.

Nothaft, H. and C.M. Szymanski. 2013. Bacterial protein N-glycosylation: new perspectives and applications. J. Biol. Chem. 288(10): 6912–6920.

Nouaille, S., L. Rault, S. Jeanson, P. Loubiere, Y. Le Loir and S. Even. 2014. Contribution of *Lactococcus lactis* reducing properties to the downregulation of a major virulence regulator in *Staphylococcus aureus*, the agr system. Appl. Environ. Microbiol. 80(22): 7028–7035.

Nyati, K.K. and R. Nyati. 2013. Role of *Campylobacter jejuni* Infection in the pathogenesis of Guillain-Barre Syndrome: An Update. Biomed Res. Int. 852195.

O'Loughlin, J.L., T.P. Eucker, J.D. Chavez, D.R. Samuelson, J. Neal-McKinney, C.R. Gourley, J.E. Bruce and M.E. Konkel. 2015. Analysis of the *Campylobacter jejuni*genome by SMRT DNA sequencing identifies restriction-modification motifs. Plos One 10(2): 18.

Ong, S.E., G. Mittler and M. Mann. 2004. Identifying and quantifying *in vivo* methylation sites by heavy methyl SILAC. Nat. Methods 1(2): 119–126.

Ouidir, T., T. Kentache and J. Hardouin. 2016. Protein lysine acetylation in bacteria: Current state of the art. Proteomics 16(2): 301–309.

Oyola, S.O., T.D. Otto, Y. Gu, G. Maslen, M. Manske, S. Campino, D.J. Turner, B. MacInnis, D.P. Kwiatkowski, H.P. Swerdlow and M.A. Quail. 2012. Optimizing Illumina next-generation sequencing library preparation for extremely AT-biased genomes. BMC Genomics 13: 1.

Palyada, K., Y.Q. Sun, A. Flint, J. Butcher, H. Naikare and A. Stintzi. 2009. Characterization of the oxidative stress stimulon and PerR regulon of *Campylobacter jejuni*. BMC Genomics 10: 481.

Pan, J., Z. Ye, Z. Cheng, X. Peng, L. Wen and F. Zhao. 2014. Systematic analysis of the lysine acetylome in *Vibrio parahemolyticus*. J. Proteome Res. 13(7): 3294–302.

Parkhill, J., B.W. Wren, K. Mungall, J.M. Ketley, C. Churcher, D. Basham, T. Chillingworth, R.M. Davies, T. Feltwell, S. Holroyd, K. Jagels, A.V. Karlyshev, S. Moule, M.J. Pallen, C.W. Penn, M.A. Quail, M.A. Rajandream, K.M. Rutherford, A.H. van Vliet, S. Whitehead and B.G. Barrell. 2000. The genome sequence of the food-borne pathogen *Campylobacter jejuni* reveals hypervariable sequences. Nature 403(6770): 665–668.

Pendleton, S., I. Hanning, D. Biswas and S.C. Ricke. 2013. Evaluation of whole-genome sequencing as a genotyping tool for *Campylobacter jejuni* in comparison with pulsed-field gel electrophoresis and *flaA* typing. Poult. Sci. 92(2): 573–580.

Pinto, A. Di, V.T. Forte, M. Corsignano Guastadisegni, C. Martino, F.P. Schena and G. Tantillo. 2007. A comparison of DNA extraction methods for food analysis. Food Control 18(1): 76–80.

Prokhorova, T.A., P.N. Nielsen, J. Petersen, T. Kofoed, J.S. Crawford, C. Morsczeck, A. Boysen, P. Schrotz-King. 2006. Novel surface polypeptides of *Campylobacter jejuni* as traveller's diarrhoea vaccine candidates discovered by proteomics. Vaccine 24(40-41): 6446–6455.

Quail, M.A., M. Smith, P. Coupland, T.D. Otto, S.R. Harris, T.R. Connor, A. Bertoni, H.P. Swerdlow and Y. Gu. 2012. A tale of three next generation sequencing platforms: Comparison of Ion Torrent, Pacific Biosciences and Illumina MiSeq sequencers. BMC Genomics 13: 341.

Quast, C., E. Pruesse, P. Yilmaz, J. Gerken, T. Schweer, P. Yarza, J. Peplies and F.O. Glöckner. 2013. The SILVA ribosomal RNA gene database project: Improved data processing and web-based tools. Nucleic Acids Res. 41(Database issue): D590–D596.

Quick, J., A.R. Quinlan and N.J. Loman. 2014. A reference bacterial genome dataset generated on the MinION (TM) portable single-molecule nanopore sequencer. Gigascience 3: 6.

Rathbun, K.M., J.E. Hall and S.A. Thompson. 2009. Cj0596 is a periplasmic peptidyl prolyl cis-trans isomerase involved in *Campylobacter jejuni* motility, invasion, and colonization. BMC Microbiol. 9: 160.

Reid, A.N., R. Pandey, K. Palyada, H. Naikare and A. Stintzi. 2008. Identification of *Campylobacter jejuni* genes involved in the response to acidic pH and stomach transit. Appl. Environ. Microbiol. 74(5): 1583–1597.

Reid, A.N., R. Pandey, K. Palyada, L. Whitworth, E. Doukhanine and A. Stintzi. 2008. Identification of *Campylobacter jejuni* genes contributing to acid adaptation by transcriptional profiling and genome-wide mutagenesis. Appl. Environ. Microbiol. 74(5): 1598–1612.

Reisinger, V. and L.A. Eichacker. 2008a. Isolation of membrane protein complexes by blue native electrophoresis. Methods Mol. Biol. 424: 423–431.

Reisinger, V. and L.A. Eichacker. 2008b. Solubilization of membrane protein complexes for blue native PAGE. J. Proteomics 71(3): 277–283.

Remenant, B., E. Jaffrès, X. Dousset, M.-F. Pilet and M. Zagorec. 2015. Bacterial spoilers of food: Behavior, fitness and functional properties. Food Microbiol. 45(Part A): 45–53.

Revez, J., T. Schott, M. Rossi and M.L. Hanninen. 2012. Complete genome sequence of a variant of *Campylobacter jejuni* NCTC 11168. J. Bacteriol. 194(22): 6298–6299.

Revez, J., J. Zhang, S.A. Thomas, R. Kivisto, M. Rossi and M.L. Hanninen. 2014. Genomic variation between *Campylobacter jejuni* isolates associated with milk-borne-disease outbreaks. J. Clin. Microbiol. 52(8): 2782–2786.

Rezanka, T., Z. Kresinova, I. Kolouchova and K. Sigler. 2012. Lipidomic analysis of bacterial plasmalogens. Folia Microbiol. 57(5): 463–472.

Rodrigues, R.C., N. Haddad, D. Chevret, J.M. Cappelier and O. Tresse. 2016. Comparison of proteomics profiles of *Campylobacter jejuni* strain Bf under microaerobic and aerobic conditions. Front. Microbiol. 7: 1596.

Roh, S.W., K.H. Kim, Y.D. Nam, H.W. Chang, E.J. Park and J.W. Bae. 2010. Investigation of archaeal and bacterial diversity in fermented seafood using barcoded pyrosequencing. ISME J. 4(1): 1–16.

Rossen, L., P. Nørskov, K. Holmstrøm and O.F. Rasmussen. 1992. Inhibition of PCR by components of food samples, microbial diagnostic assays and DNA-extraction solutions. Int. J. Food Microbiol. 17(1): 37–45.

Rouger, A., O. Tresse and M. Zagorec. 2017a. Bacterial contaminants of poultry meat: Sources, species, and dynamics. Microorganisms 5: 50.

Rouger, A., B. Remenant, H. Prévost and M. Zagorec. 2017b. A method to isolate bacterial communities and characterize ecosystems from food products: Validation and utilization as a reproducible chicken meat model. Int. J. Food Microbiol. 247: 38–47.

Rouger, A., N. Moriceau, H. Prevost, B. Remenant and M. Zagorec. 2018. Diversity of bacterial communities in French chicken cuts stored under modified atmosphere packaging. Food Microbiol. 70: 7–16.

Rul, F. and V. Monnet. 2015. How microbes communicate in food: A review of signaling molecules and their impact on food quality. Curr. Opin. Food Sci. 2: 100–105.

Sanger, F., S. Nicklen and A.R. Coulson. 1977. DNA sequencing with chain-terminating inhibitors. Proc. Natl. Acad. Sci. USA 74(12): 5463–5467.

Scott, N.E., B.L. Parker, A.M. Connolly, J. Paulech, A.V. Edwards, B. Crossett, L. Falconer, D. Kolarich, S.P. Djordjevic, P. Højrup, N.H. Packer, M.R. Larsen and S.J. Cordwell. 2011. Simultaneous glycan-peptide characterization using hydrophilic interaction chromatography and parallel fragmentation by CID, higher energy collisional dissociation, and electron transfer dissociation MS applied to the N-linked glycoproteome of *Campylobacter jejuni*. Mol. Cell. Proteomics 10(2): M000031–MCP201.

Scott, N.E., N.B. Marzook, J.A. Cain, N. Solis, M. Thaysen-Andersen, S.P. Djordjevic, N.H. Packer, M.R. Larsen and S.J. Cordwell. 2014. Comparative proteomics and glycoproteomics reveal increased N-linked glycosylation and relaxed sequon specificity in *Campylobacter jejuni* NCTC11168 O. J. Proteome Res. 13(11): 5136–5150.

Shamshiev, A., A. Donda, T.I. Prigozy, L. Mori, V. Chigorno, C.A. Benedict, L. Kappos, S. Sonnino, M. Kronenberg and G. De Libero. 2000. The alphabeta T cell response to self-glycolipids shows a novel mechanism of CD1b loading and a requirement for complex oligosaccharides. Immunity 13(2): 255–264.

Shendure, J. and H.L. Ji. 2008. Next-generation DNA sequencing. Nature Biotechnol 26(10): 1135–1145.

Shendure, J., S. Balasubramanian, G.M. Church, W. Gilbert, J. Rogers, J.A. Schloss and R.H. Waterston. 2017. DNA sequencing at 40: Past, present and future. Nature 550(7676): 345–353.

Sheppard, S.K., K.A. Jolley and M.C.J. Maiden. 2012. A gene-by-gene approach to bacterial population genomics: Whole genome MLST of *Campylobacter*. Genes 3(2): 261–277.

Sieuwerts, S., F.A.M. de Bok, J. Hugenholtz, and J.E.T. van Hylckama Vlieg. 2008. Unraveling microbial interactions in food fermentations: From classical to genomics approaches. Appl. Environ. Microbiol. 74(16): 4997–5007.

Sieuwerts, S., D. Molenaar, S.A.F.T. van Hijum, M. Beerthuyzen, M.J.A. Stevens, P.W.M. Janssen, C.J. Ingham, F.A.M. de Bok, W.M. de Vos and J.E.T. van Hylckama Vlieg. 2010. Mixed-culture transcriptome analysis reveals the molecular basis of mixed-culture growth in *Streptococcus thermophilus* and *Lactobacillus bulgaricus*. Appl. Environ. Microbiol. 76(23): 7775–7784.

Silva, J., D. Leite, M. Fernandes, C. Mena, P.A. Gibbs and P. Teixeira. 2011. *Campylobacter spp.* as a foodborne pathogen: A review. Front. Microbiol. 2: 200.

Sivadon, V., D. Orlikowski, F. Rozenberg, J.C. Quincampoix, C. Caudie, M.C. Durand, J.L. Fauchère, T. Sharshar, J.C. Raphaël and J.L. Gaillard. 2005. Prevalence and characteristics of Guillain-Barre syndromes associated with *Campylobacter jejuni* and cytomegalovirus in greater Paris. Pathol. Biol. (Paris) 53(8-9): 536–538.

Skandamis, P.N. and G.J.E. Nychas. 2012. Quorum sensing in the context of food microbiology. Appl. Environ. Microbiol. 78(16): 5473–5482.

Stahl, M. and A. Stintzi. 2011. Identification of essential genes in *C. jejuni* genome highlights hyper-variable plasticity regions. Funct. Integr. Genomics 11(2): 241–257.

Stanley, K. and K. Jones. 2003. Cattle and sheep farms as reservoirs of *Campylobacter*. J. Appl. Microbiol. 94: 104S–113S.

Strom, K., J. Schnurer and P. Melin. 2005. Co-cultivation of antifungal *Lactobacillus plantarum* MiLAB 393 and *Aspergillus nidulans*, evaluation of effects on fungal growth and protein expression. FEMS Microbiol. Lett. 246(1): 119–124.

Sulaeman, S., M. Hernould, A. Schaumann, L. Coquet, J.-M. Bolla, E. Dé and O. Tresse. 2012. Enhanced adhesion of *Campylobacter jejuni* to abiotic surfaces is mediated by membrane proteins in oxygen-enriched conditions. PLoS One 7(9): e46402.

Taberlet, P., N.E. Zimmermann, T. Englisch et al. 2012. Genetic diversity in widespread species is not congruent with species richness in alpine plant communities. Ecol. Lett. 15(12): 1439–1448.

Tam, V.C. 2013. Lipidomic profiling of bioactive lipids by mass spectrometry during microbial infections. Semin. Immunol. 25(3): 240–248.

Taylor, A.J., S.A.I. Zakai and D.J. Kelly. 2017. The periplasmic chaperone network of *Campylobacter jejuni*: Evidence that SalC (Cj1289) and PpiD (Cj0694) are involved in maintaining outer membrane integrity. Front. Microbiol. 8:531.

Tian, R.J. 2014. Exploring intercellular signaling by proteomic approaches. Proteomics 14(4-5): 498–512.

Tyler, A.D., S. Christianson, N.C. Knox et al. 2016. Comparison of Sample Preparation Methods Used for the Next-Generation Sequencing of *Mycobacterium tuberculosis*. PLoS ONE 11(2): e0148676.

van Dijk, E.L., Y. Jaszczyszyn and C. Thermes. 2014. Library preparation methods for next-generation sequencing: Tone down the bias. Exp. Cell. Res. 322(1): 12–20.

Van Houdt, R., P. Moons, A. Aertsen, A. Jansen, K. Vanoirbeek, M. Daykin, P. Williams and C.W. Michiels. 2007. Characterization of a luxI/luxR-type quorum sensing system and N-acyl-homoserine lactone-dependent regulation of exo-enzyme and antibacterial component production in *Serratia plymuthica* RVH1. Res. Microbiol.158(2): 150–158.

van Rensburg, M.J.J., C. Swift, A.J. Cody, C. Jenkins and M.C.J. Maiden. 2016. Exploiting bacterial whole-genome sequencing data for evaluation of diagnostic assays: *Campylobacter* species identification as a case study. J. Clin. Microbiol. 54(12): 2882–2890.

Varsaki, A., C. Murphy, A. Barczynska, K. Jordan and C. Carroll. 2015. The acid adaptive tolerance response in *Campylobacter jejuni* induces a global response, as suggested by proteomics and microarrays. Microb. Biotechnol. 8(6): 974–988.

Vermassen, A., A. de la Foye, V. Loux, R. Talon and S. Leroy. 2014. Transcriptomic analysis of *Staphylococcus xylosus* in the presence of nitrate and nitrite in meat reveals its response to nitrosative stress. Front. Microbiol. 5: 691.

Villani, F., A. Casaburi, C. Pennacchia, L. Filosa, F. Russo and D. Ercolini. 2007. Microbial ecology of the soppressata of Vallo di Diano, a traditional dry fermented sausage from southern Italy, and *in vitro* and in situ selection of autochthonous starter cultures. Appl. Environ. Microbiol. 73(17): 5453–5463.

Voisin, S., D.C. Watson, L. Tessier, W. Ding, S. Foote, S. Bhatia, J.F. Kelly and N.M. Young. 2007. The cytoplasmic phosphoproteome of the Gram-negative bacterium *Campylobacter jejuni*: evidence for modification by unidentified protein kinases. Proteomics 7(23): 4338–4348.

Walsh, A.M., F. Crispie, K. Kilcawley, O. O'Sullivan, M.G. O'Sullivan, M.J. Claesson and P.D. Cotter. 2016. Microbial succession and flavor production in the fermented dairy beverage kefir. mSystems 1(5): e00052–16.

Wang, Q., Y. Zhang, C. Yang, H. Xiong, Y. Lin, J. Yao, H. Li, L. Xie, W. Zhao, Y. Yao, Z.B. Ning, R. Zeng, Y. Xiong, K.L. Guan, S. Zhao and G.P. Zhao. 2010. Acetylation of metabolic enzymes coordinates carbon source utilization and metabolic flux. Science 327(5968): 1004–1007.

Watson, E., M.P. Alberdi, N.F. Inglis, A. Lainson, M.E. Porter, E. Manson, L. Imrie, K. Mclean and D.G.E. Smith. 2014. Proteomic analysis of *Lawsonia intracellularis* reveals expression of outer membrane proteins during infection. Vet. Microbiol. 174(3-4): 448–455.

Wilson, I.G. 1997. Inhibition and facilitation of nucleic acid amplification. Appl. Environ. Microbiol. 63(10): 3741–3751.

Wohlbrand, L., H.S. Ruppersberg, C. Feenders, B. Blasius, H.P. Braun and R. Rabus. 2016. Analysis of membrane-protein complexes of the marine sulfate reducer *Desulfobacula toluolica* Tol2 by 1D blue native-PAGE complexome profiling and 2D blue native-/SDS-PAGE. Proteomics 16(6): 973–988.

Young, N.M., J.R. Brisson, J. Kelly, D.C. Watson, L. Tessier, P.H. Lanthier, H.C. Jarrell, N. Cadotte, F. St Michael, E. Aberg and C.M. Szymanski. 2002. Structure of the N-linked glycan present on multiple glycoproteins in the Gram-negative bacterium, *Campylobacter jejuni*. J. Biol. Chem. 277(45): 42530–42539.

Zautner, A.E., A.M. Goldschmidt, A. Thurmer, J. Schuldes, O. Bader, R. Lugert, U. Gross, K. Stingl, G. Salinas and T. Lingner. 2015. SMRT sequencing of the *Campylobacter coli* BfR-CA-9557 genome sequence reveals unique methylation motifs. BMC Genomics 16: 1088.

Zdenkova, K., B. Alibayov, L. Karamonova, S. Purkrtova, R. Karpiskova and K. Demnerova. 2016. Transcriptomic and metabolic responses of *Staphylococcus aureus* in mixed culture with *Lactobacillus plantarum*, *Streptococcus thermophilus* and *Enterococcus durans* in milk. J. Ind. Microbiol. Biotechnol. 43(9): 1237–1247.

Zhang, M.J., D. Xiao, F. Zhao, Y.X. Gu, F.L. Meng, L.H. He, G.Y. Ma and J.Z. Zhang. 2009. Comparative proteomic analysis of *Campylobacter jejuni* cultured at 37 degrees C and 42 degrees C. Jpn. J. Infect. Dis. 62(5): 356–361.

Zhao, S., G.H. Tyson, Y. Chen, C. Li, S. Mukherjee, S. Young, C. Lam, J.P. Folster, J.M. Whichard and P.F. McDermott. 2016. Whole-Genome sequencing analysis accurately predicts antimicrobial resistance phenotypes in *Campylobacter* spp. Appl. Environ. Microbiol. 82(2): 459–466.

7

Genomics and Proteomics Features of *Listeria monocytogenes*

Sahoo Moumita,[1] Bhaskar Das,[2] Abhinandan Patnaik,[1] Paramasivan Balasubramanian[2] and Rasu Jayabalan[1,*]

1. Introduction

Foodborne illness has become the major concern for the public health sector due to the significant rise of the globalization of food trade, leading to the consumption of contaminated foods every day. With the change in the lifestyle, humans are more prone to consumption of minimally processed, contaminated, ready to eat convenience foods. In most of the cases, pathogenic bacterial contamination has played the eminent role in epidemic outbreaks of foodborne illness. *Listeria monocytogenes* has been distinguished as one of the major foodborne pathogens since the early 1980s, causing several outbreaks of foodborne listeriosis with high mortality in immunocompromised hosts. The presence of *L. monocytogenes* as pathogenic bacteria for animals has been reported by Murray et al. (1926), where it was first isolated from rabbits, causing "circling disease". *L. monocytogenes* is a small, Gram-positive, non-sporulating, catalase-positive, facultative anaerobe, flagellated, rod-shaped bacterium and is mainly classified in the Firmicutes division (Donnelly 2001, Donnelly and Diez-Gonzalez 2013). Genus *Listeria* consists of both pathogenic and nonpathogenic species, such as *L. monocytogenes, L. innocua, L. seeligeri, L. welshimeri, L. grayi, L. ivanovii* (subspecies *ivanovi* and subspecies *londoniensis*), *L. marthii* and *L. rocourtiae* (Seeliger 1986, Boerlin 1992, Leclercq 2010, Tham and Danielsson-Tham

[1] Food Microbiology and Bioprocess Laboratory, Department of Life Science, National Institute of Technology, Rourkela 769 008, Odisha, India.
[2] Agriculture and Environmental Laboratory, Department of Biotechnology and Medical Engineering, National Institute of Technology, Rourkela 769 008, Odisha, India.
* Corresponding author: jayabalanr@nitrkl.ac.in

2013). The differentiation between various species of genus *Listeria* is made based on the biochemical reactions, such as beta hemolysis, reduction of nitrates to nitrites, production of acid from mannitol, L-rhamnose, and D-xylose, etc. Out of the twenty species of *Listeria* genus, *L. monocytogenes* is considered as a potential candidate for causing illness in humans. While very few reports of disease outbreaks caused by *L. seeligeri* are available, *L. innocua* and *L. welshimeri* were reported as being incapable of causing illness in humans or animals (Donnelly and Diez-Gonzalez 2013). These nonpathogenic *Listeria* species were used as a species of interest instudying the detailed pathogenesis of *Listeria* infection. Due to the presence of diversified genomic characteristics and intricate surface proteome, *Listeria* is capable of surviving diverse environmental conditions including food and the cytosol of eukaryotic cells. Of the 13 serovars of *L. monocytogenes* (i.e., 1/2a, 1/2b, 1/2c, 3a, 3b, 3c, 4a, 4b, 4c, 4d, 4e, 4ab and 7), only 3 (i.e., 1/2a, 1/2b, and 4b) are responsible for 98% of human listeriosis outbreaks (Roche et al. 2008). Listeriosis, a life-threatening disease caused by *L. monocytogenes*, favourably affects the host with compromised immune conditions, leading to meningitis, meningoencephalitis, and fetus infection, mostly due to the consumption of contaminated foods. Although fewer cases of listeriosis outbreaks have been reported, it has a higher mortality rate than cases of *Clostridium botulinum* (Huang et al. 2014*)*. A recent investigation of the total genome of *L. monocytogenes* has reported the prediction of a total of 2,853 proteins, of which 133 are surface proteins. These surface proteins are classified according to their different anchoring systems with several structural domains (Bierne and Cossart 2007). Additionally, the presence of these exclusive proteomic features made *L. monocytogenes* capable of having a protective cell-mediated immune response in mammalian host cells (phagocytes). The role of different genes and proteins involved in the pathogenesis of *L. monocytogenes*, along with their regulation, is discussed in the following sections. Furthermore, possible techniques for the detection of *L. monocytogenes* in contaminated food are also discussed.

2. Listeriosis

L. monocytogenes is the primary causative organism of human listeriosis, which follows an intracellular mode of infection. The incidence of *L. monocytogenes* as a foodborne pathogen is lower as compared to other notorious pathogens, such as *Campylobacter* and *Salmonella*, but the food is the primary root of infection for *Listeria*. The contamination may occur at any stage of food processing and, eventually, the pathogen could acquire a sporadic or an epidemic stage. *Listeria* produces enzymes like superoxide dismutase that are responsible for the catalytic dismutation of superoxides produced by phagocytes, which in turn favours the survival of the pathogen in the host system (Vasconcelos and Deneer 1994). The pathogen adapts to a wide range of environmental conditions and is known to affect both healthy and immunodeficient population, although the symptoms and conditions differ. *Listeria* shows an invasive mechanism in affecting dendritic cells and macrophages, the immune cells which provide the primary defence against foreign bodies and are responsible for presenting processed antigens to T-cells in order to generate a secondary response to infection (Drevets and Bronze 2008). The non-invasive pathway causes feverish gastroenteritis (Schlech 1997) which is associated with symptoms like headaches, nausea, fatigue and diarrhea, to name a few. The non-invasive infection manifests the disease symptoms within hours as the incubation period for the pathogen, in this case, is around 18–20 hours, whereas the invasive infection has an incubation period of 30 days and shows more acute symptoms, such as conjunctivitis, meningitis, pneumonia and cardiac lesions (Bundrant et al. 2011).

Upon ingestion of the contaminated food product, the pathogen must first survive through the physiological conditions in the body before finally reaching the incubation period which may vary from days to months, altogether depending on the number of colonies initially taken in with the food. *Listeria,* after entering into the body, sustains itself through the unfavourable conditions of the alimentary canal, e.g., low pH conditions of the stomach and high salinity in the gastrointestinal (GI) tract. Once the pursuit of survival is over, the pathogen colonizes the GI tract and multiplies until it permeates through the intestinal layer barrier and seeps into the circulatory and the lymphatic system, eventually reaching other vital organs where the systemic phase of infection takes over, leading to the diseased condition. The virulence factors responsible for the pathogenesis of *L. monocytogenes* are present on a *prfA*-dependent region, located on the chromosome which comprises of a cluster of genes. The product of *prfA* is an upregulator of the virulence genes present in the gene cluster. The virulence genes *plcA* and *plcB* code for certain phospholipases which function in vacuole escape from macrophages and epithelial cells, whereas the product of *mpl* gene, which is a metalloprotease, ensures the cell to cell spreading of the bacteria by facilitating localization to the host surfaces from the cytosol (Bhunia 2018). Also, listeriolysin O, a hemolysin present in *Listeria*, is an important factor responsible for virulence. Phagosomes are lysed by listeriolysin O and, as a result, *Listeria* is released into the cytoplasm; hence, the bacteria proliferate causing further infection (Geoffroy et al. 1987). Loss of functioning of any of these factors results in loss of pathogenicity. Other species of *Listeria*, which include *L. ivanovii* and *L. seeligeri*, possess a special virulence factor which helps them utilize the phospholipids that are available in erythrocytes of ruminant animals (Barbuddhe et al. 2008).

3. Routes of Listeriosis

3.1 Gastrointestinal mode of Infection-Survival and Pathogenesis

L. monocytogenes has to overcome many stress conditions within the body of the host in order to ensure that it survives until it elicits its pathogenic properties. For the pathogen to begin colonizing the GI tract before it invades the system, it must first sustain the acidic conditions of the stomach, the variations in osmotic pressure in different regions of the alimentary canal and also the bile salts that are released into the small intestine. The survival of *Listeria* under low pH conditions is to be credited to the glutamate decarboxylase (GAD) system. Uptake of glutamate through antiporters, and its decarboxylation to gamma-aminobutyrate, is catalyzed by three enzymes of *Listeria*'s GAD system (Cotter et al. 2001). This decarboxylation reaction consumes the cytoplasmic protons of the bacteria and, hence, the pH of cytoplasm increases. This mechanism is of paramount importance to the survival of *Listeria* under acidic conditions and works efficiently only when the surrounding environment provides a sufficient amount of glutamate. Hence, glutamate-rich foods are a boon for the survival and passage of *Listeria* through the stomach. There is an increase in osmolarity of GI tract as the region has high salinity. This increase in osmotic pressure is known to instigate the expression of genes important for colonization and further proliferation of *Listeria*. Membrane transporters like BetL (Sleator et al. 1999) and Gbu (Ko and Smith 1999) that are present in *Listeria* are associated with the uptake of osmolytes, which in turn maintain the water level in the cytoplasm. *Listeria* escapes the pressure of bile salts in the small intestine with the help of an operon which is homologous to the osmoprotectant uptake (Opu) system present in other Gram-positive bacteria (Sleator et al. 1999). The products of this gene play a role in excluding the bile salts from the system.

The true mechanism of the pathogenic manifestation of *Listeria* in GI tract is not yet known, but the organism tends to effect villi responsible for nutrient uptake and secretions. Persons under medication for gastric acid problems are usually at greater risk. The early symptoms are experienced within 24 hours of incubation, these may include a mild fever along with vomiting and diarrhoea (Bortolussi 2008).

3.2 Listeriosis Linked to Disorders in Pregnancy

Pregnant women have a decreased cell-mediated immune response in order to prevent the rejection of the developing fetus. As a result of this natural immunosuppressive action of the system, the body is left vulnerable to several infections and diseases, listeriosis being one of them. Early symptoms may include mild fever, but prolonged sustenance may lead to severe outcomes, such as the child being born with the infection, and there are even cases of stillbirths and spontaneous abortion (Gibbs 2002). The localisation of the bacterium to the placenta is relatively less likely to occur but if even a single bacterium finds its way it would find a suitable environment and begin to proliferate, thereby spreading the infection to the mother's body. Immediate abortion followed by infection is very likely a defence mechanism of the body to ensure the survival of the mother.

3.3 Systemic Listeriosis

Listeria has been known to infect immunocompromised individuals. The individuals who are already in a diseased condition are more susceptible to the infection. In systemic infection, the bacterium attaches to and colonizes the intestine for a short period and then infects other major organs by crossing the intestinal layer and spreading through the blood circulatory system and the lymphatic system. The liver, which is the major drug metabolising organ, receives around 90% of the bacterial population. In the liver, the virulence factor InternalinB (InlB) is responsible for the entry of the pathogen into the hepatocytes (Parida et al. 1998). Thus, the hepatic cycle of infection begins, which gives rise to a condition called liver abscess. The systemic mode of infection of *Listeria* also includes organs like the spleen, lymph nodes, and gall bladder which are infected by a comparatively smaller population of the bacterium. *Listeria* continues to spread through the system until it crosses the blood-brain barrier (Lecuit 2005) and infects the brain, which leads to meningitis and encephalitis of the brain stem. Fever, bacteremia and ataxia are symptoms associated with the systemic spread of listeriosis.

4. Mechanism of Pathogenesis

Of the 13 serovars of *L. monocytogenes* (i.e., 1/2a, 1/2b, 1/2c, 3a, 3b, 3c, 4a, 4b, 4c, 4d, 4e, 4ab and 7), only 3 (i.e., 1/2a, 1/2b, and 4b) are responsible for 98% of human listeriosis outbreaks (Roche et al. 2008). It is an intracellular pathogen that causes gastroenteritis and fever in addition to β-hemolysis. Its pathogenicity is brought about by several virulence factors which are either surface bound or secreted proteins (Table 1) aiding in its entry, persistence, spread and evasion. This is a complex two-stage process, involving an intestinal and systemic phase of infection. In the intestinal phase, it colonizes the intestine and subsequently breaches the epithelial barrier for transportation by the blood or lymph (Bhunia 2018). During the systemic dissemination, it is translocated by dendritic cells and macrophages to the liver, lymph nodes, spleen, brain and placenta (in case of pregnancy). The crucial virulence genes, in its 3 Mb chromosome, include a major virulence gene cluster (*vgc*) in the *Listeria* pathogenicity island 1 (LIPI1; 9 Kb; *prfA-plcA-hly-mpl-actA-plcB*), internalin genes (*inlA* and *inlB*), *hpt*, *bsh* and *bilE*, which are regulated by *prfA* gene, the major regulator of *Listeria* virulence gene expression (Fig. 1). The last five genes along with *gadA* are co-regulated by an alternative sigma factor, Sigma B (*σB*).

Upon consumption, GAD (glutamate decarboxylase) helps in acid resistance of *L. monocytogenes* during the gastric passage, while BilE, BSH and OpuC impart bile salt and osmotic tolerance in the intestinal lumen. Then it is translocated from the intestinal lumen to the basolateral side of the mucosal epithelium via three possible ways (Fig. 2): (1) through phagocytic M-cells (Microfold cells found in the gut-associated lymphoid tissue (GALT) of the Peyer's patches in the small intestine (passive process), (2) through dendritic cells and macrophages (in the lamina propria of intestinal mucosa, are the antigen-presenting cells which either present antigens to T-cells of the immune system, or, help in systemic spread) (passive process), and (3) through enterocytes (intestinal absorptive cells found in the small intestine) (active process through internalin/E-cadherin pathway). In the case of the active processes,

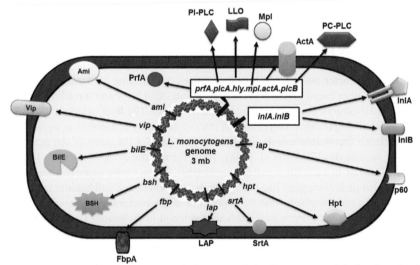

Fig. 1. Arrangement of *L. monocytogenes* virulence genes in its chromosome and the location of their gene products (Adapted and revised from Bhunia 2018).

Table 1. Major virulence factors responsible for *L. monocytogenes* pathogenesis (Bhunia 2018, Cabanes et al. 2011, Bierne and Cossart 2007, Lebreton et al. 2011, Prokop et al. 2017, Roche et al. 2008).

Protein	Name	Gene	Receptor	Function
1. Surface proteins				
Lmo0433	Internalin A (InlA)	*inlA*	E-cadherin	Adhesion and invasion of enterocytes and placenta
Lmo0434	Internalin B (InlB)	*inlB*	Met (tyrosine kinase), gC1q-R/p32	Invasion of hepatocytes and endothelium
Lmo2821	Internalin J (InlJ)	*inlJ*	E-cadherin	Invasion of enterocytes
Lmo0320	Virulence-associated invasion protein (Vip)	*vip*	Gp96 (chaperone protein)	Invasion of enterocytes
Lmo1634	*Listeria* adhesion protein (LAP)	*lap*	Hsp60 (heat shock protein)	Adhesion to enterocytes
Lmo2558	Autolysin amidase (Ami)	*ami*	Peptidoglycans	Adhesion to host cells
Lmo0582	p60 (a cell wall hydrolase, invasion-associated protein)	*iap*	Peptidoglycans	Adhesion and invasion of host cells
Lmo0204	Actin polymerization protein (ActA)	*actA*	Heparan sulphate	Actin tail formation for movement in host cell cytoplasm
Lmo0929	Sortase A (SrtA)	*srtA*	-	InlA, InlJ and Vip anchoring on *Listeria* surface
Lmo1829	Fibronectin-binding protein (FbpA)	*fbpA*	Fibronectin	Adhesion to host cells
Lmo1847	lipoprotein promoting entry (LpeA)	*lpeA*	-	Invasion of host cells and metal transport
Lmo2196	OppA (surface bound protein of ABC transporter system)	*oppA*	-	Low temperature tolerance and oligopeptide transport
Lmo1076	Autolysin (Auto)	*aut*	-	Invasion of host cells
-	Hexose phosphate translocase (Hpt)	*hpt*	-	Multiplication in host cells
2. Secreted and intracellular proteins				
Lmo0200	Protein regulatory factor (PrfA)	*prfA*	-	Central virulence regulator
Lmo0895	Sigma B (σB)	*sigB*	-	Stress transcription factor
Lmo0202	Listeriolysin O (LLO)	*hly*	Cholesterol	Hemolysin aiding lysis of phagosomes
Lmo0203	Zinc metalloprotease (Mpl)	*mpl*	-	Maturation of PC-PLC
Lmo0201	Phosphatidylinositol-specific phospholipase C (PI-PLC)	*plcA*	-	Lysis of phagosome membrane
Lmo0205	Phosphatidylcholine-specific phospholipase C (PC-PLC)	*plcB*	-	Lysis of double membrane vacuole (cell-to-cell spread)
-	Bile exclusion system (BilE)	*bilE*	-	Bile salt tolerance (gut)
Lmo2067	Bile salt hydrolase (BSH)	*bsh*	-	Bile salt tolerance (gut)
Lmo1426	OpuC	*opuC*	-	Osmotic tolerance (gut)
Lmo0206	OrfX	*orfX*	-	Nucleomodulin (interacts with RybP)
Lmo0438	Listeria nuclear targeted protein (LntA)	*lntA*	-	Nucleomodulin (targets the chromatin repressor BAHD1)

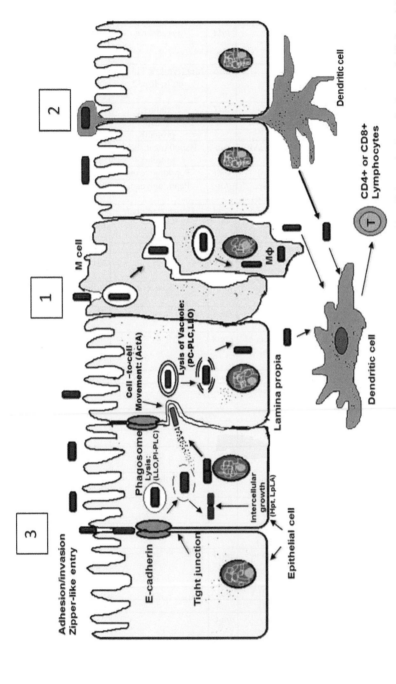

Fig. 2. Three possible translocation pathways of *L. monocytogenes* through the intestinal epithelial lining and its cellular mechanism of pathogenesis (1) through M-cells, (2) by dendritic cells, and (3) active invasion through epithelial cells by E-cadherin signalling pathway (Adapted and revised from Bhunia 2018).

bacteria interact with the receptors on the host cell surface, giving rise to a signalling cascade which facilitates the host cell invasion (Bhunia 2018). The cellular mechanism of *L. monocytogenes* infection cycle encompasses the following four fundamental steps (Fig. 2).

4.1 Attachment and Invasion of Host Cell

Adhesins are specialized surface proteins which recognize specific receptors on the host cell surface, thereby mediating bacterial adhesion. These host receptors are the components of extracellular matrix, like structural proteins (collagen and elastin), specialized proteins (fibrillin, fibronectin, and laminin), proteoglycans and mucins, which play a crucial role in bacterial adhesion to the epithelium (Bierne and Cossart 2007). Several adhesion factors act in unison for the attachment of *L. monocytogenes* bacteria onto the host cells. InlA interacts with host cell receptors, called epithelial cadherin (E-cadherin), and initiates the intestinal and placental infection. InlB interacts with Met/gC1q-R/proteoglycan (GAGs) receptors in order to instigate hepatic and endothelial infection. Other major virulence factors involved are InlJ, Vip, LAP, FbpA, Ami, p60, LpeA and Auto (Bhunia 2018).

4.1.1. InlA. Encoded by *inlA* and a part of the internalin multigene family, it shares the same locus with *inlB*. InlA (88 kDa, 800 amino acids) is a surface protein bound by covalent anchoring to cell wall peptidoglycan (*meso*-diaminopimelic acid residues) through its C-terminal LPXTG (Leu-Pro-X-Thr-Gly) sorting motif-mediated by transpeptidase sortase (SrtA). Presence of N-terminal leucine-rich repeats (LRRs) in InlA is critical for the invasion of intestinal and placental cells subsequent to interaction with the receptor, E-cadherin (Ecad, a glycoprotein, cell adhesion molecule which maintains the integrity of adherens junction of polarized epithelium and stays connected to cytoskeletal proteins). The Pro residue at position 16 in human Ecad is responsible for InlA recognition (Bierne and Cossart 2007). This leads to actin cytoskeleton rearrangement for bacterial entry though the recruitment of β- and α-catenin, vezatin and myosin VIII. A full-length and un-mutated InlA is necessary for *Listeria* pathogenicity and crossing of the blood-brain barrier (Bhunia 2018).

4.1.2. InlB. Encoded by *inlB* and part of the internalin multigene family, InlB (65 kDa, 630 amino acids) is a surface protein bound by non-covalent interaction with the lipoteichoic acid of the cell wall through its C-terminal GW module (three repeats of 80 amino acid long clusters, each starting with GW, i.e., Gly-Trp repeats). It also has N-terminal LRRs which help with its entry into hepatocytes and endothelial cells via interaction with the hepatocytes growth factor receptor, Met (a tyrosine kinase) and a coreceptor, gC1q-R/p32 (globular part of complement cascade protein C1q, it interacts with GW module). This leads to actin cytoskeletal rearrangement for bacterial entry through the recruitment of Cb1, Shc, Gab1, which in turn activates PI-3 kinase, Rac (a GTPase) and Arp2/3 (actin nucleator protein) (Bhunia 2018).

4.1.3. InlJ. Encoded by *inlJ* and part of internalin multigene family, InlJ (92 kDa, 851 amino acids) is a surface protein bound by covalent anchoring to cell wall peptidoglycan through its C-terminal LPXTG sorting motif (similar to InlA) mediated by transpeptidase sortase (SrtA). Presence of N-terminal leucine-rich repeats (LRRs) in InlJ that are similar to InlA is critical for the invasion of enterocytes subsequent to interaction with the receptor, E-cadherin (a glycoprotein) (Cabanes et al. 2011).

4.1.4. Vip. Vip (43 KDa, 399 amino acids) is a surface protein bound by covalent anchoring to cell wall peptidoglycan through its C-terminal LPXTG sorting motif-mediated by transpeptidase sortase (SrtA). Its N-terminal signal sequence helps in the invasion of enterocytes subsequent to interaction with the receptor, endoplasmic reticulum resident chaperone Gp 96 (a 96 kDa chaperone protein belonging to the Hsp 90 family) (Cabanes et al. 2011).

4.1.5. LAP. LAP (104 kDa, 866 amino acids) is a bifunctional membrane-bound enzyme having two major domains, i.e., the N-terminal acetaldehyde, dehydrogenase, and the C-terminal alcohol, dehydrogenase. It is secreted and translocated onto the cell surface for interaction with the Hsp60 receptor on the host cell in order to enhance its affinity towards the epithelium (Bhunia 2018).

4.1.6. FbpA. FbpA (55.3 kDa, 570 amino acids) is a novel multifunctional virulence factor of *L. monocytogenes*. It binds to fibronectin receptor (450 kDa glycoprotein in plasma, extracellular fluid and cell surface of the host), thereby contributing in adherence and colonization of the epithelium. It also behaves as a chaperone protein or escort protein for LLO and InlB, the two key virulence factors (Barbuddhe et al. 2008). It does not possess a characteristic surface-exposed domain. Rather, it is exported via the auxiliary secretory protein (SecA2)—dependent secretion pathway (Bierne and Cossart 2007).

4.1.7. Ami. Ami is a surface-associated protein bound to the lipoteichoicacid of bacterial cell wall via non-covalent interaction with the GW modules present in its C-terminal domain (Barbuddhe et al. 2008). It is an autolytic enzyme also serving as an adhesin. Its size varies between the serovars, in 1/2a it is 917 amino acids long while in 4b it has 770 amino acids. Ami has three domains, the N-terminal alanine amidase, followed by a 30 amino acid long signal sequence, and the C-terminal cell wall anchoring (CWA) domain containing GW modules. This CWA domain helps it in anchoring to the cell wall (here it exerts autolytic action), as well as in adherence to the host cell surface. This CWA domain is highly variable among the two serovars, 1/2a contains eight GW modules while 4b consists of six GW modules (Bhunia 2018).

4.1.8. p60. Autolysin p60 (60 kDa, 484 amino acids) is a major extracellular protein, better known as cell wall hydrolase. It contains LysM (lysine motif) domains responsible for its non-covalent anchoring to the cell wall (here it exerts murein hydrolysis for proper cell division). They are secreted out of the bacteria cell via SecA2-dependent secretion pathway (Barbuddhe et al. 2008). It is an invasion-associated protein that helps with entry into hepatocytes and macrophages.

4.1.9. LpeA. LpeA belongs to a family of surface-exposed proteins called the lipoprotein-associated antigen I (LraI). It is a lipoprotein and a component of the ABC transporter system. It also consists of a metal binding site composed of His-67, His-139, Glu-205 and Asp-280, thus helping in metal transport (Zn/Mn ions). It is necessary for adherence and invasion into epithelium and hepatocytes. LpeA might act directly through interaction with cell receptor or indirectly being an ABC transporter, thereby influencing the signalling cascade mediated invasion (Bierne and Cossart 2007).

4.1.10. Auto. Auto encoded by *aut* gene is a surface protein having autolytic activity. It is necessary for invasion of host cells by controlling the bacterial surface architecture or by peptidoglycan hydrolysis modulating host immune response (Cabanes et al. 2011). Auto contains an N-terminal autolysin domain and a C-terminal CWA domain comprising of four GW modules (Barbuddhe et al. 2008).

4.2 Lysis of Phagosomes

In the process of invasion into the host cell, *L. monocytogenes* gets trapped inside primary single membrane vacuoles, called the phagosomes, by the process called phagocytosis. Before being digested by the hydrolytic enzymes from lysosomal fusion, *L. monocytogenes* escapes by disrupting the phagosomes with its enzymes, namely listeriolysin O (LLO) and phosphatidylinositol-specific phospholipase C (PI-PLC) (Barbuddhe et al. 2008).

4.2.1. LLO. LLO (60 kDa) is a sulfhydryl (SH)-activated and cholesterol-binding pore-forming hemolysin. Multiple numbers of LLO molecules (monomers) oligomerize in the vacuole membrane and lyse it by pore formation. The maximum lytic activity of LLO is seen at pH 5.5, same as that inside the phagosomes, thereby aiding in its lysis and release of *L. monocytogenes* into host cell cytoplasm. LLO has a PEST-like sequence (Pro-Glu-Ser-Thr), which helps the bacteria to persist in the host cell cytoplasm by acting as the target for degradation. LLO lacking this sequence induces host cell apoptosis and death, thereby hampering bacterial spread. Therefore, LLO plays a key role in both phagosomal lysis as well as cell-to-cell spread during the *L. monocytogenes* infection-cycle. LLO is also known to kill erythrocytes, hepatocytes, dendritic cells and B-lymphocytes by membrane lysis, intracellular release of hydrolytic enzymes and DNA degradation (Bhunia 2018).

4.2.2. PI-PLC. Encoded by *plcA*, PI-PLC (36 kDa) works synergistically with LLO to cause phagosomal lipid-bilayer membrane disruption thereby aiding in the escape of *L. monocytogenes*. Phosphatidylinositol acts as the substrate for PI-PLC, and it has absolutely no activity on phosphatidylcholine (Bhunia 2018).

4.3 Intracellular Growth and Multiplication

After escaping into the cytoplasm, *L. monocytogenes* replicates rapidly inside the host cell by fuelling its organelles using energy from the host cell-mediated via Hpt and lipoate protein ligase (LpLA1) (Bhunia 2018).

4.3.1. Hpt. Hpt helps *L. monocytogenes* exploit hexose phosphates from the host cell as a source of carbon and energy to fuel their growth and multiplication in a host cell. Hpt scavenges host-derived glucose-1-phosphate, glucose-6-phosphate, fructose-6-phosphate and mannose-6-phosphate from host cell cytoplasm. Hpt is a translocase which, upon entering the host cell, induces a set of virulence factors required for virulence and intracellular proliferation. Hpt is the first virulence factor specifically involved in the replication phase of pathogenesis (Barbuddhe et al. 2008).

4.3.2. LpLA1. *L. monocytogenes* produces LpLA1 in order to scavenge lipoic acid from the host cell cytoplasm. It ligates this lipoic acid to the E2 subunit of the pyruvate dehydrogenase (acts as a cofactor for the enzyme activity) forming E2-lipoamide, having an important role in the aerobic metabolism of bacteria (Bhunia 2018).

4.4 Cell-to-Cell Systemic Spread

4.4.1. ActA. ActA (90 kDa, 639 amino acids) endows *L. monocytogenes* with an actin-assembly capability in order to steer intracellular bacterial movement (actin-based motility) in the host cell cytoplasm (Barbuddhe et al. 2008). It is a bacterial cell surface protein anchored to the cell membrane bilayer via hydrophobic tail of the protein present at the C-terminal (a carboxy-terminal stretch of hydrophobic residues followed by a few charged residues serving as a stop-transfer signal) (Bierne and Cossart 2007). ActA along with PC-PLC initiates the process of cell-to-cell spread via actin polymerization forming structures called actin-comet tails, which propel *L. monocytogenes* through the cytoplasm towards the cell membrane to infect the adjacent cells. Then, *L. monocytogenes* is enveloped by a filopodia-like protruding structure which is engulfed by the adjacent cell, thereby forming a secondary double-membrane vacuole facilitating cell-to-cell bacterial spread. The outer membrane is from the newly infected cell and the inner membrane is from the previously infected cell. Finally, this secondary vacuole is lysed by LLO and PC-PLC in order to aid in bacterial escape, which in turn initiates a new infection-cycle (Roche et al. 2008). ActA is the first protein identified as actin nucleation-promoting factor. It comprises of 3 domains, the N-terminal domains which interact with Arp2/3 complex, thereby initiating actin accumulation. The proline-rich domain located at the centre interacts with a special family of proteins called Enabled (Ena)/ vasodilator-stimulated phosphoprotein (VASP), thereby accelerating directional actin assembly for actin-based motility. The C-terminal domain aids in its anchorage to the *Listeria* cell wall, as already discussed. Detection of actin filaments on *Listeria* surface attracts the microfilament crosslinking proteins (α-actinin, villin and fimbrin) around the bacteria, in addition to the association of other proteins like tropomyosin, talin, vinculin, profilin and plastin (Bhunia 2018).

4.4.2. PC-PLC. PC-PLC (29 kDa) exhibits both lecithinase and sphingomyelinase activity around neutral pH. Phosphatidylcholine acts as the substrate for PC-PLC,and it shows a weak hemolytic activity. The *plcB* gene encodes a 33 kDa proenzyme precursor protein, which then requires a zinc metalloprotease, Mpl (encoded by *mpl* gene of lecithinase operon), for its maturation in order to form the active PC-PLC enzyme (needs zinc as the cofactor) (Bhunia 2018).

5. General Features of the *L. monocytogenes* Genome

L. monocytogenes have a circular chromosome of 3 Mb length and low G+C content (around 38%). The genome sequence encodes approximately 2900 protein-coding genes. There are around 517 polycistronic operons coding for 1719 genes. A large number of genes encode transport systems, transcriptional regulators, surface-bound proteins and secreted proteins associated with colonization of *L. monocytogenes* on a wide range of niches. 331 genes are reported to encode transport proteins, out of which 39 were supposed to be phosphotransferase sugar-uptake systems (Cabanes et al. 2011), while 209 genes (7% of total genome) encode transcriptional regulators responsible for coordinated gene expression for bacterial adaptation and virulence. Also, 133 surface proteins and 86 secreted proteins are encoded by the *L. monocytogenes* genome. Among the 133 surface proteins, 41 proteins belong to the LPXTG family (Bierne and Cossart 2007). Several studies have reported three distinct phylogenetic lineages of *L. monocytogenes*, based on their ribopatterns and association with outbreaks. Lineage I strains have highest pathogenic potential,and they are involved in most epidemic outbreaks in humans (1/2b, 4b, 3b, 4d and 4e). Lineage II strains show intermediate pathogenicity and are responsible for sporadic outbreaks in humans (1/2a, 1/2c, 3c and 3a). Lineage III strains comprise mostly of animal pathogens (4a and 4c). Of these serovars 1/2a, 1/2b and 4b are responsible for 98% of the disease outbreaks in humans, while 4b is the most virulent (Bhunia 2018).

6. Regulation of Virulence-Associated Genes

Expression of the *L. monocytogenes* genome is orchestrated by a complex regulatory network, where PrfA-σB interplay holds the central role. While PrfA is the major virulence regulator, σBplays a crucial role in the intestinal adaptation of *L. monocytogenes*. PrfA is the main switch of a regulon comprising of several virulence-associated loci scattered all over the bacterial chromosome, including the internalin-multigene family (Barbuddhe et al. 2008). The genes coding for the virulence factors is mainly concentrated in the vgc in the LIPI1. The vgc comprises of three transcriptional units, where the *hly* monocistron occupies the central position coding for a pore-forming hemolysin LLO (vital virulence factor). Upstream from *hly*, there lies the *plcA-prfA* bicistronic operon, where the *prfA* gene is placed immediately downstream of *plcA*, and they are cotranscribed (Roche et al. 2008). *plcA* encodes PI-PLC which acts synergistically with LLO and PC-PLC (encoded by *plcB* is a zinc-metalloprotease) in lysis of the primary and secondary vacuoles in the infection cycle. Downstream from *hly*, there lies the *mpl-actA-plcB* operon. Where, *mpl* (coding for Mpl) is responsible for maturation of PC-PLC, while actA endows *L. monocytogenes* with actin-driven motility inside the host cells (Barbuddhe et al. 2008). PrfA-dependent genes are induced only when *L. monocytogenes* grows in the host cell cytoplasm. Transcription of some virulence genes, such as *inlA* and *inlB*, are co-regulated by the stress-responsive σB. Regulators of two-component systems such as VirR and DegU are also associated with the pathogenesis of *L. monocytogenes*. Moreover, the synthesis of some virulence factors, like PrfA, InlA, ActA, and LLO, is post-transcriptionally regulated (Sabet et al. 2008). Some of the major virulence regulators are described in the subsequent paragraphs.

6.1 PrfA

It is the direct, best characterized and pleiotropic transcriptional regulator of *L. monocytogenes* virulence. It belongs to the Crp/Fnr family of transcriptional activators (Sabet et al. 2008). *prfA* gene contains three separate transcription binding sites in its promoter region responsible for its expression. First is *prfA1*, which is a 14-bp sequence (i.e., -TTAACANNTGTTAA-) called PrfA-box. The PrfA protein activates most of the virulence genes by recognizing and binding to the PrfA-box located upstream of their transcription start site. All the *prfA* regulated virulence genes contain the PrfA-box. Second is *prfA2* (i.e., TTGTTACT-N$_{14}$-GGGTAT), which highly resembles the consensus sequence of σB-dependent promoters. Thus, in addition to its self-regulation, the *prfA* gene is also partially regulated by σB via this *prfA2* (Roche et al. 2008). The two promoters *prfA1* and *prfA2* are present immediately upstream of the *prfA* coding region. While the third promoter, i.e., *prfA3*, is located upstream of the *plcA* gene. Thus, self-regulation of *prfA* is mediated by the activation of *plcA* transcription (Bhunia 2018). Different genes are regulated differently by PrfA regulator, and they are divided into three groups accordingly. Group I consists of 11 positively regulated genes which are preceded by PrfA box. Group II consists of eight genes which are negatively regulated, while Group III is comprised of 53 genes (only two having PrfA-box). Therefore, PrfA can either up-regulate or down-regulate different genes and can directly or indirectly activate different genes by associating with different sigma factors (Cabanes et al. 2011).

6.2 σ^B

Upon ingestion, *L. monocytogenes* has to first withstand the host's proteolytic enzymes, high acidity, bile stress, and inflammatory immune response in the stomach and intestine. A protein subunit of RNA polymerase (RNAP), i.e., alternate sigma factor σ^B (encoded by *sigB*), is responsible for its survival during passage through the stress environment (Roche et al. 2008). The alternative sigma factor σ^B regulates stress response genes, such as those necessary for growth during oxidative and osmotic stress, low pH stress, low temperature and carbon starvation. It regulates 55 genes, including several virulences and stress response genes like *bsh, gadA, opuC, inlA* and *prfA* (Bhunia 2018). Both σ^B and PrfA are critical for transcription of *inlA* and *inlB*. There are 105 positively regulated genes, while 111 genes are negatively regulated by σ^B. The σ^B operon comprises of σ^B itself, along with seven regulators of σ^B genes (*rsb*) (Barbuddhe et al. 2008). The σB regulon consists of genes coding for solute transporters, novel cell wall proteins, universal stress proteins, transcriptional regulators, and those involved in osmoregulation, carbon metabolism, virulence, motility, niche-specific survival and adaptation, and chemotaxis (Cabanes et al. 2011).

6.3 VirR

A novel virulence regulatory factor VirR (encoded by *virR*) is a response regulator of a two-component system. It is a second key virulence regulon after the *prfA* regulon, and it controls the virulence by regulating surface component modifications. These modifications may affect the interaction of *L. monocytogenes* with host cells and components of the immune system. There are 12 VirR-regulated genes like *dltA* and *mprF* (Cabanes et al. 2011).

6.4 DegU

DegU (encoded by *degU*) is a pleiotropic response regulator involved in the expression of virulence genes and motility at low temperature. Two *L. monocytogenes* operons are positively regulated by DegU and the flagella-specific genes, and monocistronically transcribed *flaA* genes are expressed only at low temperature (Cabanes et al. 2011).

7. Detection of *Listeria monocytogenes*

7.1 Enrichment and Enumeration

For the prevention of sporadic outbreaks of listeriosis caused by *L. monocytogenes*, early detection is one of the major concerns for the food processing industries. Detection of *Listeria* contaminants in various food products or the food processing environments is achieved by the application of various standard and rapid microbiological procedures. Among the different detection procedures, the most widely used protocols for the detection of *Listeria* in food products, such as poultry, meat, dairy products, vegetables, fruits, and seafood products, are accomplished by United States Department of Agriculture-Food Safety Inspection Service (USDA-FSIS) and Food and Drug Administration (FDA). In European countries, an extensively accepted method for the detection of *Listeria* in food products is the Netherland Government Food Inspection Service (NGFIS), developed by Van Netten et al. (1989). Additionally, along with the classical conventional microbiological techniques for *Listeria* detection, there are different, widely-applied, rapid techniques available which have been recognized for regulatory screening.

L. monocytogenes can be easily cultured from most of the clinical samples as they can be identified easily in the sterile sites, such as in blood or cerebrospinal fluid. *Listeria* is usually capable of growing on the unselective media such as blood agar plates, following incubation at 35–37°C for 24–48 h. However, in comparison with the clinical samples, detection of *Listeria* in food and faecal samples is more difficult due to the complex matrix and competing for microflora present in the food and faecal samples. Generally, most of the food testing protocol especially for the detection of *Listeria* spp. includes the use of an enrichment broth, where samples are completely mixed with the broth in order to prepare a homogenized solution following with the incubation for 24–48 hours. After incubation with the enrichment broth, a small portion of the sample is mixed with enrichment broth for further incubation, following this, it is plated in a selective agar medium which is specific for *Listeria* spp. The composition of the enrichment broth may vary with the specific strains of *Listeria* species, which consists of a different combination of antimicrobial agents. In specific cases, antimicrobial agents, to which *L. monocytogenes* are resistant, are used and can provide the detection of *L. monocytogenes* by restricting the growth of other unwanted microflora.

The most commonly-used antimicrobial agents include nalidixic acid, cycloheximide and acriflavin. Specific agar media are used for isolation of *L. monocytogenes* through direct plating methods, although agar media with less selective property have also been utilized successfully. According to the FDA protocol, buffered *Listeria* enrichment broth (BLEB) is used as an enrichment broth for mixing the food samples followed with the plating onto specific agar, such as Oxford, MOX or PALCAM agar plates. Additionally, the USDA method employs the use of University of Vermont medium (UVM) as an enrichment medium in the initial step, with simultaneous transfer to MOX medium and slightly modified BLEB as the secondary step after 24 hours of incubation. The USDA-FSIS protocol verifies the quality of the hazardous analysis and critical control point (HACCP) systems used by the food processing industries, such as poultry and meat products, by collecting the samples from ready to eat food products for pathogenic microorganisms like *L. monocytogenes*. Although the current protocols rely on highly specific enrichment medium for identification of *L. monocytogenes* in food samples, these media are less sensitive towards the low level of *L. monocytogenes* contamination at the initial stage (Donnelly and Diez-Gonzalez 2013). Additionally, food products consisting of low levels of *L. monocytogenes* in early stages can outgrow in multiplied numbers during storage conditions, leading towards the outbreaks of listeriosis in humans. Hence, enrichment of the contaminated samples alone for the detection of the low level of *L. monocytogenes* contamination is not sufficient. Regarding enumeration, the described procedures in FDA's Bacteriological Analytical manual consist of two different methodologies: (1) direct plating method and (2) most probable number (MPN) technique. The MPN method implies the use of modified BLEB medium in a nine-tube series for the enumeration process with the higher efficacy and sensitivity towards the low level of *L. monocytogenes* contamination detecting 100 CFU/g or less. On the other hand, the direct plating method involves the use of UVM as the primary diluent of the food samples for preparation of a homogenized solution, followed by plating it directly on a selective agar media such as modified Oxford medium (MOX). Additionally, the direct plating method has been considered as less selective and sensitive as compared to MPN method, due to its inferior sensitivity in detecting the actual amount of contaminant (pathogenic bacterial cell) present in food samples. Hence, direct plating methods are possibly considered for those samples containing a high level of *L. monocytogenes* contamination. Additionally, newly developed FDA protocol provides application of MPN filter and DNA probed colony hybridization techniques for enumeration methodologies. The addition of chromogenic substances in agar media can immensely improve the ability to discriminate among *Listeria* spp. Based on the identification of single virulence factor, such as phosphatidylinositol-specific phospholipase C, few commercial chromogenic substances are capable of detecting *L. monocytogenes* in food samples (Donnelly and Diez-Gonzalez 2013).

7.2 Advanced Molecular Techniques

Rapid advancement in the molecular techniques used for the detection and quantification of pathogenic microbial contaminants in food samples has been proved as a promising technology in the aspect of food safety in food processing industries. There are several commercial methodologies available for the detection of *L. monocytogenes*. Polymerase chain reaction (PCR)-based techniques have gained a considerable amount of attention due to their ability to utilize primer-based technology in targeting the virulence and non-virulence based factors, such as invasion-associated protein (*iap*), hemolysin (*hly*) and 16S rRNA genes in the case of *Listeria* detection (Tham and Danielsson-Tham 2013). Additionally, real-time PCR or quantitative real-time PCR (qPCR), which helps to quantify the amount of DNA or RNA amplified during the PCR reaction with the help of some fluorescence molecules, has reduced the time required for acquiring the results of the *Listeria* detection by taking only 2 days, as compared to the conventional technologies which used to take 7 days (Postollec et al. 2011). Detection of *L. monocytogenes* using enzyme-linked immunosorbent assay (ELISA) in food samples such as poultry, processed meat and ready to eat food products has been standardized. A comparison between two different methodologies, such as ELISA (specific for *L. monocytogenes* detection) and ISO 11290-1:1996, for isolation and identification of *L. monocytogenes* contamination in food samples produced an immense concordance index (Portanti et al. 2011).

Conclusions

Understanding of the molecular mechanisms of the pathogenesis is of critical importance for the development of prevention and detection strategies of a pathogen. This chapter attempted to highlight the proteins involved in the pathogenesis of *L. monocytogenes*, along with the genes coding them and their regulation. With the use of advanced techniques like comparative genomics, proteomics, transcriptomics and phenotypic arrays, it will be possible to predict the genes and proteins of different strains of *L. monocytogenes*, which are essential in encountering the stresses in different food environments and pathogenicity.

Acknowledgement

The authors acknowledge the support of PhD fellowships by National Institute of Technology, Rourkela, Odisha, India.

References

Barbuddhe, S., T. Hain and T. Chakraborty. 2008. Comparative Genomics and Evolution of Virulence. pp. 311–336. *In*: D. Liu (ed.). Handbook of *Listeria monocytogenes*. CRC press, Boca Raton, Florida.

Bell, C. and A. Kyriakides. 2009. *Listeria monocytogenes*. pp. 675–717. *In*: C.D. Blackburn and P. McClure (eds.). Foodborne Pathogens: Hazards, Risk Analysis and Control. Woodhead Publishing, Cambridge.

Bhunia, A.K. 2018. Foodborne Microbial Pathogens: Mechanisms and Pathogenesis. Springer-Verlag, New York.

Bierne, H. and P. Cossart. 2007. *Listeria monocytogenes* surface proteins: From genome predictions to function. Microbiol.Mol. Biol. Rev. 71(2): 377–397.

Boerlin, P., J. Rocourt, F. Grimont, P.A. Grimont, C. Jacquet and J.C. Piffaretti. 1992. *Listeria ivanovii* subsp. *londoniensis* subsp. *nov*. Int. J. Syst. Evol. Microbiol. 42(1): 69–73.

Bortolussi, R. 2008. Listeriosis: A primer. CMAJ. 179(8): 795–797.

Bundrant, B.N., T. Hutchins, H.C. den Bakker, E. Fortes and M. Wiedmann. 2011. Listeriosis outbreak in dairy cattle caused by an unusual *Listeria monocytogenes* serotype 4b strain. J. Vet. Diagn. Invest. 23(1): 155–158.

Cabanes, D., S. Sousa and P. Cossart. 2011. *Listeria* Genomics. pp. 141–170. *In*: M. Wiedmann and W. Zhang (eds.). Genomics of Foodborne Bacterial Pathogens. Springer-Verlag, New York.

Cotter, P.D., C.G. Gahan and C. Hill. 2001. A glutamate decarboxylase system protects *Listeria monocytogenes* in gastric fluid. Mol. Microbiol. 40(2): 465–475.

Donnelly, C.W. 2001. *Listeria monocytogenes*: A continuing challenge. Nutr. Rev. 59(6): 183–194.

Donnelly, C.W. and F. Diez-Gonzalez. 2013. *Listeria monocytogenes*. pp. 45–74. *In*: R.G. Labbé and S. García (eds.). Guide to Foodborne Pathogens. John Wiley & Sons, Ltd., Chichester.

Drevets, D.A. and M.S. Bronze. 2008. *Listeria monocytogenes*: epidemiology, human disease, and mechanisms of brain invasion. FEMS Immunol. Med. Microbiol. 53(2): 151–165.

Geoffroy, C.H., J.L. Gaillard, J.E. Alouf and P.A. Berche. 1987. Purification, characterization, and toxicity of the sulfhydryl-activated hemolysin listeriolysin O from *Listeria monocytogenes*. Infec. Immun. 55(7): 1641–1646.

Gibbs, R.S. 2002. The origins of stillbirth: Infectious diseases.Semin. Perinatol. 26(1): 75–78.

Huang, G., S.L. Mason, J.A. Hudson, S. Clerens, J.E. Plowman and M.A. Hussain. 2014. Proteomic differences between *Listeria monocytogenes* isolates from food and clinical environments. Pathogens 3(4): 920–933.

ISO. 11290-1 (1996). Microbiology of food and animal feeding stuffs. Horizontal method for the detection and enumeration of *Listeria monocytogenes*. Part, 1.

Ko, R. and L.T. Smith. 1999. Identification of an ATP-driven, osmoregulated glycine betaine transport system in *Listeria monocytogenes*. Appl. Environ. Microbiol. 65(9): 4040–4048.

Lebreton, A., G. Lakisic, V. Job, L. Fritsch, T.N. Tham, A. Camejo, P.J. Mattei, B. Regnault, M.a.Nahori, D. Cabanes and A. Gautreau. 2011. A bacterial protein targets the BAHD1 chromatin complex to stimulate type III interferon response. Science 331(6022): 1319-1321.

Leclercq, A., D. Clermont, C. Bizet, P.A. Grimont, A. Le Flèche-Matéos, S.M. Roche, C. Buchrieser, V. Cadet-Daniel, A. le Monnier, M. Lecuit and F. Allerberger. 2010. *Listeria rocourtiae* sp. nov. Int. J. Syst. Evol. Microbiol. 60(9): 2210–2214.

Lecuit, M. 2005. Understanding how *Listeria monocytogenes* targets and crosses host barriers. Clin. Microbiol. Infect. 11(6): 430–436.

Murray, E.G.D., R.A. Webb and M.B.R. Swann. 1926. A disease of rabbits characterised by a large mononuclear leucocytosis, caused by a hitherto undescribed bacillus *Bacterium monocytogenes* (n. sp.). J. Pathol. Bacteriol. 29(4): 407–439.

Parida, S.K., E. Domann, M. Rohde, S. Müller, A. Darji, T. Hain and T. Chakraborty. 1998. Internalin B is essential for adhesion and mediates the invasion of *Listeria monocytogenes* into human endothelial cells. Mol. Microbiol. 28(1): 81–93.

Portanti, O., T. Di Febo, M. Luciani, C. Pompilii, R. Lelli and P. Semprini. 2011. Development and validation of an antigen capture ELISA based on monoclonal antibodies specific for *Listeria monocytogenes* in food. Vet. Ital. 47(3): 281–290.

Postollec, F., H. Falentin, S. Pavan, J. Combrisson and D. Sohier. 2011. Recent advances in quantitative PCR (qPCR) applications in food microbiology. Food Microbiol 28(5): 848–861.

Prokop, A., E. Gouin, V. Villiers, M.A. Nahori, R. Vincentelli, M. Duval and O. Dussurget. 2017. OrfX, a nucleomodulin required for *Listeria monocytogenes* virulence. mBio8(5): e01550–17.

Roche, S.M., P. Velge and D. Liu. 2008. Virulence determination. pp. 241–266. *In*: D. Liu (ed.). Handbook of *Listeria monocytogenes*. CRC press, Boca Raton, Florida.

Sabet, C., A. Toledo-Arana, N. Personnic, M. Lecuit, S. Dubrac, O. Poupel and H. Bierne. 2008. The *Listeria monocytogenes* virulence factor InlJ is specifically expressed *in vivo* and behaves as an adhesin. Infect. Immun. 76(4): 1368–1378.

Schlech3rd, W.F. 1997. *Listeria* gastroenteritis—old syndrome, new pathogen. N. Engl. J. Med. 336(2): 130–132.

Seeliger, H.P.R. 1986. Listeria. Bergey's manual of systematic bacteriology 2: 1235–1245.

Sleator, R.D., C.G. Gahan, T. Abee and C. Hill. 1999. Identification and disruption of BetL, a secondary glycine betaine transport system linked to the salt tolerance of *Listeria monocytogenes* LO28. Appl. Environ. Microbiol. 65(5): 2078–2083.

Tham, W. and M.L. Danielsson-Tham. 2013. *Listeria monocytogenes*. pp. 124–140. *In*: W. Tham and M.L. Danielsson-Tham (eds.). Food Associated Pathogens. CRC Press, Boca Raton, Florida.

Van Netten, P., I. Perales, A. Van de Moosdijk, G.D.W. Curtis and D.A.A. Mossel. 1989. Liquid and solid selective differential media for the detection and enumeration of *L. monocytogenes* and other *Listeria* spp. Int. J. Food Microbiol. 8(4): 299–316.

Vasconcelos, J.A. and H.G. Deneer. 1994. Expression of superoxide dismutase in *Listeria monocytogenes*. Appl. Environ. Microbiol. 60(7): 2360–2366.

8

The Structural and Functional Analysis of *Escherichia coli* Genome

Shikha Gupta and *Sangeeta Pandey**

1. Introduction

The presence of a large and wide range of microbes in and around the earth forms anintegral part of the biosphere (Fraser et al. 2000). The microbial genomics analysis provides insights into the genome of organisms (such as genomic size, mRNA and genetic organization, G+C content, topology) and help researchers to explore the microbial diversity and basis of evolution in microbes present in the environment. In addition to this, the functional genomics approaches also evaluate the factors or genes of microbes that are responsible for their adaption in a wide range of environments, such as high temperature, pH, salinity, oxygen concentration. On the other hand, the comparative genomics approaches have the potential to identify virulent factors, associated pathogenicity islands, mechanisms of pathogenesis and also to examine putative drug or antibiotics targets, new vaccine candidates and, thereby, strategies for diagnosis and treatment of disease (Chan 2006, Zhou and Miller 2002).

Escherichia coli is the most significant and extensively studied as model prokaryotic organism in the field of biotechnology, molecular biology for various biochemical, metabolic and genetics studies (Han and Lee 2006). It was first discovered as a *Bacterium coli commune* in 1884 by the German microbiologist and paediatrician Theodor Escherich in the gut of the neonates. He studied its characteristics and examined its role in metabolism and in pathogenesis of gastroenteritis in infants. Later on, the name of the bacterium was changed to *Escherichia coli* in honour of its discoverer (Escherich 1988, Shulman et al. 2007).

Amity Institute of Organic Agriculture, Amity University Uttar Pradesh, Sector 125, Noida, Uttar Pradesh 201313.
* Corresponding author: spandey5@amity.edu

It is a facultative anaerobic, Gram-negative, non-sporing, rod-shaped, gamma-proteobacteria that ubiquitously adhered to the epithelial layer of the gastrointestinal tract of mammals with caecum and the colon as its main commensal niches (Tenaillon et al. 2010). It aids in the breakdown of food, vitamin K2 production and also acts as a bacteriostatic agent against harmful bacteria by producing bacteriocins and other organic acids (Delmas et al. 2015).

The cells are about 2.0 micrometres (µm) in length and 0.25–1.0 µm in diameter with some strains being either motile (due to the presence of flagella with the peritrichous arrangement) or non-motile. The optimum temperature is 37°C and pH is between 6.0 to 8.0 for growth of *E. coli* (Welch 2006, Percival and Williams 2014). The cell wall is composed of a mesh-like network of peptidoglycan (N-acetylmuramic acid is cross-linked with L-alanine, D-glutamic acid, meso-diaminopimelic acid and D-alanine) of size 2–7 nm, found within the periplasm between the outer and inner membrane. (Welch 2006, Boags et al. 2017, Gumbart et al. 2014). Generally, 1–3 layers of peptidoglycan are present in Gram-negative bacteria but, in *E. coli*, only a single layer of peptidoglycan is prevalent (Huang et al. 2008). The physiological and genetic characteristics of *E. coli* are shown in Fig. 1 (Percival and Williams 2014, Welch et al. 2002).

E. coli is a large and diverse group of bacteria belonging to family *Enterobacteriaceae*, in which some are beneficial, non-pathogenic, commensal strains while some are responsible for causing diseases in humans and animals, such as urinary tract infection, respiratory problems, pneumonia, diarrhoea and food poisoning in humans (Kaper et al. 2004, Kohler and Dobrindt 2011). Another classification system is based on the presence of surface-associated antigens (serotypes): O antigens in which lipopolysaccharide molecules are present on the surface; K antigen is the acidic capsular polysaccharide, while H antigen strains have flagellin (involved in the motility) on the surface. Pathogenic strains are further classified into enteropathogenic, uropathogenic and extra intestinal. Enteropathogenic strains are responsible for causing diarrhoea and acute gastroenteritis in infants and adults. Extra intestinal pathogenic *E. coli* (ExPEC) are facultative pathogens that can cause diseases such as UTI, meningitis, sepsis, pneumonia at sites outside the gastrointestinal tract while living harmlessly in the gut. The strains of *E. coli* that are responsible for infecting the urinary tract are termed as Uropathogenic *E. coli* (UPEC). Another classification of pathogenic strains is based on their pathogenesis: Enterohemorrhagic (EHEC),

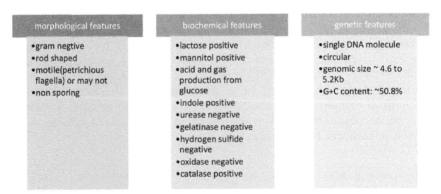

morphological features	biochemical features	genetic features
•gram negtive •rod shaped •motile(petrichious flagella) or may not •non sporing	•lactose positive •mannitol positive •acid and gas production from glucose •indole positive •urease negative •gelatinase negative •hydrogen sulfide negative •oxidase negative •catalase positive	•single DNA molecule •circular •genomic size ~ 4.6 to 5.2Kb •G+C content: ~50.8%

Fig. 1. Characteristics of *Escherichia coli.*

enterotoxigenic (ETEC), enteroinvasive (EIEC), enteropathogenic (EPEC), diffusely adherent (DAEC) and entero aggregative (EAEC). Enterohemorrhagic *E. coli* strains are zoonotic pathogens which cause hemorrhagic colitis, gradually leading to hemolytic uremic syndrome (HUS) andacute renal failure in humans. The enterotoxigenic strains produce heat labile and stable toxins that allow secretion of excessive fluid from the intestinal epithelial cell layer which leads to diarrhoeal disease, whereas the enteroinvasive strains invade the intestine resulting in necrosis and causing dysentery (less severe than *Shigella* dysentery). The enteroaggregative *E. coli* are newly discovered strains whose pathogenesis is unclear but are found to be responsible for chronic diarrhoea in children (Garmendia et al. 2005, Smith et al. 2007, Lukjancenko et al. 2010, Kohler and Dobrindt 2011, Sperandio and Nguyen 2012).

It is the most extensively studied prokaryotic organism in the field of molecular biology and biotechnology by virtue of its easy availability, simpler cell morphology, small genetic composition, rapid growth rate with the generation time of 20–30 minutes and it is also safe to handle and easy to culture in the laboratory (Cronan 2014).

E. coli is the first choice for researchers who wish to study the molecular cloning-based experiments because of its fully sequenced genome, ability to easily transform the cells with recombinant plasmids and efficiently expressed, produced and characterized the desired proteins and several industrially important metabolites, food additives, vitamins, etc., encoded by the recombinant DNA (Idalia and Bernardo 2017).

E. coli is the most significant organism for bacterial conjugation due to the presence of double-stranded, extra chromosomal circular plasmid called F episome or fertility factor in some strains. These strains possess pili which permit the transfer of foreign DNA/plasmid to and from another bacterium (Griffiths et al. 1999). Moriguchi et al. (2013) reported the transfer of DNA from *E. coli* to yeast *Saccharomyces cerevisiae* which results in the genetic exchange between to different species.

These organisms have the ability to withstand elevated temperatures and are also commonly found in faeces of animal and humans. They generally act as indicator organisms with respect to faecal pollution in water and indicate the amount of faecal material present in water (Odonkor and Ampofo 2013).

In this chapter, we present the genomic sequence of four *E. coli* strains, one commensal *E. coli* strain K-12 MG1655, two enterohemorrhagic strains of O157:H7 (EDL933 and Sakai) and one uropathogenic strain O6: K2:H1 strain CFT073 and the function of the gene products encoded by the genome of *E. coli*. Here, we also discuss the techniques of proteomics in order to analyse the *E. coli* proteome.

2. Structural genomics

2.1 Commensal E. coli Strain K-12

The complete sequence of the genome of a non-pathogenic laboratory strain of *E. coli* (K-12) was reported in the 5 September 1997 issue of Science (Blattner et al. 1997).

The genome of *E. coli* K-12 consists of circular double-stranded DNA of size 4.6 Mb,in which around 4 Mb consists of protein-coding genes, 0.8% and 0.7% accounts for stable RNA and non-coding repeats, respectively, while the remaining ~ 10.7% encode genes of regulatory and other functions. As stated by Blattner et al. (1997), there are 2 replichores in the genome representing origin and termination point of chromosome replication. There are 7 rRNA operons (*rrnA, rrnB, rrnC, rrnD, rrnE, rrnG, rrnH*) which transcribe ribosomal RNA (16S rRNA gene, a 23S rRNA gene, and a 5S rRNA gene) in the direction of chromosome replication (Condon et al. 1995, Blattner et al. 1997).

There are a total 86 tRNA genes in the genome, of which 56 transcribe in the direction of replication. The authors also determined 6 additional new genes for tRNA-*valZ, lysY, lysZ* and *lysQ*, transcribed by *lysT* operon, and the other two are asnW and ilex. The analysis of the base composition of genome indicates that the content of guanine nucleobase is higher as compared to its corresponding cytosine and other nucleobase pair adenine-thymine. There is an operon (as illustrated in Fig. 2) consisting of 6 genes encoding enzymes (monooxygenase (*mhpA*), dioxygenase (*mhpB*), the hydrolase (*mhpC*), the hydratase (*mhpD*), the dehydrogenase (*mhpF*) and 4-hydroxy2-oxovalerate aldolase (*mhpE*)) for degradation of aromatic compounds such as phenylpropionate, phenylacetic acid, 3- and 4-hydroxyphenylacetic acid, phenylpropionic acid, 3-hydroxyphenylpropionic acid (HPP), 3-hydroxycinnamic acid, phenylethylamine, tyramine and dopamine (Blattner et al. 1997, Diaz et al. 2001).

There are 14 flagellar protein-encoding genes in the genome of *E. coli* enlisted in Table 1, in which most of the genes are identical to *Salmonella Typhimurium* (Jones et al. 1989, Blattner et al. 1997).

There is a total of 4,288 ORFs (open reading frames) in the genomic sequence of *E. coli*, with an average size of 317 amino acids. The longest ORF codes for a protein of 2,383 amino acids were found to be associated with pathogenicity of certain strains of *E. coli* and other enteric bacteria (Blattner et al. 1997).

There are a total of 2,584 operons in the genome, of which the functions of 2,192 operons have been characterized and predicted. There are ~1,600 operons consisting of 1 gene, ~364 operons have 2 genes, ~100 operons have 3 genes and the remaining 6% operons transcribed 4 or more genes. Among 2,405 operons there are 68% that contain 1 promoter, 20% contain 2 promoters, while 12% contain 3 or more promoters, which act as a recognition site for RNA polymerase to initiate the transcription of genes encoded by the operons. There are also regulatory sites present in the chromosome, consisting of genes encoding regulatory proteins which in turn inhibit the gene expression. It has been estimated that there are 15 to 25% regulatory proteins found in the *E. coli* genome. There are certain repeated sequences found throughout in the genome that are present in multiple copies as tandem repeats. The *rhs* elements are the largest repeated sequences in *Escherichia coli* genome, which present in 5 sets (*rhsA, rhsB, rhsC, rhsD, rhsE*). They act as accessory elements present in the clockwise orientation, which accounts for ~1% of the genome (Zhao et al. 1993, Blattner et al. 1997).

There are 581 REP (repetitive extragenic) sequences present, referred to as mobile elements; these are the target for transposase enzymes. In addition to this, there are

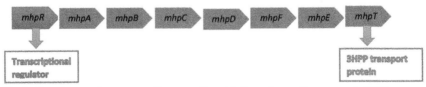

Fig. 2. Genetic map of operon encoding genes for catabolism of aromatic compounds (HPP).

Table 1. Genes of Flagellar operon encoding for flagella synthesis.

Genes	Function
flgN	Initiate filament assembly
flgM	Repressor of flagellin synthesis
flgA	Encodes P-ring protein of basal body
flgB	Involved in synthesis of proteins of basal body
flgC	Involved in synthesis of proteins of basal body
flgD	Basal body rod protein
flgE	Hook protein
flgF	Encodes M-ring protein of basal body
flgG	Encode rod protein of basal body
flgH	Encodes precursor for L ring protein of basal body
flgI	Encodes precursor for P ring protein of basal body
flgJ	Flagellar protein
flgK	Hook associated protein 1
flgL	Hook associated protein 3

other repeated sequences dispersed throughout the genome, such as BIME (Bacterial interspersed mosaic elements), IRU or ERIC (enterobacterial repetitive intergenic consensus sequence/intergenic repeat unit) elements, Box C and RSA. Another repetitive sequence, *Ter* sequence (10 copies),is found throughout the genome and blocks the progression of the replication fork beyond the termination point (Blattner et al. 1997, Mori 2004, Tobes and Pareja 2006).

2.2 Enterohemorrhagic E. coli O157:H7 Strain Sakai

It is one of the Shiga toxin-producing enterohemorrhagic strains of *E. coli* isolated from the Sakai outbreak in 1996 and its genomic sequence was studied and reported by the University of Wisconsin and Kazua Institute of Japan.

The genome of *E. coli* O157:H7 Sakai is 5.4 Mb in size, greater than that of *Escherichia coli* K-12 strain by 859Kb. It comprises of plasmid pO157 (92,721 bp) and a cryptic plasmid of 3,306 bp (pOSAK1), which contributes to the pathogenesis of strain. Unlike *E. coli* K-12, there are 2 asymmetrical replichores representing origin and termination point of replication in this strain (replichore 2 is greater than replichore 1). There are 296 strain-specific loops called S-loops of size 1,393 bp, almost 25% of the entire genome, with GC content 48.2%, mainly composed of prophage and prophage-like elements. There are a number of mobile elements present in the chromosome, which includes 80 sets of 20 types of insertion sequences (IS), for example, IS629, IS679-related elements (ISEc8, 682 and 683), 18 sets of prophages (Sp1-18) and lambda-like phages (Stx-1 and Stx-2 transducing phage), Mu-like phage, P4 like phages and other remnants of phages. It was suggested that half of the prophage are of bacteriophage origin, which reflects that bacteriophage plays an important role in O157:H7 strain evolution and diversity. There are also 6 prophage-like elements (Sakai prophage-like elements, SpLE1–6) apart from the

18 prophages, which are similar to the cryptic prophage of strain K-12 and also not evolved from bacteriophage. There is another mobile element present in the genome, known as *rhs* element, in 9 copies; out of which, four (*rhsA, rhsC, rhsD, rhsE*) are common in *E. coli* K-12 strain, two (*rhsF, rhsG*) are also found in other strains, but three (*rhsI, rhsJ, rhsK*) are specific to O157:H7 Sakai strain. There are seven sets of *rrn operons* encoding ribosomal RNA found in the genome that is identical to that of *E. coli* K-12. There are 102 tRNA genes identified in the genome, out of which 86 are common in K-12 strain while 20 are specific to O157:H7 Sakai strain. Moreover, out of 86 tRNA genes that are conserved in K-12 strain, 4 tRNA genes (*leuPV* genes and *lysQ*) are absent in O157:H7 Sakai strain. A total of 5,361 ORFs or protein coding regions, either known or predicted, are identified in the entire chromosome, of which 3,729 are common in K-12 strain (known as conserved ORFs), while the remaining are O157:H7 Sakai specific ORFs. There are a total of 14 loci identified in the chromosome associated with the genes encoding proteins for fimbrial biosynthesis, of which five loci are conserved, five are partially conserved with the K-12 strain, while four are O157:H7 Sakai specific (Hayashi et al. 2001).

The LIM locus of the O157:H7 Sakai chromosomes encodes genes for 2 proteins, namely ECs1825 and ECs3485, belonging to the TrcA chaperon family involved in the pili formation and development of adherence phenetic character (Hayashi et al. 2001). Another locus named LEE (locus of enterocyte effacement) encodes genes for proteins such as adhesin intimin, a type III secretion system (T3SS), translocated intimin receptor (*tir*) which are involved in the attachment and effacement of A/E lesions on the epithelial layer of intestines, a prerequisite for EHEC mediated infection (Cepeda-Molero et al. 2017).

The lambda phage genome encodes Stx1 and Stx2 which play a major role in EHEC mediating HUS (Hemolytic-uremic syndrome). It also encodes one *lom*-like gene (*lomX*), one copper/zinc-superoxide dismutase (SOD) gene (*sodC*), which increases their ability to survive in the phagosome. The plasmid pO157 genome encodes genes for enterohemolysin protein, catalase protein (*katP*) and proteins which are similar to LCT (large clostridial toxins), a major virulent factor of *Clostridium difficile* associated colitis (Hayashi et al. 2001, Pruitt et al. 2010). The genes for iron/ siderophore uptake are conserved in both strains of *E. coli* K-12 and O157:H7 Sakai but *fec* operon (*fecABCDE*) which encodes genes for the ferric citrate transport system is absent in the O157:H7 Sakai genome (Hayashi et al. 2001, Enz et al. 2003). In addition to this, the O157:H7 Sakai genome consists of another iron transport system encoded by genes of TonB-dependent transporters (TBDTs) that enables the bacteria to bind and utilize iron chelating compounds called siderophores (Hayashi et al. 2001, Noinaj et al. 2010).

The O157:H7 Sakai chromosome contains genes of the transportation system, for example, ABC transporter family, phosphoenolpyruvate: Carbohydrate phosphotransferase (PTS) transport system. It also contains genes for the metabolism of sucrose, urease and sorbose. Moreover, it possesses the genes for a glutamate fermentation system, as well as for biodegradation of aromatic compounds (Hayashi et al. 2001).

2.3 Uropathogenic E. coli Strain CFT073

The UPEC strain CFT073 was first isolated from the blood of a patient suffering from pyelonephritis and sequenced by combined approaches of shotgun and primer walking experiments.

The size of the genome is 5.2 Mb with 5,533 protein-coding genes; no virulent plasmid has been reported. There are 5 cryptic prophage genes present in the genome of the bacterium with no ability to produce viable phage. The virulent factors are present in the horizontally acquired pathogenicity islands (code for genes that contribute to the pathogenesis of the strain), which include genes *pheV, pheU,* and *asnT* located at tRNAs. PAI-*pheV* contains genes for aerobactin synthesis, haemolysin, capsule synthesis and two autotransporters, while *pheU* contains pili associated with pyelonephritis, as well as siderophore receptor genes. The gene products of *asnT* gene resemble the pathogenesis of *Yersinia pestis* and contain *Yersinia* bactin genes (Welch et al. 2002).

On comparing its genome with the other three *E. coli* genomes, i.e., EDL933, Sakai and strain K12 MG1655, it was found that 432 genes were specific to UPEC strain CFT073 that may play a role in bacterial virulence in the urogenital tract. There are 12 sets of fimbriae or pili, of which 10 fimbriae belong to the family of chaperone-usher (responsible for biosynthesis of pili) and the rest are of type IV pili (responsible for twitching motility). These genes facilitate the colonization of the urinary tract in hosts, a prerequisite for the development of urinary tract pathogenesis (Hooton and Stamm 1997).

FimB and FimE are two recombinases encoded by genes *fimB* and *fimE* (5 copies) and are responsible for phase switching phenomenon where a ~300 bp DNA fragment consisting promoter region of fimbriae protein *fimA* was inverted. This will, in turn, mediate the on/off expression of fimbriae and confer virulence of bacteria (Klemm 1986). There are at least 7 autotransporters of type V secreted proteins in the genome which may contribute to the pathogenesis of UTI. The most well-characterized example is secreted autotransporter toxin (Sat) belonging to the SPATE (serine protease autotransporters) of *Enterobacteriaceae* family (Henderson and Nataro 2001, Guyer et al. 2002).

Comparison of the full *E. coli* genome of all four strains, i.e., the commensal K-12 MG1655 and the three pathotypes- 2 O157:H7 strains EDL933 and Sakai and one uropathogenic strain CFT073, revealed that *E. coli* pangenome consists of core genome and flexible genome. Core genome accounts for nearly 4 Mb genetic data that consist of genes present in all strains and referred as conserved backbone essential for maintaining cellular life, while flexible genome consists of some strain-specific genes or elements, such as plasmids, transposons, insertion sequence elements, prophages, genomic islands (GEIs), pathogenicity islands which are associated with pathogenicity and also with the ability to survive in wide range of environments (Table 2).

3. Transcriptomics analysis

After determination of genome sequence of *E. coli* strains, including K-12 (Blattner et al. 1997), the haemorrhagic strains O157:H7 EDL933 (Perna et al. 2001), O157:H7

Table 2. The genomic content of all 4 strains of *E. coli* (Blattner et al. 1997, Perna et al. 2001, Welch et al. 2002, Hayashi et al. 2001).

E. coli Strains	Type	Genomic Size (Mb)	G+C Content (%)	Number of ORFs Present	Number of IS Elements/ Transposable Elements	Number of Prophages/Cryptic Prophages	Plasmids
K-12 MG1655	Commensal, non-pathogenic	4.6	50.8	4,294	42	10	-
O157:H7 EDL933	Enterohemorrhagic	5.5	50.5	5,361	60	16	pO157
O157:H7 Sakai	Enterohemorrhagic	5.5	50.5	5,361	80	24 (18 prophages + 6 prophage like elements)	pO157, pOSAK1
O6: K2:H1 CFT073	Uropathogenic	5.2	50.47	5,533	49	5	-

ORFs: Open Reading Frames; IS: Insertion Sequences

Sakai (Hayashi et al. 2001) and the uropathogenic strain CFT073 (Welch et al. 2002), the next task is to characterize different gene products encoded by the genome of *E. coli* using functional genomics approaches,such as DNA microarray, ORF analysis (ORFeome), DNA microarray, RNAseq and construction of mutants by Gene knockout (Griffiths et al. 2000, Mori et al. 2000, Matte et al. 2003, Barsy and Greub 2013, Bunnik and Le Roch 2013).

3.1 Translation apparatus

The translation machinery of *E. coli* comprises rRNA, ribosomal proteins, tRNA, aminoacyl-tRNA synthetases and other translation factors (Laursen et al. 2005).

There are 37 known genes encoding the tRNA synthetases and other modification enzymes, 17 genes encoding translation factors and 55 genes for ribosomal proteins. Some are listed in Table 3 (Riley 1993, Tao et al. 1999, Soye et al. 2015).

Table 3. Genes encoding translation system components.

Gene	Gene Product
alaS	Alanyl-tRNA synthetase
pheST	Operon encodingphenylalanyl-tRNA synthetase pheS and pheT co-transcribed encodes large and small subunit, respectively of enzyme
thrS	Threonyl-tRNA synthetase
lysS *lysU*	Two lysyl-tRNA synthetases *lysS:* Housekeeping synthetases; constitutively expressed *lysU:* Exhibits stimulation induced expression
prfB	Peptide chain release factor
gltX	Glutamyl-tRNA Synthetase
metG	Methionyl-tRNA Synthetase
efp	Elongation factor
infB	Protein chain initiation factor 2
rpsA	30S ribosomal subunit protein S1
rplW	50S ribosomal subunit protein L23
rpsS	30S ribosomal subunit protein S19
rplB	50S ribosomal subunit protein L2

3.2 Metabolism of Nitrogen

Ammonia (inorganic nitrogen) is metabolized into glutamate which is ultimately the precursor of amino acids such as arginine, glutamine, proline and the polyamines (Reitzer 2004).

In *E. coli* there are 2 enzymes,glutamate dehydrogenase (GDH, encoded by *gdhA*) and glutamate synthetase (also called glutamine oxoglutarate aminotransferase [GOGAT] encoded by *gltBD*), required for the synthesis of glutamate under ammonia limiting conditions (Helling 1998, Tao et al. 1999, Riley 1993). The DNA microarray-based genomic technology allowed for the identification the 75 NtrC/Nac (Nitrogen regulatory protein C) regulated genes in the genome of *Escherichia coli* that will regulate the physiological state of the cell under nitrogen limiting conditions. NtrC is a 2-component system consisting of NtrB (*glnL,* encodes sensor kinase) and NtrC (*glnG,* encodes response regulator) which activates the transcription of σ54-dependent genes that include: Ammonia transporter (*AmtB*), glutamine synthetase (encoded by *glnA*), *glnK* (N-regulation), and *nac* (N assimilation control) amino acid permeases and catabolic enzymes. The nitrogen assimilation control (*Nac*) protein activates transcription of σ70-dependent genes in order to attenuate the effects of the low availability of ammonia (nitrogen source) (Thompson and Zhou 2004).

3.3 *Formation of Amino Acids*

Arginine and its precursor ornithine are significant nitrogen sources for *E. coli* under nitrogen limiting conditions (Tao et al. 1999, Caldara et al. 2008).

The genes required for de novo synthesis of amino acids include the *argA* gene encoding N-Acetylglutamate synthase (EC 2.3.1.1) responsible for biosynthesis of arginine from glutamate. The genes of the *ilvGMEDA* operon encode for biosynthesis of three branch chain amino acids, isoleucine, leucine and valine. It is a repressible operon: Turns off when all three amino acids are present while expressed in the absence of any of the three amino acids (Lawther et al. 1990, Parekh and Hatfield 1997). Chorismate synthase, encoded by *aroC* gene, is the rate-limiting enzyme of shikimate pathway, catalysing the conversion of 5-enolpyruvylshikimate-3-phosphate 1 (EPSP) to chorismate (Bornemann et al. 1995, Tao et al. 1999). For tyrosine synthesis, the first step is the condensation reaction catalysed by 3-deoxy-d-arabino-heptulosonate-7-phosphate synthase (DAHPS), encoded by *aroF* gene, between phosphoenolpyruvate (PEP) and erythrose-4-phosphate (E4P). The chorismate is the major intermediate for the synthesis of aromatic amino acids, as elucidated in the Fig. 3. The *tyrA* gene encodes for a combined protein for chorismate mutase and prephenate dehydrogenase. Moreover, *tyrA* and *aroC* genes reside very closely on the chromosome and, hence, form an operon (Mattern and Pittard 1971, Tao et al. 1999, Chávez-Béjar et al. 2008).

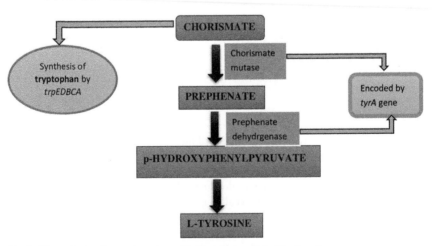

Fig. 3. Biosynthesis of aromatic amino acids from intermediate, Chorismate.

For tryptophan biosynthesis *trp* operon, *trpEDBCA* (position in the same way)encode enzymes for the conversion of chorismate to tryptophan. The *trp* operon is regulated by TrpR repressor protein at the transcription level.

3.4 Vitamins and Cofactors

There are 106 genes reported in the genome that are involved in the biosynthesis of vitamins, cofactor, prosthetic groups and carriers. Some are listed in Table 4, including *hemC*, structural gene of heme biosynthetic, another *entCEBA* operon consists of genes encoding enzymes of biosynthesis of enterobactin siderophore (Kwon et al. 1996). There are three kinds of Glutaredoxins (Grx1, Grx2, Grx3) found, with Grx2 being the most abundant in the cell. They are small (~12KDa) redox enzymes with two cysteine residues in the active site and are responsible for glutathione-/NADPH-dependent redox reactions and, thereby, maintain the reduced state of cytoplasm and alleviate the harmful effects of oxidative stress. Another group of enzymes present in *E. coli* is called Glutathione S-transferase (*gst*).This group also provides protection against chemical and oxidative stress mediated by xenobiotic compounds (Allocati et al. 2009). Tetrahydrofolate serves as a carbon donor in many anabolic reactions, such as *de novo* synthesis of nucleotides (RNA/DNA) and amino acids. It is synthesized by allosteric enzyme GTP cyclohydrolase I (EC 3.5.4.16, also known as GTP 7,8-8,9-dihydrolase) (Illarionova et al. 2002). The periplasmic enzyme γ-glutamyl transpeptidase (GGT; EC 2.3.2.2) of *Escherichia coli* catalyses the hydrolysis of glutathione and releases free glutamate and dipeptide of cysteinyl-glycine which can act as cysteine and glycine source (Suzuki et al. 1989, Okada et al. 2006).

Table 4. Genes for synthesis of vitamins and cofactors.

Gene	Geneproduct
hemC	Porphobilinogen deaminase
entC	Isochorismate synthase 1
entB	Isochorismatase
entA	2,3-dihydro-2,3-dihydroxybenzoate dehydrogenase
entE	2,3-Dihydroxybenzoate-AMP ligase
entF	Apo-serine activating enzyme
entD	phosphopantetheinyl transferase
entH	proofreading thioesterase
folE	GTP cyclohydrolase 1
thiH	tyrosine lyase, involved in thiamine-thiazole moiety synthesis
ggt	γ-glutamyltranspeptidase
grxB	GlutaredoxinB
trx	Thioredoxins
ribC	Riboflavin synthase
pdxA	Pyridoxine synthetase
bioB	Biotin synthetase

3.5 Nucleotides Anabolism

E. coli has 2 operons, *nrdAB* and *nrdHIEF*, that consist of genes encoding ribonucleotide reductases enzyme that catalyse the conversion from ribonucleotides to deoxyribonucleotides (Table 5) (Riley 1993, Tao et al. 1999).

3.6 Metabolism of Fatty Acids and Lipids

The genes encoding the enzymes carrying out different steps of the pathway of metabolism of fatty acids are present throughout the genome. They are listed in Table 6 with their function (Magnuson et al. 1993, Riley 1993, Tao et al. 1999, Cho et al. 2006).

Table 5. Genes for biosynthesis of nucleotides.

Gene	Gene Product
nrdA	NrdAB class Ia ribonucleotidereductase encoding subunit B1 and subunit B2
nrdB	
nrdE *nrdF*	NrdEF class Ib ribonucleotidereductase
nrdH	NrdH redoxin with thioredoxin like activity and glutaredoxin like sequence.
nrdI	NrdI for stimulatory effecton ribonucleotide reduction
prsA	Ribose-phosphate pyrophosphokinase
upp	Uracil phosphoribosyltransferase for synthesis of Uridine monophosphate UMP
gmk	Guanylate kinase / GMP kinase
pfs (mtn)	5'-methylthioadenosine/S-adenosylhomocysteine nucleosidase
ndk	Nucleoside diphosphate kinases
adk	Adenylate kinase

3.7 Carbohydrate Metabolism

DNA microarray also exhibited the following genes encoding enzymes for carbohydrate and energy metabolism in *E. coli*, as listed in Table 7 (Riley 1993, Tao et al. 1999).

3.8 Virulent Factors

The genes encoding virulent factors listed in Table 8 are acquired by *E. coli* through horizontal transfer of either plasmid, bacteriophage, genetic elements or genome from another bacterium (Donnenberg and Whittam 2001, Chapman et al. 2006).

3.9 Stress Tolerance Mechanism

The stress response is mediated by both pathogenic and commensal strains of *Escherichia coli* in order to cope with the variety of rapidly changing environmental factors, such as temperature, pH, water activity, nutrient deprivation, etc.

Table 6. Genes for metabolism of fatty acids.

Gene	Gene Product
panF	Pantothenate permease catalyse the Na$^+$ mediated uptake of pantothenate from outside
coaA	Pantothenate kinase; phosphorylate pantothenate
aas	Acyl-ACP synthetase
accA	trans carboxylase- alpha subunit of Acetyl -CoA carboxylase
accB	Biotin carboxyl carrier protein of Acetyl -CoA carboxylase
accC	Biotin carboxylase of Acetyl -CoA carboxylase
accD	trans carboxylase- beta subunit of Acetyl -CoA carboxylase
acpP	Acyl carrier protein
acpS	ACP synthetase
cfa	Cyclopropane fatty acid synthase
fabA	Beta- hydroxydecanoyl ACP dehydrase; unsaturated fatty acid biosynthesis
fabB	Beta- ketoacyl-ACP synthase I
fabF	Beta-Ketoacyl-ACP synthase II (3-Oxoacyl-[acyl carrier protein] synthase II)
fabD	Malonyl-CoA: ACP transacylase; initiate the fatty acid synthesis.
fabE	Acetyl CoA carboxylase
fabG	Beta-Ketoacyl-ACP reductase
fabH	Beta-Ketoacyl-ACP synthase III (acetoacetyl-ACP synthase); the condensation of acetyl-CoA
fadR	Transcriptional regulator of fabA
tes	thioesterase
ato	Degradation of 4-carbon fatty acids
fad regulon	Consist of genes encoding enzymes for the transport, activation and b-oxidation of fatty acids
fadL	outer-membrane-bound fatty acid transport protein; Uptake of exogenous fatty acid into the cell
fadD	inner-membrane-associated acyl-CoA synthetase
fadE	acyl-CoA dehydrogenase of beta- oxidation pathway
fadB	3-hydroxyacyl-CoA dehydrogenase of beta- oxidation pathway
fadR	Encode for FadR repressor that deactivates the genes of Fad regulon in the absence of long-chain fatty acids (LCFAs)

It is controlled by alternative sigma factors *rpoH, rpoE* enlisted in the Table 9 at transcription level to activate stress related proteins such as heat shock proteins (DnaK (HSP70), DnaJ, GrpE, GroEL (HSP60), and GroES), cold shock proteins (CspA), acidic resistance (AR) system (acid tolerance), starvation proteins encoded by *cst* genes for carbon starvation and the *pex* genes under carbon, nitrogen, or phosphorus depletion condition, osmoregulatory protein such as ProP and ProU (osmoregulant and compatible solute betaine and proline transporter) (Chung et al. 2006).

Table 7. Genes for carbohydrate metabolism.

Gene	Gene Product
cyoABCDE operon	Gene encode for cytochrome O complex (ubiquinol oxidase); an aerobic enzyme
cyoA	cytochrome oxidase C
cyoE	heme O synthase
dld	D-lactate dehydrogenase
frsA	fermentation-respiration switch protein
ndh	NADH: quinone oxidoreductase II
nuoA	NADH: quinone oxidoreductase subunit A
gnd	6-phosphogluconate dehydrogenase
zwf	$NADP^+$-dependent glucose-6-phosphate dehydrogenase
adhP	Ethanol dehydrogenase / alcohol dehydrogenase
atpA	ATP synthase F1 complex subunit α
atpB	ATP synthase Fo complex subunit α
atpC	ATP synthase F1 complex subunit ε
atpD	ATP synthase F1 complex subunit β
atpE	ATP synthase Fo complex subunit
atpF	ATP synthase Fo complex subunit b
atpG	ATP synthase F1 complex subunit γ
atpH	(ATP synthase F1 complex subunit
csrA	carbon storage regulator
tpiA	triose-phosphate isomerase
pykF	pyruvate kinase I
pykA	pyruvate kinase II
pgk	phosphoglycerate kinase

Table 8. Genes encoding virulent factors of *E. coli*.

Gene	Gene Product	Description
pap	P-fimbriae	alpha-D-Gal (1-4)-Beta-D-Gal
sfaS	S-fimbrial adhesin protein	Affinity for sialic acid; responsible for neonatal meningitis
fimH	type 1 fimbriae	Affinity for alpha-D-mannosides on host glycoproteins (Prasadarao et al. 1993)
hlyA	α-hemolysin	Exotoxin; causes hemolysis of red blood cells; peritonitis, meningitis (Goni and Ostolaza 1998)
cnf1	cytotoxic necrotizing factor type 1	Activates GTP binding protein such as Rho, Rac, and Cdc42 GTPases; disturbs signal transduction in cell (Fabbri et al. 2008)
ireA	Iron responsive element	Serve as siderophore receptor for acquiring iron (Russo et al. 2001)
iutA	ferric aerobactin receptor	For iron uptake (Murakami et al. 2000)

Table 9. Alternative sigma factors of *E. coli.*

Sigma Factors	Gene	Function
σ^{70}	*rpoD*	Heat shock genes
σ^{54} (σ^{N})	*glnF, nrtA, rpoN*	Nitrogen-regulated genes
σ^{32}	*htpT, rpoH*	Heat-shock genes
σ^{24}(σ^{E})	*rpoE*	Heat-shock genes
σ^{28}	*flbB, flaI, rpoF*	for flagella biosynthesis
σ^{38}(σ^{S})	*rpoS, katF*	mediate stress response against starvation

4. Proteomics

Proteomics is the emerging discipline of omics research, which allows better understanding of function, structure, localization of protein in cell and also of its post-translation modification, such as processing, phosphorylation, glycosylation and level of expression (quantitative analysis) which is impossible to deduce from transcriptomics studies, which mainly analyse gene expression rather than protein. The term 'proteomics' was first coined in 1997 in parallel to genomics (the study of genome). The term proteome, coined by Australian geneticist Mark Wilkins in 1994, stands for an array of well-defined proteins that are expressed by the genome of an organism, tissue or cell (Shah and Mishra 2011). It also allows researchers and scientists to decipher the changes in protein expression under different genotypic and phenetic (includes environmental stress) conditions (Han and Lee 2006, Radhouani et al. 2012).

4.1 Proteomic Methods

The techniques for proteomics studies are mainly categorized into gel based approaches, non-gel-based approaches and predictive proteomics.

The gel and non-gel-based strategies employ the separation of entire proteome on the gel and non-gel matrices, including 2-DE, 2-DIGE, MALDI-TOF MS, respectively, while the predictive proteomics use *in silico* or computational methods to identify proteins (Han and Lee 2006, Radhouani et al. 2012).

Two-dimensional electrophoresis (2-DE) is the conventional approach for proteomic analyses ans is also known as blue-collar proteomics. It was first developed by O'Farrell and Klose in 1975 for functional studies of the genome (Klose and Kobalz 1995), while Van Bogelen and his colleagues used 2-DE for analysis of *Escherichia coli* proteome (Han and Lee 2006).

2-DE is a separation technique which involves exploiting two properties of proteins in 2 dimensions: First is isoelectric focusing (IEF), in which proteins are separated according to their isoelectric point (pI), the pH at which net charge and solubility of protein becomes zero and the second dimension is size or mass of protein by SDS-polyacrylamide gel electrophoresis (SDS-PAGE). So, it is the sequential separation of proteins on the basis of charge and mass for better proteome analyses (Monribot and Boucherie 2000).

There are some factors that limit the efficiency of 2-DE, such as arduousness, time-consuming and more error-prone protocol, loss and contamination of protein sample and producing non-reproducible data (Han and Lee 2006, Petriz and Franco

2014). It is ineffective in distinguishing hydrophobic membrane proteins since they are difficult to solubilize and extract (Magdeldin et al. 2012, Petriz and Franco 2014). Moreover, the protein spots of lower abundant proteins are often masked by those of high abundance protein, so it is difficult to detect these proteins on the gel and, therefore, impossible to assign the function to such proteins. This technique is restricted to proteins of molecular weight in the range of 5–150 kDa and pH range (<3.5 to >10) (Petriz and Franco 2014, Joshi and Patil 2017).

Various modifications have been proposed in order to enhance the separation efficiency of 2-DE for better visualization and resolution of protein spots on the gel, such as theuse of IPG strips of narrow-pH-range (pH 4 to 5, 4.5 to 5.5, 5 to 6, 5.5 to 6.7, 6 to 9, and 6 to 11) which form more stable pH gradient,use of prefractionation strategies prior to electrophoresis, which allows the proper resolution and enrichment of less abundant proteins and also proteins of subcellular compartments, for example cytosol, periplasm, inner or outer membrane. This includes sequential and selective extraction and precipitation methods, use of isoelectric focusing (IEF separation), chromatographic methods, such as size exclusion, affinity, ion exchange, and reverse phase resins, differential and density gradient centrifugation (Corthals et al. 1997, Herbert et al. 2007). Several researchers have extracted and recovered more than 200 membrane proteins of *E. coli*, such as inner membrane proteins MCPs, AtpA, AtpB, YiaF and AcrA, the membrane-associated FtsZ protein and the outer membrane proteins YbhC, OmpW, Tsx, Pal, FadL, OmpT and BtuB, by employing prefractionation methods (Molloy et al. 2000, Lai et al. 2004). In order to visualize the proteins spots more clearly on the gel, there are various staining procedures employed either before or after electrophoresis,such as organic dyes (Coomassie blue), silver staining (silver nitrate), radiolabelling, reverse staining, fluorescent staining (Sypros and Ruthenium red-based dyes) and chemiluminescent staining (Chevalier 2010).

There is another gel-based technique in the proteomic research,known as differential electrophoresis (DIGE), which has not only improved reproducibility, but also allowed the researchers to quantitatively compare two different protein samples (proteome) at the same time. In this method, different proteome samples are covalently labelled with fluorophore agents, cyanine dye 1-(5-carboxypentyl)-1′-propylindocarbocyanine halide (Cy3) N-hydroxysuccinimidyl ester and 1-(5-carboxypentyl)-1′-methylindodicarbocyanine halide (Cy5) N-hydroxysuccinimidyl ester (Radhouani et al. 2012), and then subjected to two-dimensional gels for separation. After electrophoresis, protein spots are independently visualized using different wavelengths and then corresponding gel images are normalized by utilizing appropriate computer/software programs to significantly quantify protein spots and, thus, characterize proteome expression (Conrads et al. 2003). This technique allows for the direct scanning and imaging of gel after electrophoresis rather than the manual detection of protein spots in the 2-D gel, thereby reducing the amount of time needed and the number of gels to be run (Han and Lee 2006).

After detecting the relevant protein spot on the gel, excision and sequence-specific digestion using enzymes, mainly trypsin, elution of peptides and subsequent analysis of peptides using non-gel approaches, such as peptide fingerprinting approach, by matrix-assisted laser desorption/ ionization (MALDI), time-of-flight (TOF), MS or liquid chromatography-tandem mass spectrometry,was done as shown in Fig. 4 (Han and Lee 2006, Radhouani et al. 2012).

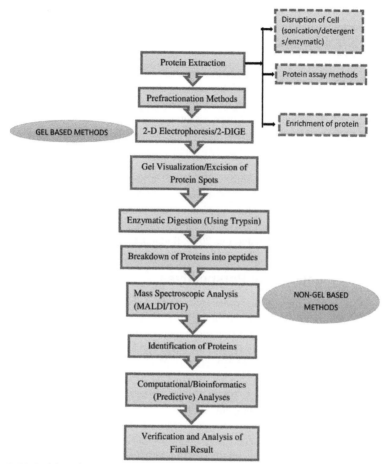

Fig. 4. Methodology for proteomic analyses.

The multiplexed proteomics is another proteomic method that allows researchers to simultaneously determine protein expression as well as other characteristics of protein, such as post-translational modifications, drug-binding capability, etc., in two different proteome samples (Schulenberg and Patton 2003).

Gel-free proteomics method includes MS based systematic, high throughput techniques which analyse peptides rather than whole proteome complex for better insights of protein expression, and allow quantification of protein (Scherp et al. 2011). It involves digestion of proteins sample obtained directly or resolved by gel-based method (such as 2-DG, 2-DIGE) into peptides, followed by chromatography-based separation of mixed peptides (such as affinity, covalent chromatography, strong anion/cation exchange, size exclusion or the use of packed reactive dye compound or reverse-phase columns). These separated peptides are then subjected to mass spectroscopy techniques, such as MALDI-TOFMS, ESI-MS, or MS/MS, for high throughput determination of their expression and abundance in the proteome (Monteoliva and Albar 2004).

MudPIT is the most robust and popular technique, in which two-dimensional chromatography is coupled with tandem mass spectroscopy in order to analyse complex of peptides mixture generated from proteome of organism/cell/organelle (Florens and Washburn 2006).

The ultimate objective of proteomics analyses is to validate and relate the results of genome sequencing and transcriptomics analysis with the physiological functions of the cell (Thompson and Zhou 2004). The proteomic characterization of *E. coli* proteome will be beneficial to future research because the knowledge of *E. coli* proteome will provide abetter understanding of various biological and physiological processes in prokaryotes and thereby can be applied in case of eukaryotic studies. It will help metabolomic studies to increase the production of recombinant desired proteins and industrially important metabolites. The *E. coli* proteome is the composite of a total of 4,285 proteins of molecular weight ranging from 1.59 to 248 kDa; isoelectric point between 3.38 to 13.0. The protein composition is identified in different organelles of *E. coli:* 2,885 in the cytoplasm, 670 in the inner membrane, 87 in the outer membrane and 138 in the periplasm (Han and Lee 2006, Radhouani et al. 2012).

Conclusion and Future Perspectives

Escherichia coli is one of the most characterized, simple and ideal prokaryotic organisms in various disciplines of biotechnology, such as genetics, biochemistry, microbiology, structural biology, cell and molecular biology. The determination of complete genomic structure of *E. coli* variants, K-12 MG1655 (reference strain), O157: H7 EDL933 and Sakai O6: K2: H1 CFT073, form the basis of understanding the evolutionary relationship which has revealed the presence of conserved, ancestral genomic backbone of size ~ 4 Mb and other strain-specific elements accounting for beneficial traits by non-pathogenic or human infection by potential pathogenic and drug-resistant *E. coli* strains. This has significant medical applications because it provides an opportunity to accurately diagnose the wide range of extraintestinal infections, as well as to develop putative antimicrobial agents and advanced vaccines targeting specific virulent factors involved in pathogenicity. Furthermore, the proteomics (structure to function approach) characterization of *E. coli* proteome will shed light on development and regulatory processes not only of prokaryotes but also the cellular activities of the eukaryotic organism. In addition to this, the knowledge of *E. coli* proteome might be further useful in Recombinant DNA technology and metabolomic research fields.

Acknowledgements

The authors thank DST-SERB for providing financial support to the research grant ECR/2017/000080 and financially supporting Shikha Gupta to pursue phD. The authors are also thankful to the Amity University Uttar Pradesh for providing infrastructural support.

References

Allocati, N., L. Federici, M. Masulli and C.D. Ilio. 2009. Glutathione transferases in bacteria. FEBS J. 276(1): 58–75.

Barsy, M.D. and G. Greub. 2013. Functional genomics of intracellular bacteria. Brief. Funct. Genomics 12(4): 341–353.

Blattner, F.R.G., C.A. Plunkett III, T.A. Perna, V. Burland, M. Riley, J. Collado-Vides, J.D. Glasner, C.K. Rode, G.F. Mayhew, J. Gregor, N.W. Davis, H.A. Kirkpatrrick, M.A. Goeden, D.J. Rose, B. Mau and J. Shao. 1997. The complete genome sequence of *Escherichia coli* K-12. Science 277: 1453–1474.

Boags, A., P.C. Hsu, F. Samsudin, P.J. Bond and S. Khalid. 2017. Progress in molecular dynamics simulations of Gram-negative bacterial cell envelopes. J. Phys. Chem. Lett. 8(11): 2513–2518.

Bornemann, S., M.K. Ramjee, S. Balasubramanian, C. Abell, J.R. Coggins, D.J. Lowe and R.N. Thorneley. 1995. *Escherichia coli* chorismate synthase catalyzes the conversion of (6S)-6-fluoro-5-enolpyruvylshikimate-3-phosphate to 6-fluorochorismate. Implications for the enzyme mechanism and the antimicrobial action of (6S)-6-fluoroshikimate. J. Biol. Chem. 270(39): 22811–22815.

Bunnik, M.E. and G.K. Le Roch. 2013. An introduction to functional genomics and system biology. Adv. Wound Care (New Rochelle). 2(9): 490–498.

Caldara, M., G. Dupont, F. Leroy, A. Goldbeter, L.D. Vuyst and R. Cunin. 2008. Arginine biosynthesis in *Escherichia coli*: Experimental perturbation and mathematical modeling. J. Biol. Chem. 283(10): 6347–58.

Cepeda-Molero M., C.N. Berger, A.D.S. Walsham, S.J. Ellis, S. Wemyss-Holden, S. Schuller, G. Frankel and L.A. Fernandez. 2017. Attaching and effacing (A/E) lesion formation by enteropathogenic *Escherichia coli* on human intestinal mucosa is dependent on non-LEE effectors. PLoS Pathog. 13(10): e1006706.

Chan, V.L. 2006. Microbial Genomes. pp. 1–19. *In*: V.L. Chan, P.M. Sherman, B. Bourke (eds.). Bacterial Genomes and Infectious Diseases. Humana Press, Totowa, NJ.

Chapman, T.A., X.Y. Wu, I. Barchia, K.A. Bettelheim, S. Driesen, D. Trott, M. Wilson and J.J.C. Chin. 2006. Comparison of virulence gene profiles of *Escherichia coli* strains isolated from healthy and diarrheic swine. Appl. Environ. Microbiol. 72(7): 4782–4795.

Chavez-Bejar, M.I., A.R. Lara, H. Lopez, G. Hernández-Chávez, A. Martinez, O.T. Ramirez, F. Bolívar and G. Gosset. 2008. Metabolic engineering of *Escherichia coli* for l-tyrosine production by expression of genes coding for the chorismate mutase domain of the native chorismate mutase-prephenate dehydratase and a cyclohexadienyl dehydrogenase from *Zymomonas mobilis*. Appl. Environ. Microbiol. 74(10): 3284–3290.

Chevalier, F. 2010. Standard dyes for total protein staining in gel-based proteomic analysis. Materials 3(10): 4784–4792.

Cho, B., E. Knight and B. Palsson. 2006. Transcriptional regulation of the fad regulon genes of *Escherichia coli* by ArcA. Microbiology 152(8): 2207–2219.

Chung, H.J., W. Bang and M.A. Drake. 2006. Stress response of *Escherichia coli*. Compr. Rev. Food Sci. Food Saf. 5: 52–64.

Condon, C., D. Liveris, C. Squires, I. Schwartz and C.L. Squires. 1995. rRNA operon multiplicity in *Escherichia coli* and the physiological implications of *rrn* inactivation. J. Bacteriol. 177(14): 4152–4156.

Conrads, T.P., H.J. Issaq and V.M. Hoang. 2003. Current strategies for quantitative proteomics. pp. 133–159. *In*: R.D. Smith and T.D. Veenstra (eds.). Advances in Protein Chemistry. Academic Press, Boston.

Corthals, G.L., M.P. Molloy, B.R. Herbert, K.L. Williams and A.A. Gooley. 1997. Prefractionation of protein samples prior to two-dimensional electrophoresis. Electrophoresis 18: 317–323.

Cronan, J.E. 2014. *Escherichia coli* as an experimental organism. In: eLS. John Wiley & Sons Ltd, Chichester. https://doi.org/10.1002/9780470015902.a0002026.pub2.

Delmas, J., G. Dalmassoand R. Bonnet. 2015. *Escherichia coli*: The good, the bad and the ugly. Clin. Microbiol. 4: 195.

Diaz, E., A. Ferrandez, M.A. Prieto and J.L. Garcia. 2001. Biodegradation of aromatic compounds by *Escherichia coli*. Microbiol. Mol. Biol. Rev. 65(4): 523–569.

Donnenberg, M.S and T.S. Whittam. 2001. Pathogenesis and evolution of virulence in enteropathogenic and enterohemorrhagic *Escherichia coli*. J. Clin. Invest. 107(5): 539–548.

Enz, S., S. Mahren, C. Menzel and V. Braun. 2003. Analysis of the ferric citrate transport gene promoter of *Escherichia coli*. J. Bacteriol. 185(7): 2387–2391.

Escherich, T. 1988. The intestinal bacteria of the neonate and breast-fed infant (1885). Rev. Infect. Dis. 10: 1220–1225.

Fabbri A., S. Travaglione, L. Falzano and C. Fiorentini. 2008. Bacterial protein toxins: Current and potential clinical use. Curr. Med. Chem. 15: 1116–1125.

Florens, L. and M.P. Washburn. 2006. Proteomic analysis by multidimensional protein identification technology. pp. 159–175. *In*: D. Nedelkov and R.W. Nelson (eds.). New and Emerging Proteomic Techniques. Methods in Molecular Biology. Humana Press, Springer, New York.

Fraser, C.M., J.A. Eisen and S.L. Salzberg. 2000. Microbial genome sequencing. Nature 406: 799–803.

Garmendia, J., G. Frankel and V.F. Crepin. 2005. Enteropathogenic and enterohemorrhagic *Escherichia coli* infections: Translocation, translocation, translocation. Infect. Immun. 73(5): 2573–2585.

Goni F.M. and H. Ostolaza. 1998. *E. coli* alpha-hemolysin: A membrane-active protein toxin. Braz. J. Med. Biol. Res. 31(8): 1019–1034.

Griffiths, A.J.F., W.M. Gelbart, J.H. Miller and R.C. Lewontin. 1999. The genetics of bacteria and phages. *In*: Modern Genetic Analysis. W.H. Freeman, New York.

Griffiths, A.J.F., J.H. Miller, D.T. Suzuki, R.C. Lewontin and W.M. Gelbart. 2000. Genomics. *In*: An Introduction to Genetic Analysis.W. H. Freeman,New York.

Gumbart, J.C., M. Beeby, G.J. Jensen and B. Roux. 2014. *Escherichia coli* peptidoglycan structure and mechanics as predicted by atomic-scale simulations. PLoS Comput. Biol. 10(2): e1003475.

Guyer, D.M., S. Radulovic, F.E. Jones and H.L.T. Mobley. 2002. Sat, the secreted autotransporter toxin of uropathogenic *Escherichia coli*, is a vacuolating cytotoxin for bladder and kidney epithelial cells. Infect Immun. 70(8): 4539–4546.

Han, M.J. and S.Y. Lee. 2006. The *Escherichia coli* proteome: Past, present, and future prospects. Microbiol. Mol. Biol. Rev. 70(2): 362–439.

Hayashi, T., K. Makino, M. Ohnishi, K. Kurokawa, K. Ishii, K. Yokoyama, C.G. Han, E. Ohtsubo, K. Nakayama, T. Murata, M. Tanaka, T. Tobe, T. Iida, H. Takami, T. Honda, C. Sasakawa, N. Ogasawara, T. Yasunaga, S. Kuhara, T. Shiba, M. Hattori and H. Shinagawa. 2001. Complete genome sequence of enterohemorrhagic *Escherichia coli* O157:H7 and genomic comparison with a laboratory strain K-12. DNA Res. 8: 11–22.

Helling, R.B. 1998. Pathway choice in glutamate synthesis in *Escherichia coli*. J. Bacteriol. 180(17): 4571–4575.

Henderson, I.R. and J.P. Nataro. 2001. Virulence functions of autotransporter proteins. Infect. Immun. 69(3): 1231–1243.

Herbert, B.R., P.G. Righetti, A. Citterio and E. Boschetti. 2007. Sample preparation and prefractionation techniques for electrophoresis-based proteomics. pp. 15–40. *In*: M.R. Wilkins, R.D. Appel, K.L. Williams and D.F. Hochstrasser (eds.). Proteome Research. Principles and Practice. Springer, Heidelberg.

Hooton, T.M. and W.E. Stamm. 1997. Diagnosis and treatment of uncomplicated urinary tract infection. Infect. Dis. Clin. North Am. 11: 551–581.

Huang, K.C., R. Mukhopadhyay, B. Wen, Z. Gitai and N.S. Wingreen. 2008. Cell shape and cell-wall organization in Gram-negative bacteria. Proc. Natl. Acad. Sci. USA. 105(49): 19282–19287.

Idalia, V.M.N. and F. Bernardo. 2017. *Escherichia coli* as a model organism and its application in biotechnology, *In*: A. Samie (ed.). Recent Advances on Physiology, Pathogenesis and Biotechnological Applications. InTech.

Illarionova, V., W. Eisenreich, M. Fischer, C. Haussmann, W. Romisch, G. Richter and A. Bacher. 2002. Biosynthesis of tetrahydrofolate. Stereochemistry of dihydroneopterin aldolase. J. Biol. Chem. 277(32): 28841–28847.

Jones, C.J., M. Homma and R.M. Macnab. 1989. L-, P-, and M-ring proteins of the flagellar basal body of *Salmonella*Typhimurium: Gene sequences and deduced protein sequences. J. Bacteriol. 171(7): 3890–3900.

Joshi, K. and D. Patil. 2017. Proteomics. pp. 273–294. *In*: B. Patwardhan and R. Chaguturu (eds.). Innovative Approaches in Drug Discovery. Academic Press, Boston.

Kaper, J.B., J.P. Nataro and H.L. Mobley. 2004. Pathogenic *Escherichia coli*. Nat. Rev. Microbiol. 2: 123–140.

Klemm, P. 1986. Two regulatory *fim* genes, *fimB* and *fimE*, control the phase variation of type 1 fimbriae in *Escherichia coli*. EMBO J. 5(6): 1389–1393.

Klose, J. and U. Kobalz. 1995. Two-dimensional electrophoresis of proteins: An updated protocol and implications for a functional analysis of the genome. Electrophoresis 16(6): 1034–59.

Kohler, C.D. and U. Dobrindt. 2011. What defines extraintestinal pathogenic *Escherichia coli*? Int. J. Med. Microbiol. 301: 642–647.

Kwon, O., M.E. Hudspeth and R. Meganathan. 1996. Anaerobic biosynthesis of enterobactin *Escherichia coli*: Regulation of *entC* gene expression and evidence against its involvement in menaquinone (vitamin K2) biosynthesis. J. Bacteriol. 178(11): 3252–3259.

Lai, E.M., U. Nair, N.D. Phadke and J.R. Maddock. 2004. Proteomic screening and identification of differentially distributed membrane proteins in *Escherichia coli*. Mol. Microbiol. 52: 1029–1044.

Laursen, B.S., H.P. Sørensen, K.K. Mortensen and H.U. Sperling-Petersen. 2005. Initiation of protein synthesis in bacteria. Microbiol. Mol. Biol. Rev. 69(1): 101–23.

Lawther, R.P., J.M. Lopes, M.J. Ortuno and M.C. White. 1990. Analysis of regulation of the *ilvGMEDA* operon by using leader-attenuator-*galK* gene fusions. J. Bacteriol. 172(5): 2320–2327.

Lukjancenko, O., T.M. Wassenaar and D.W. Ussery. 2010. Comparison of 61 sequenced *Escherichia coli* genomes. Microb. Ecol. 60: 708–720.

Magdeldin, S., Y. Zhang, B. Xu, Y. Yoshida and T. Yamamoto. 2012. Two-dimensional polyacrylamide gel electrophoresis—A practical perspective. *In*: S. Magdeldin (ed.). Gel Electrophoresis—Principles and Basics. InTech.

Magnuson, K., S. Jackowski, C.O. Rock and J.E. Cronan Jr. 1993. Regulation of fatty acid biosynthesis in *Escherichia coli*. Microbiol. Rev. 57(3): 522–542.

Matte, A., J. Sivaraman, I. Ekiel, K. Gehring, Z. Jia and M. Cygler. 2003. Contribution of structural genomics to understanding the biology of *Escherichia coli*. J. Bacteriol. 185(14): 3994–4002.

Mattern, I.E. and J. Pittard. 1971. Regulation of tyrosine biosynthesis in *Escherichia coli* K-12: isolation and characterization of operator mutants. J. Bacteriol. 107(1): 8–15.

Molloy, M.P., B.R. Herbert, M.B. Slade, T. Rabilloud, A.S. Nouwens, K.L. Williams and A.A. Gooley. 2000. Proteomic analysis of the *Escherichia coli* outer membrane. Eur. J. Biochem. 267: 2871–2881.

Monribot, C. and H. Boucherie. 2000. Two-dimensional electrophoresis with carrier ampholytes. pp. 31–55. *In*: T. Rabilloud (ed.). Proteome Research: Two-Dimensional Gel Electrophoresis and Identification Methods. Principles and Practice. Springer, Heidelberg.

Monteoliva, L. and J.P. Albar. 2004. Differential proteomics: An overview of gel and non-gel-based approaches. Brief. Funct. Genomic. Proteomic. 3(3): 220–39.

Mori, H. 2004. From the sequence to cell modeling: Comprehensive functional genomics in *Escherichia coli*. J. Biochem. Mol. Biol. 37(1): 83–92.

Mori, H.K.I., T. Horiuchi and T. Miki. 2000. Functional genomics of *Escherichia coli* in Japan. Res. Microbiol. 151(2): 121–128.

Moriguchi, K., N. Edahiro, S. Yamamoto, K. Tanaka, N. Kurata and K. Suzuki. 2013. Transkingdom genetic transfer from *Escherichia coli* to *Saccharomyces cerevisiae* as a simple gene introduction tool. Appl. Environ. Microbiol. 79(14): 4393–4400.

Murakami K., H. Fuse, O. Takimura, H. Inoue and Y. Yamaoka. 2000. Cloning and characterization of the *iutA* gene which encodes ferric aerobactin receptor from marine *Vibrio* species. Microbios. 101(400): 137–46.

Noinaj, N., M. Guillier, T.J. Barnard and S.K. Buchanan. 2010. TonB-dependent transporters: Regulation, structure, and function. Annu. Rev. Microbiol. 64: 43–60.

Odonkor, S.T. and J.K. Ampofo. 2013. *Escherichia coli* as an indicator of bacteriological quality of water: an overview. Microbiol. Res. 4(1).

Okada, T., H. Suzuki, K. Wada, H. Kumagai and K. Fukuyama. 2006. Crystal structures of γ-glutamyltranspeptidase from *Escherichia coli*, a key enzyme in glutathione metabolism, and its reaction intermediate. Proc. Natl. Acad. Sci. USA 103(17): 6471–6476.

Parekh, B.S. and G.W. Hatfield. 1997. Growth rate-related regulation of the *ilvGMEDA* operon of *Escherichia coli* K-12 is a consequence of the polar frameshift mutation in the *ilvG* gene of this strain. J. Bacteriol.179(6): 2086-2088.

Percival, S.L. and D.W. Williams. 2014. *Escherichia coli*. pp. 89–117. *In*: S.L. Percival, M. Yates, D. Williams, R. Chalmers and N. Gray (eds.). Microbiology of Waterborne Diseases (Second Edition). Academic Press, London.

Perna, N.T., G. Plunkett III, V. Burland, B. Mau, J.D. Glasner, D.J. Rose, G.F. Mayhew, P.S. Evans, J. Greor, H.A. Kirkpatrick, G. Posfai, J. Hackett, S. Klink, A. Boutin, Y. Shao, L. Miller, E.J. Grotbeck, N.W. Davis, A. Lim, E.T. Dimalanta, K.D. Potamousis, J. Apodaca, T.S. Anantharaman, J. Lin, G. Yen, D.C. Schwartz, R.A. Welch and F.R. Balttner. 2001. Genome sequence of the enterohaemorrhagic *Escherichia coli* O157:H7. Nature 409: 529–533.

Petriz, B.A. and O.L. Franco. 2014. Application of cutting-edge proteomics technologies for elucidating host–bacteria interaction. pp. 1–24. *In*: R. Donev (ed.). Advances in Protein Chemistry and Structural Biology. Academic Press, London.

Prasadarao, N.V., C.A. Wass, J. Hacker, K. Jann and K.S. Kim. 1993. Adhesion of S-fimbriated *Escherichia coli* to brain glycolipids mediated by *sfaA* gene encoded protein of S-fimbriae. J. Biol. Chem. 268: 10356–10363.

Pruitt, R.N., M.G. Chambers, K.K.-S.Ng, M.D. Ohi and D.B. Lacy. 2010. Structural organization of the functional domains of *Clostridium difficile* toxins A and B. Proc. Natl. Acad. Sci. USA 107(30): 13467–13472.

Radhouani, H., L. Pinto, P. Poeta and G. Igrejas. 2012. After genomics, what proteomics tools could help us understand the antimicrobial resistance of *Escherichia coli*? J. Proteomics. 75(10): 2773–2789.

Reitzer, L. 2004. Biosynthesis of glutamate, aspartate, asparagine, L-alanine, and D-alanine. EcoSal Plus 1(1).

Riley, M. 1993. Functions of the gene products of *Escherichia coli*. Microbiol. Rev. 57(4): 862–952.

Russo, T.A., U.B. Carlino and J.R. Johnson. 2001. Identification of a new iron-regulated virulence gene, *ireA*, in an extraintestinal pathogenic isolate of *Escherichia coli*. Infect. Immun. 69(10): 6209–6216.

Scherp P., G. Ku, L. Coleman and I. Kheterpal. 2011. Gel-based and gel-free proteomic technologies. pp. 163–190. *In*: J. Gimble and B. Bunnell (eds.). Adipose-Derived Stem Cells. Methods in Molecular Biology (Methods and Protocols). Humana Press, Totowa, NJ.

Schulenberg, B. and W.F. Patton. 2003. Multiplexed proteomics. pp 107–115. *In*: P.M. Conn (ed.). Handbook of Proteomic Methods. Humana Press, Totowa, NJ.

Shah, T.R. and A. Misra. 2011. Proteomics. pp. 387–427. *In*: A. Misra (ed.). Challenges in Delivery of Therapeutic Genomics and Proteomics. Elsevier, London.

Shulman, S.T., H.C. Friedmann and R.H. Sims. 2007. Theodor Escherich: The first paediatric infectious diseases physician? Clin. Infect. Dis. 45:1025–1029.

Smith, J.L., P.M. Fratamico and N.W. Gunther. 2007. Extraintestinal pathogenic *Escherichia coli*. Foodborne Pathog. Dis. 4(2): 134–163.

Soye, B.J.D., J.R. Patel, F.J. Isaacs and M.C. Jewett. 2015. Repurposing the translation apparatus for synthetic biology. Curr. Opin. Chem. Biol. 28: 83–90.

Sperandio, V. and Y. Nguyen. 2012. Enterohemorrhagic *Escherichia coli* (EHEC) pathogenesis. Front. Cell. Infect. Microbiol. 2: 90.

Suzuki, H., H. Kumagai, T. Echigo and T. Tochikura. 1989. DNA sequence of the *Escherichia coli* K-12 gamma-glutamyltranspeptidase gene, *ggt*. J. Bacteriol. 171(9): 5169–5172.

Tao, H., C. Bausch, C. Richmond, F.R. Blattner and T. Conway. 1999. Functional genomics: Expression analysis of *Escherichia coli* growing on minimal and rich media. J. Bacteriol. 181(20): 6425–6440.

Tenaillon, O., D. Skurnik, B. Picard and E. Denamur. 2010. The population genetics of commensal *Escherichia coli*. Nat. Rev. Microbiol. 8: 207–217.

Thompson, D.K. and J. Zhou. 2004. The functional genomics of model organisms: Addressing old questions from a new perspective. pp. 325–375. *In*: J. Zhou, D.K. Thompson, Y. Xu and

J.M. Tiedje (eds.). Microbial Functional Genomics. John Wiley & Sons, Inc., Hoboken, NJ, USA.

Tobes, R. and E. Pareja. 2006. Bacterial repetitive extragenic palindromic sequences are DNA targets for Insertion Sequence elements. BMC Genomics 7: 62.

Welch, R.A., V. Burland, G. Plunkett, P. Redford, P. Roesch, D. Rasko, E.L. Buckles, S.-R. Liou, A. Boutin, J. Hackett, D. Stroud, G.F. Mayhew, D.J. Rose, S. Zhou, D.C. Schwartz, N.T. Perna, H.L.T. Mobley, M.S. Donnenberg and F.R. Blattner. 2002. Extensive mosaic structure revealed by the complete genome sequence of uropathogenic *Escherichia coli*. Proc. Natl. Acad. Sci. USA 99(26): 17020–17024.

Welch, R.A. 2006. The genus *Escherichia*. pp. 60–71. *In*: M. Dworkin, S. Falkow, E. Rosenberg, K.H. Schleifer and E. Stackebrandt (eds.). The Prokaryotes. Springer, New York.

Zhao, S., C.H. Sandt, G. Feulner, D.A. Vlazny, J.A. Gray and C.W. Hill. 1993. *Rhs* elements of *Escherichia coli* K-12: complex composites of shared and unique components that have different evolutionary histories. J. Bacteriol. 175(10): 2799–2808.

Zhou, J. and J.H. Miller. 2002. Microbial genomics—challenges and opportunities: The 9th international conference on microbial genomes. J. Bacteriol. 184(16): 4327–4333.

9

Stress Responses of LAB

*Angela Longo** and *Giuseppe Spano*

1. Introduction

Lactic acid bacteria (LAB) are characterized as a heterogeneous group of Gram-positive, low-GC (< 55 mol%) generally nonsporulating and nonmotile, catalase-negative (with the exception of some species of the genus *Pediococcus*), rod (*Lactobacillus, Bifidobacterium*) or cocci (*Streptococcus, Pediococcus, Leuconostoc*) shaped bacteria,characterized by the ability to produce lactic acid as the major metabolic end-product of carbohydrate fermentation (Carr et al. 2002). LAB include both homofermenters, producing mainly lactic acid, and heterofermenters, which, apart from lactic acid, yield a large variety of fermentation products, such as acetic acid, ethanol, carbon dioxide, and formic acid (Mozzi et al. 2010). Like all Gram-positive bacteria, their cell envelope is a multilayered structure, which is mainly composed of peptidoglycan with embedded teichoic acids, proteins, and polysaccharides and which is essential for the cellular integrity and shape (Silhavy et al. 2010). LAB are extremely widespread in nature, preferring nutrient-rich habitats. They are part of the normal microflora of the mouth, intestines and human vagina and are also indigenous to food-related habitats, including milk, plants (vegetables, cereal grains), wine and meat. LAB are involved in numerous industrial applications, ranging from starter cultures, to drive food and beverage fermentations, exercising a preservative effect on the fermented product, to bioconversion agents; specific LAB are commonly used in the formulation of functional probiotic foods for their health-promoting effects in consumers (Table 1) (Wu et al. 2011, Giraffa 2012, Wu et al. 2014). LAB fermentation is a safe, economical, and traditional method of food processing, used all over the world,to improve the microbial safety andto offer technological, nutritional and health benefits. Except for a few species (such as *Streptococcus pneumoniae* or *Streptococcus pyogenes*), LAB are nonpathogenic organisms, which is why they have the "Generally Recognized as Safe (GRAS)" and "Qualified Presumption of Safety (QPS)" states (Bernardeau et al. 2008, Giraffa

Department of Agriculture, Food and Environment Sciences, University of Foggia, Italy.
* Corresponding author: angela.longo@unifg.it

Table 1. Examples of application of lactic acid bacteria.

Applications	Products	Strains	References
Dairy products	Cheese	*L. lactis, Leuconostoc* spp.	(Zhu et al. 2009) (Smit et al. 2005)
	Butter and buttermilk	*L. lactis* subsp. *lactis, L. lactis* subsp. *lactis* var. *diacetylactis, L. lactis* subsp. *cremoris, Leuc. mesenteroides* subsp. *cremoris*	(Leroy and De Vuyst 2004), (Saraoui et al. 2017)
	Yoghurt	*Lb. delbrueckii* subsp. *bulgaricus,* *Streptococcus thermophilus*	(Zhu et al. 2009)
	Fermented, probiotic milk	*Lb. casei, Lb. acidophilus,* *Lb. rhamnosus, Lb. johnsonii,* *B. lactis, B. bifidum, B. breve*	(Furet et al. 2004)
	Kefir	*Lb. kefir, Lb. kefiranofaciens,* *Lb. brevis*	(Simova et al. 2002), (Chen et al. 2008)
Fermented foods	Meat	*Lb. sakei, Lb. curvatus* *P. acidilactici, P. pentosaceus*	(Chaillou et al. 2005), (Vermeiren et al. 2004), (Ammor and Mayo 2007)
	Vegetables	*P. acidilactici, P. pentosaceus,* *Lb. plantarum, Lb. fermentum* *Leuc. mesenteroides, P. cerevisiae, Lb. brevis*	(Xiong et al. 2012), (Salminen and Wright 2004), (Trias et al. 2008), (Tamang et al. 2005)
	Cereals	*Lb. sanfransiscensis, Lb. farciminis, Lb. fermentum,* *Lb. brevis, Lb. plantarum, Lb. amylovorus*	(Charalampopoulos et al. 2002), (Capozzi et al. 2012), (Coda et al. 2011)
Alcoholic beverages	Wine	*O. oeni, Lb. acetolerans*	(D.A. Mills et al. 2005)

2012). Furthermore, during the fermentation process, these microorganisms produce important aroma and flavor compounds (such as diacetyl or small peptides) through their metabolic activities (e.g., lipolysis and proteolysis), and they contribute to the texture (e.g., through the production of exopolysaccharides [EPS]) in milk products (yogurt, cheese, butter, kefir and koumiss) but also in vegetables (sauerkraut), meat products (sausages) and wine (Papadimitriou et al. 2016). LAB play an important role also in bread production, especially for their ability to produce antimicrobial compounds, like bacteriocins, and to inhibit the main bread contaminants, such as *Aspergillus*, *Fusarium*, and *Penicillium* (Gerez et al. 2009).

Today, the main LAB genera extensively exploited in the food industry are: *Lactobacillus* (*Lb. sakei* important in meat and fermented meat products or *Lb. bulgaricus* a starter culture in the manufacturing of yogurt) (Champomier Vergès et al. 1999, Hao et al. 2011), *Lactococcus* (*L. lactis* converts glucose to lactate, leading to a gradual reduction in pH of the growth medium) (Frees et al. 2003), *Leucocostoc* (milk, vegetables), *Pediococcus* (meat, vegetables), *Oenococcus* (in wine-making, *O. oeni* is the main species which induces malolactic fermentation) (Tononand Lonvaud-Funel 2000), *Enterococcus* (milk) and *Streptococcus* (milk) (Schroeter and Klaenhammer 2009, Papadimitriou et al. 2016).

The industrialization of food transformations increased the economic importance of LAB, as they play a crucial role in the development of the organoleptic and hygienic quality of fermented products. Lactic acid bacteria, like other microorganisms, must face adverse environmental conditions during steps of technological processing, storage and consumption which include up or downshifts in temperature, sudden changes in pH value and osmolarity, or nutrient limitations. Research in this direction would promote better understanding of their responses to stresses. To monitor population dynamics during food fermentation and better understand the spoilage process of LAB, several molecular techniques have been applied for their identification (Cocolin et al. 2000, Yost and Nattress 2000, Andrighetto et al. 2001).

2. Stress Physiology of Lactic Acid Bacteria

LAB are widely applied in the food industry, agricultural production, animal husbandry and pharmaceutical engineering, in which they encounter a wide range of stresses in their constantly changing environments, both abiotic (during food production, manipulation of starter or probiotic cultures) and biotic (in the host or in complex ecosystems) (Papadimitriou et al. 2016, Wang et al. 2018). Variations in temperature, pH, solute concentrations, nutrients, and oxygen levels can inhibit cell growth and lead to cell death. In order for the LAB to perform their activity, it is necessary to survive, maintaining high viable counts, the technological stress encountered during food manufacture, as well as the hostile environments found within the product to which they are added and the host intestinal flora (van de Guchte et al. 2002, Tsakalidou and Papadimitriou 2011). Then, cellular robustness plays a key role, especially in view of the development of new applications, such as pharmaceutical preparations, live vaccines and probiotic foods. Several studies show that LAB are particularly robust:

S. Mills et al. (2011) reported that *Lactobacillus* spp. survived the low pH of the stomach (pH 2.0 to 4.0), while *O. oeni* strains were able to proliferate in the presence of 13% ethanol at pH 3.2 and 18°C.

Booth (2002) defined the stress as "any change in the genome, proteome or environment that imposes either reduced growth or survival potential. Such changes lead to attempts by a cell to restore a pattern of metabolism that either fits it for survival or for faster growth". Bacteria have developed defense mechanisms against stress that allow them to survive in a hostile environment with different ways to sense changes and trigger a cascade of alterations in gene expression and protein activity (Hoe et al. 2013, Wu et al. 2014). Indeed, a stressful environment usually affects the microbial cell physiology and,in response to environmental stresses, microorganisms have developed signal transduction systems that control the coordinated expression of genes involved in different cellular processes, such as cell division, DNA metabolism, membrane composition and transportprotein (Ljungh and Wadström 2009, Hoe et al. 2013) (Fig. 1).

Many studies conducted on lactobacilli and bifidobacteria showed that exposure to acidic environments leads to changes in cell membrane lipid composition (Taranto et al. 2003, Taranto et al. 2006, Ruiz et al. 2007). Broadbent et al. (2010) found a dramatic increase in the ratio of saturated to unsaturated fatty acid (FA) and cyclopropane FA content in the membranes of acid-adapted *Lb. casei* ATCC 334. These changes modulate membrane features, including fluidity, hydrophobicity and proton-permeability (Corcoran et al. 2007, Muller et al. 2011). In *O. oeni*, the main bacterial starter used for secondary fermentation in wine, under stress conditions, changes in the composition of fatty acids of cell membranes usually occurred, as well as the synthesis of stress proteins (Wen-ying and Zhen-kui 2013). During industrial processing (starter handling and storage) and in the passage through the gastrointestinal tract (acidity and bile salt), bacteria find several environmental stress conditions which can cause structural and physiological injury to the cells, resulting in loss of viability. For that, most of bacteria react by blocking the replication processes and activating adaptation mechanisms (Prasad et al. 2003).

The current knowledge on the environmental stress responses in LAB varies between species and depending on the type of stress (Arena et al. 2007). Within the stress responsive mechanisms, knowledge is accumulating on stress such as acid, heat, cold, osmotic and oxidative stress.

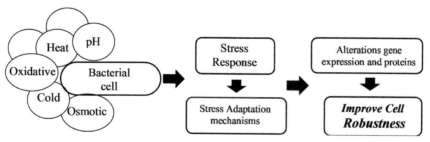

Fig. 1. Examples of stress response mechanism in lactic acid bacteria.

2.1 Stress Response Mechanisms in LAB

The ability of lactic acid bacteria to tolerate industrial stress is essential,considering their economic importance for food fermentation and their health-related implications as probiotics (Konings 2002). To survive these harsh conditions, LAB have evolved both physiological and genetic mechanisms (Beales 2004) including:

- The production (i.e., up-regulation or *de novo* synthesis) of proteins involved in damage restoring, in the preservation of cell homeostasis, and/or in the eradication of the stressing cause (i.e., achieved by modulation of stress regulons);
- The modifications of cellular structures resulting in a temporary enhanced resistance or tolerance to the stress (i.e., changes in cell morphology, alterations in membrane fatty acids composition and content);
- The cell entrance into a quiescent physiological condition;
- The evasion of host defenses, in case of host-microbe interaction (Nezhad et al 2015).

Below the main stress factors encountered by LAB in their industrial applications and their adaptation mechanisms are discussed.

2.1.1. Mechanisms of acid resistance. LAB are generally neutrophils (optimal pH for growth between 5 and 9), with the exception of some species of the genera *Lactobacillus, Leuconostoc* and *Oenococcus*. They are usually faced with acidic stress conditions due to the generation of acidic end products of fermentation,associated with specific fermented foods/beverages (i.e., sauerkraut, green olives, pickles, sausages, baked goods, cheeses, fermented milk drinks) as well as with transition through the gastrointestinal tract. The resistance to acids is fundamental for LAB growth, for their application in fermentation processes and for the production and functionality of a probiotic culture. LAB, as well as other bacteria, developed sophisticated adaptation mechanisms, which are able to increase their cell robustness (Wang et al. 2018). Acid stress can be described as the combined biological effect of low pH and weak (organic) acids, such as acetate, propionate and lactate, present in the environment (food) as a result of fermentation, or alternatively, when added as preservatives. Weak acids in their protonated form can diffuse into the cell and dissociate, thereby lowering the intracellular pH (pHi), resulting in the inhibition of various essential metabolic and anabolic processes, and influencing the transmembrane ΔpH, impairing not only the pH homeostasis but also the transport systems that depend on the proton-motive force (Corcoran et al. 2008). In LAB, acid tolerance (AT) increases in at least two distinct physiological states: (i) during logarithmic growth an adaptive response referred to as L-ATR (lactic acid tolerance response), can be induced by incubation at a non-lethal acidic pH; (ii) after entry into the stationary phase, AT increases as a result of the induction of a general stress response (van de Guchte et al. 2002).

Several mechanisms are involved in the acid resistance regulation of LAB. For example, the induction of cytoplasmic (DnaK and GroEL) and periplasmic (HdeA and HdeB) chaperones, as well as Clp protease complex, able to remove damaged proteins (due to low pH) participate in protein homoeostasis (Lund et al. 2014) or the production of a new set of shock proteins, known as stress proteins (Gandhi and Shah 2016). Moreover, the development of biological mechanisms, such as the Arginine Dihydrolase (ADS) pathway, an increased activity and amount of the F_0F_1-

ATPase, the amino acid decarboxylation-antiporter reactions, the pre-adaptation and cross-protection systems, are all important in dealing with acidic stress.

Arginine dihydrolase system (ADS). When microbial cells are exposed to an acidic environment, they produce alkaline compounds which, when combined with intracellular protons, neutralize the internal pH. There are two major substrates for alkali production: (i) urea, which is rapidly hydrolyzed in ammonia and CO_2 by the ureases, and (ii) arginine that is catabolized by the ADS system. Activity of this defense mechanism depends on the combination of different factors/environmental conditions, such as arginine availability, energy depletion, low oxygen concentration rather than low pH and species (van de Guchte et al. 2002). ADS system, which leads to ATP and ammonia production, involves three enzymatic reactions: In the first reaction, catalyzed by arginine deiminase (AD, encoded by the *arc*A gene), arginine is converted to citrulline with the production of by-product NH_3; then, citrulline is transformed into ornithine and carbamoyl phosphate by ornithine transcarbamoylase (OTC, encoded by *arc*B); and, finally, carbamoyl phosphate is catabolized in order to generate ammonia, carbon dioxide and ATP by carbamate kinase (CK, encoded by *arc*C) (Cotter and Hill 2003). A fourth protein that completes this pathway is arginine/ornithine antiporter (encoded by *arc*D gene) (Gobbetti et al. 2005) (Fig. 2). The resulting NH_3 reacts with H+ and helps to alkalize the environment, while the generated ATP can enable extrusion of cytoplasmic protons by the F_0F_1-ATPase (R. Wu et al. 2011, C. Wu et al. 2014).

Several studies show that arginine and aspartate are connected to the regulation of acid resistance in some bacterial species/strains (C. Wu et al. 2012, 2013, Zhang et al. 2012). For example, C. Wu et al. (2012) demonstrated that the addition of 50 mM arginine or aspartate improve the resistance to acids in *L. casei*.

The F_1F_0-ATPase is a membrane-located proton pump that can hydrolyze or synthesize intracellular ATP. This proton pump consists of two main portions, a hydrophilic enzyme (F1), peripherally bound, composed of α, β, γ, δ, and ε subunits and an hydrophobic transmembrane complex (F0), composed three subunits, that mediate proton translocation between two compartments of the organelle, thereby maintaining pH homeostasis. Proton passage through Fo is regulated by the F1 portion (van de Guchte et al. 2002, Cotter and Hill 2003, Kajfasz and Quivey 2011, Wang et al. 2018). Kullen and Klaenhammer (1999) reported high transcriptional levels of *atp* gene (coding F_1-F_0-ATPase) in *L. acidophilus* inoculated in acidic environments. The **Amino acid decarboxylation-antiporter reactions (proton pump)** is an additional mechanism of acidic stress tolerance in LAB. It can maintain intracellular pH homeostasis via a decarboxylation reaction. A proton is consumed in the reaction, and the product is exported from the cell via an antiporter (Azcarate-Peril et al.

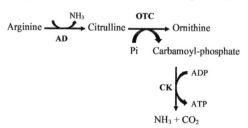

Fig. 2. ADI deiminase pathway.

2004). For example, glutamate decarboxylase (GAD) can catalyze the conversion of glutamate to γ-aminobutyrate (GABA), and results in the raising of the intracellular pH (Feehily and Karatzas 2012). Among Gram-positive bacteria, a GAD system has been described in *Lactococcus lactis* (Azcarate-Peril et al. 2004). Cotter and Hill (2003) found, at low pH values and at the beginning of the stationary phase in the presence of NaCl, an over-expression of the *gad*C and *gad*B genes, important for the lactococcal survival during cheese production.

Pre-adaptation and cross-protection. These are effective methods for strengthening the resistance of LAB against acidic environments. The pre-exposure of LAB to a sublethal stress can increase their resistance to the same type and/or other kinds of lethal stress. According to van de Guchte et al. (2002),bacteria cells develop stress-sensing systems, which allow them to better resist harsh conditions and sudden environmental changes. At the basis of cross-protection, different stimuli, such as heat, oxygen, cold and low pH, can be involved. For example, heat pre-treatment provokes an acid resistance response (ATR) in *Lb. plantarum*, promoting its growth under low pH (Angelis et al. 2004), while Chu-Ky et al. (2014) showed how acid adaptation enhanced *Lb. fermentum* resistance in acidic conditions. After 180 min of acid stress exposure, in a simulated GI juice, the vitality of the acid-adapted bacterium was much higher than that of unstressed control. These systems are used by bacteria not only to counteract the acid conditions they encounter during theirlife cycle, but also as a response to stresses such as heat, cold and osmosis.

Addition of protective agents. The supplement of acid stress protectants in LAB, such as aminoacids, fatty acids or fermentable sugars, can reduce damage caused by the acidic environment. For example, arginine, aspartate andtween-80 can improve acid resistance of *Lb. casei* Zhang (Zhang et al. 2012, C. Wu et al. 2013) and increase the survival capacity of *Lb. rhamnosus* (1000 times higher than the control) (Corcoran et al. 2007). High concentrations of maltose or glucose improved the viability of *Lb. plantarum*, while the addition of tryptone and yeast extract increased the tolerance of *Lb. acidophilus* (Charalampopoulos et al. 2003); finally, glutathione carried out a protective action against LAB in acidic conditions (Pophaly et al. 2012).

2.1.2. Heat stress. Heating is one of the most serious stresses encountered by many LABs during manufacturing processes, such as food fermentation, where they are added as starters, or pasteurization, in which microorganisms face temperatures as high as 60°C. In particular, high temperatures,like the ones reached during the spray drying process, may irreversibly damage cells. In contrast, milder heat stress (between 40 and 65°C) leads to the denaturation of proteins and their subsequent aggregation, harmful to the cell (van de Guchte et al. 2002). In order to cope with heat stress, cells respond with changes in gene expression that induce synthesis of a set of proteins, known as heat-shock proteins (HSP),that are able to counteract the pleiotropic effects of heat stress (Tsakalidou and Papadimitriou 2011). Although HSPs play a crucial role in a heat stress response, they are also produced by different stress situations, e.g., acid or oxidative stress. Microbial HSPs, also known as molecular chaperones, perform physiological functions which allow cells to adapt to gradual changes in their environment and to survive or grow in lethal conditions. Several HSPs, such as DnaJ, DnaK, GrpE and GroEL chaperones, modulate the correct folding of nascent and stress-accumulated misfolded proteins, prevent their aggregation and transport proteins into correct subcellular compartments, while other

HSPs, including ClpP and ClpX, act as proteases, catalyzing the degradation of misfolded proteins generated by exposure to stress (Garrido et al. 2001, Capozzi et al. 2009). HSP have been identified and characterized in LAB, such as *Lb. acidophilus* (Altermann et al. 2005), *Lb. plantarum* (Kleerebezem et al. 2003, Fiocco et al. 2009, 2010) and *Lb. bulgaricus* (van de Guchte et al. 2006). In Gram-positive bacteria, heat shock genes are sometimes grouped into several classes based on their transcriptional regulation. For instance, Class I and class III genes are negatively regulated, while class II genes are positively regulated. Class I genes are controlled by the HrcA protein, which binds to a palindromic operator sequence named CIRCE (Controlled Invert Repeat of Chaperone Expression); in probiotic lactobacilli, the chaperone genes *gro*ESL and *dna*K are regulated by such a system. Class II genes are transcribed by involving an alternative sigma factor, i.e., σB, originally found in *Bacillus subtilis* (Petersohn et al. 2001). Class III genes are controlled by CtsR, which binds to a specific DNA direct repeat, referred to as the CtsR box. This target sequence was identified upstream of the *clp* operon of several Gram-positive bacteria. Intriguingly a CtsR-dependent regulation was proposed for *Lb. plantarum fts*H gene, encoding a membrane-bound metalloprotease (Fiocco et al. 2009), and for several *clp* genes as well as *hsp1* (Fiocco et al. 2010). The heat stress genes regulated by unknown mechanisms are grouped under class IV (Prasad et al. 2003). Corcoran et al. (2008) showed the important function of (i) chaperones such as GroESL in the refolding of denatured proteins and (ii) DnaK chaperone (also known as Hsp70) which, together with DnaJ and GrpE, assists in protein folding.

2.1.3. Oxidative stress. Aerobic bacteria produce energy primarily through oxidative phosphorylation. Although many LAB are typically microaerophiles and can produce catalase (if grown in a medium containing heme) or pseudocatalase (non-heme catalase), several species, such as *L. lactis* and *E. faecalis*, grow well anaerobically. These species lack effective oxygen scavenging cellular mechanisms, such as catalases, and are, thus, unable to synthesize ATP by respiratory means; however, they do have an exclusively fermentative metabolism. They can be susceptible to aerobic conditions during food production (e.g., after spray drying or during fermentation), because it is naturally present in food or permeates in them through packaging. For example, dairy foods such as yoghurt contain high levels of oxygen, which is incorporated during the various processing phases (homogenization, mixing and agitation) or permeated through the packaging material during the shelf life storage (van de Guchte et al. 2002). To survive these harsh conditions, bacteria have developed adaptation responses that lead to reprogramming gene expression. As reported by several studies, in *Oenococcus*, the *trx*A gene, encoding a thioredoxin, was found to be induced not only by the presence of H_2O_2 but also by heat-shock, while the *trx*B gene encoding the thioredoxin pathway enzyme, thioredoxin reductase, has been identified in *Lb. bulgaricus* (van de Guchte et al. 2002). In *L. lactis*, FlpA and FlpB regulate the uptake of Zn(II). The inactivation of *flp*A and *flp*B, resulting in exhaustion of the intracellular pool of Zn (II), increases microorganism sensitivity to H_2O_2. This finding suggests that *L. lactis* uses Zn(II) as a defense mechanism against oxidative stress (Scott et al. 2000).

Oxidative stress is caused by an imbalance between intracellular oxidant concentration, cellular antioxidant protection and oxidative change of macromolecules (such as membrane lipids, proteins) (Capozzi et al. 2009). Exposure to oxygen causes an accumulation of toxic oxygenic metabolites, which, in turn, leads to a partial or complete growth inhibition. Understanding how oxygen can perform a toxicactionon the anaerobic microorganisms requires a brief description of Reactive Oxygen Species (ROS). ROS are generated from endogenous sources, as well as from the environment; they can readily diffuse across cellular membranes and may cause oxidative damage in biomolecules such as membrane lipids, proteins and nucleic acids, impairing their biological functions or constituting one of the major causes of aging and cell death (Halliwell 2009, Furumoto et al. 2016). For example (Guerzoni et al. 2001), have shown that oxidative stress causes a change in the fatty acid composition in the cell membrane of *Lb. helveticus*. Examples of the main toxic forms of oxygen are reported below:

(i) superoxide anion (O_2^*-), whose toxicity is due to its instability, is highly reactive and can oxidize all the organic compounds of the cell. The optional aerobic and anaerobic bacteria produce the enzyme superoxide dismutase (SOD) in order to neutralize it, while bacteria that lack this enzyme, such as *Lb. plantarum*, use a high-level (20–30 mM) intracellular accumulation of Mn^{2+} ions as a scavenger for oxygen radicals (Kleerebezem et al. 2003);

(ii) hydrogen peroxide (H_2O_2) is a powerful bactericide and oxidant, able to generate other cytotoxic oxidising chemical species, such as hydroxyl radicals, in the presence of transition metals. It contains the peroxide anion, which is also toxic, and is neutralized by enzymes such as catalase or peroxidase;

(iii) hydroxyl radical (HO*), is highly reactive with biological molecules. It forms in the cell cytoplasm following the action of ionizing radiation (Storz and Imlayt 1999).

Usually, bacteria use antioxidant enzymes, in particular superoxide dismutase (SOD) and catalase (CAT), to protect themselves from ROS by eliminating superoxide and H_2O_2, respectively. Recently, An et al. (2011) reported that the cooperation between SOD and CAT could significantly enhance oxidative resistance in *Lb. rhamnosus*.

LAB produce ATP by substrate level phosphorylation, therefore the regeneration of NAD+ from NADH assumes critical importance. The organic substrate undergoes aseries of oxidative and reductive reactions mediated by pyridine nucleotides such as NADH. The simplest way to oxidize NADH is by the reduction of molecular oxygen (O_2) via the activity of NADH oxidase. Many studies on the aerotolerance of LAB suggest that the ratio and specific activities of the NADH oxidase and NADH peroxidase determine the elimination of oxygen from the cell (Higuchi et al. 2000, Talwalkar and Kailasapathy 2004). The NADH oxidase system appears to be commonly spread within LAB. Two types of NADH oxidases have been identified in LAB, the NADH-H_2O_2 oxidase and NADH-H_2O oxidase.

While the NADH-H_2O_2 oxidase catalyzes the reduction of O_2 to H_2O_2(1)

$$NADH + H+ +O_2 NAD+ + H_2O_2 (1)$$

the NADH-H_2O oxidase carries out the four-electron reduction of oxygen to water (2)

$$2NADH + 2H+ + O_2 NAD+ + 2H_2O (2)$$

The activities of NADH oxidase can also result in the incomplete reduction of oxygen, generating reactive oxygen species, such as the superoxide anion (O_2^-)

NADH +2O$_2$ NAD+ + H+ + 2O$_2$

Accordingly, some lactic acid bacteria possess NADH peroxidase that reduce H$_2$O$_2$ to H$_2$O, as shown below:

NADH + H+ + H$_2$O$_2$ NAD+ + 2 H$_2$O (Talwalkar and Kailasapathy 2004)

2.1.4. Cold stress. The industrial preservation of lactobacilli involves processes such as freezing and freeze-drying, low temperature fermentation (e.g., during cheese ripening) and refrigerated storage, in which lactic acid bacteria find temperatures below their optimal growth temperature considering that these microorganisms are either mesophilic or thermophilic (van de Guchte et al. 2002). Low temperatures are commonly used to preserve the viability and functional activity of LAB for a long time, keeping their technological properties intact (acidification activity or organoleptic properties). However, abrupt temperature downshifts can lead to undesirable side effects, such as denaturation of sensitive proteins, decreased cell viability, formation of stable secondary structures in DNA and RNA that impair replication, transcription and protein synthesis, (Derzelle et al. 2000, Carvalho et al. 2004) or reduction in membrane fluidity. In fact, as a result of a decrease in temperature, some fluidic components of the membrane become gelatinous, preventing the proteins from functioning normally. In order to increase membrane fluidity, microorganisms respond with an increase in the unsaturation of the fatty acid chains (Beales 2004). To adapt to low temperatures, microorganisms have developed a cold-shock response, during which cold-induced proteins (CIPs) are synthesized. Among these, cold-shock proteins (Csp) ($\sim$ 7.5 kDa), single-stranded DNA- and RNA-binding proteins, are the most strongly induced (Phadtare and Severinov 2010). Csp proteins are present in many species of lactic acid bacteria. In *L. lactis*, expression of *csp* genes (*csp*A, *csp*B, *csp*C, *csp*D) is strongly induced upon cold shock (Derzelle et al. 2000). Song et al. (2014) showed that in cold stress conditions (5 °C for 6 h) in *Lb. plantarum*, L67 incurs an up-regulation/ increase, at the transcription levels, of two genes (*csp*P and *csp*L). Bacterial resistance to cold stress depends on different factors related to the microorganisms and to the conditions of production (fermentation, freezing) and storage. For example, growth medium can increase the resistance of LAB at freezing (Zavaglia et al. 2000, Beal et al. 2001). Thisis the case for *Lb. delbrueckii* subsp. *bulgaricus* grown in the presence of calcium (Fonseca et al.2001), while the presence of Tween 80 leads to alterations of the fatty acid composition of LAB cells, which influence their resistance to freezing (Carvalho et al. 2004). Likewise, the cryoprotective effects of different chemicals, including glycerin and mannitol, were studied on *Lb. rhamnosus* and *Lb. paracasei,* upon storage at room and refrigerated temperature, indicating their ability to prevent cellular lesions and to prolong probiotic vitality (Savini et al. 2010). Streit et al. (2008) showed in their work how it is possible to increase *Lb. bulgaricus* CFL1 resistance by pre-adaptation, applying moderate stress before cold stress. Upon freezing, the formation of ice crystals in the external medium or inside the cells can damage the bacterial membranes, also causing cell death after freezing and thawing. The formation of large ice crystals, which could interfere with vital metabolic activities, can be inhibited by the use of a protective agent. The added compounds can have a different permeability to the cell. For instance, permeable compounds are able to bind water, then suppress excess dehydration, reduce salt toxicity and prevent formation of ice crystals, while semi-permeable compounds induce plasmolysis of cells prior to freezing. Finally, non-permeable compounds adsorb to the surface of microorganisms where they form a viscous layer (Akın et al. 2007).

2.1.5. Osmotic stress. One of the most frequently-used methods to preserve food products is increasing osmotic pressure, lowering water activity (a_w) through drying or addition of osmotically active compounds such as salts (NaCl and KCl), sugars (glucose and sucrose) or glycerol. Therefore, there will be less water available for microorganisms (Abee and Wouters 1999, Beales 2004). Osmotic stress is often encountered by LAB in industrial processes, such as cheese production and ripening where salt can reach concentrations of up to 2.8%, meat fermentation and yogurt-making process, as well as in the GIT (upper small intestine). Developing adaptive strategies to cope with osmotic stress is essential to guarantee long-term delivery of stable cultures in terms of viability and activity. At the temporary loss of turgor pressure after a hyperosmotic shock, bacteria cells respond through osmoregulation: Accumulation of osmo-protectors by activation of transport systems. LAB respond to osmotic stress by accumulating or rapidly releasing osmolytes, such as glycine-betaine (Konings 2002). *Lb. plantarum* genome coding for three systems for the uptake and biosynthesis of the osmo-protectants glycine-betaine/carnitine/choline, including two ABC transporters (opuABCD, choSQ), while in *Lb. casei* peptides can also balance the hyperosmotic stress. In their work, Piuri et al. (2003, 2005) showed that protease PrtP, the main peptide supplier, was activated during growth in a high salt medium, despite genetic expression not being altered by salt (Piuri et al. 2003, 2005).The lowering of water activity leads to changes in membrane lipid composition. The study conducted by Guillot et al. (2000) showed that the main modification in *L. lactis* was an increase in cyclopropane fatty acid C19:0, whereas the unsaturated-to-saturated ratio remained unchanged. Bacteria also respond to osmotic stress through "cross protection". For example, in *Lb. delbrueckii* the enhanced freeze-tolerance of some cells acidified at pH 5.25 for 30 min at the end of fermentation confirm a "cross talking" between cold and acid tolerance (Streit et al. 2008). Moreover, in *Lb. plantarum*, the overproduction of some small heat shock proteins, such as Hsp 18.55 and Hsp 19.3, leads to an enhanced survival in the presence of butanol (1% v/v) or ethanol (12% v/v) treatment (Fiocco et al. 2007).

3. Lab as Probiotics

Many LAB, such as *Lb. acidophilus, Lb. casei, Lb. plantarum* and *Lb. fermentum,* are used as probiotic strains, i.e. live microorganisms, as a benefit to healthin the form of a food supplement. They produce antimicrobial substances that contrast gastric and intestinal pathogens (Ljungh and Wadström 2009). To carry out their biological role, probiotics must overcome adverse conditions that they encounter in the phases of food production and crossing the gastro intestinal tract: pH and bile shift, hydrogen peroxide, oxygen levels and storage temperature (Martín et al. 2015, De Prisco and Mauriello 2016). In this regard, various strategies have been developed in orderto improve the technological and gastrointestinal robustness of probiotics, including appropriate selection of acid and bile resistant strains, use of cell protectants, stress adaptation, genetic manipulation (Gueimonde and Sánchez 2012) as well as incorporation of micronutrients, such as peptides and amino acids, and microencapsulation (Anal and Singh 2007, de Vos et al. 2010, Sarkar 2010).

Table 2. Mechanisms and genes shown to be involved in stress response in probiotic species.

Stress	Mechanisms	Protein/Genes/System Involved	Species/Strain
Acidic	• induction of chaperones and protease complex • increase in activity and amount of F0F1-ATPase • alkalization with arginine dihydrolase system(ADS) • modulation of membrane FA composition • amino acid decarboxylation-antiporter reactions • pre-adaptation and cross-protection • addition of protective agents	GroEL, GroES, DnaK, Clp, GrpE, *tdc*, AtpB, AtpA, AtpG, AtpD, HdeA and HdeB	*Lb. casei* Zhang (C. Wu et al. 2013, Zhang et al. 2012), *Lb. acidophilus* (Charalampopoulos et al. 2003), *Lb. casei* (C. Wu et al. 2014), *Lb. fermentum* (Chu-Ky et al. 2014), *Lb. plantarum* (Arena et al. 2007, Angelis et al. 2004)
Heat	synthesis of heat shock proteins (HSPs)	CtsR (transcriptional regulator) GroES, GroEL, DnaK, DnaJ, GrpE sHSP (small Heat Shock Proteins) HtrA, FtsH and Clp (proteases)	*Lb. acidophilus* (Altermann et al. 2005), *Lb. plantarum* (Kleerebezem et al.2003, Fiocco et al. 2009, 2010), *Lb. bulgaricus* (van de Guchte et al. 2006)
Cold	• induction of a set of cold shock proteins (CSPs) • medium composition • cryoprotective agents • pre-adaptation	Csp Hsp 18.5, Hsp18.55 and Hsp 19.3.	*Lb. rhamnosus* and *Lb. paracasei* (Savini et al. 2010), *Lb. plantarum* (Song et al. 2014), *Lb. bulgaricus* (Fonseca et al. 2001, Streit et al. 2007)
Oxidative	• target protection • modulation of membrane FA composition • enzymes NAD-oxidase and NADH-peroxidase	*trxA, trxB* *flpA, flpB*	*O. oeni, Lb. bulgaricus* (van de Guchte et al. 2002), *L. lactis* (Scott et al. 2000), *Lb. helveticus* (Guerzoni et al. 2001)
Osmotic	• osmoregulation • modulation of • membrane FA composition • cross-protection	PrtP protease Hsp 18.55 and Hsp 19.3	*L. lactis* (Guillot et al. 2000), *Lb. delbrueckii* (Streit et al. 2008), *Lb. plantarum* (Fiocco et al. 2007)

Microencapsulation is defined as the technology of including sensitive ingredients (solid, liquid or gaseous) within different matrices in order to protect probiotic cells in foods (Malmo et al. 2013) and gastrointestinal conditions, as well as giving protection from bacteriophages and harmful factors, increasing survival during freeze-drying, freezing and storage (Mortazavian et al. 2007). Furthermore, microencapsulation offers many advantages, such as improving the handling of probiotic cultures, as well as the masking of undesired/unrequired tastes and aromas produced by the formation of different metabolic compounds (e.g., acetic acid) during fermentation in foods (De Prisco and Mauriello 2016). Food-grade polymers, such as alginate, chitosan, carrageenan, gelatin, carboxymethyl cellulose (CMC) and pectin, are generally used for the encapsulation technology with versatile applications (Burgain et al. 2011). Currently, several encapsulation methods have been evaluated, including drying-based techniques (freeze-, spray- and fluidized bed-drying) and entrapment in gel particles by emulsion or extrusion methods(Anal and Singh 2007, Huq et al. 2013). Even prebiotics (carbohydrates not digestible by human enzymes) are widely used as stabilizing agents in order to improve the profitability of probiotics during the production and storage of microcapsules. Prebiotics are obtained from vegetable and fruit fibers and include water-soluble oligosaccharides, such as inulin, fructooligosaccharide (FOS) and xylo-oligosaccharides (XOS).Even exopolysaccharides (EPS) of microbial origin (i.e., produced by LAB) have been evaluated for prebiotic capacity (Cinquin et al. 2006, Russo et al. 2012, Caggianiello et al.2016).

4. Genetic Approaches

Gene diversity can be improved by gene mutation induction, and the likelihood of generating mutants able to survive and withstand specific environmental stress conditions can be increased.

Indeed, it is possible to select spontaneous mutant strains which are more tolerant to that kind of stress (Foster 2007, Galhardo et al. 2007). For instance, after a gradual adaptation of *Lb. acidophilus* NCFM at 65°C for 40 minutes, a derivative mutant strain was selected that showed 2 logs-fold greater survival at 65 °C when compared to the wild type. Besides, it had a higher stability at pH 2.0, maintaining unaltered probiotic properties (Streit et al. 2008, Kulkarni et al. 2017). *Lb. casei* Zhang led to the acid-resistant mutant Lbz-2, which was obtained through adaptive evolution by gradually exposing cells to acidic environments. The mutant displayed significantly higher survival ability in the presence of lactic acid, a lower permeability and a higher percentage of unsaturated fatty acids than that of the wild-type (Zhang et al. 2012, van Bokhorst-van de Veen et al. 2015).

In particular, since stress-resistant mutants carry permanent genetic modifications, the genetic selection approach can even influence the metabolic characteristics of some probiotics, and make them more resistant to the processes technologies, thus improving their use in food technology, as well as the final properties of foods (Renault 2002). It is interesting to note that the increased resistance to acids has also been associated with greater fermentative capacity (Collado and Sanz 2007).

Gene manipulation can also be used to help cells accumulate endogenous protective compounds, such as glycine/betaine, able to protect bacteria from osmotic stress (Sleator and Hill 2001).

Conclusions

LAB are a group of bacteria of extreme economic importance, both at the industrial level (food production) and in the medical field (to mucosal delivery of vaccine antigens).Therefore, their implications related to health (probiotics or pathogens), their genetics, physiology and their metabolism have been the subject of rigorous investigations in recent decades. During food processing and storage, LAB face hostile environments such as high or low temperature, sudden changes in pH and osmolarity, ROS presence or nutrient limitations. LAB employ molecular mechanisms in order to adapt and survive to these changes, including synthesis of shock/stress proteins (HSP or CSP), alterations in membrane fatty acids composition, proton pumps (ADS or F_1F_0-ATPase) or protective agents. Understanding microbial stress response mechanisms will enhance the use of preservatives and help in the control of the survival and growth of foodborne microorganisms (Beales 2004). The future challenge will be the use of acquired knowledge on stress mechanisms for biotechnology applications in order to improve the robustness and functionality of microbial starter or probiotic strains. In this regard, the use of microencapsulation techniques, as well as omics technology, is producing satisfactory results (Cook et al. 2012, Papadimitriou et al. 2016).

References

Abee, T. and J.A. Wouters. 1999. Microbial stress response in minimal processing. Int. J. Food Microbiol. 50: 65–91.

Akın, M.B., M.S. Akın and Z. Kırmacı. 2007. Effects of inulin and sugar levels on the viability of yogurt and probiotic bacteria and the physical and sensory characteristics in probiotic ice-cream. Food Chem. 104: 93–99

Altermann, E., W.M. Russell, M.A. Azcarate-Peril, R. Barrangou, B.L. Buck, O. McAuliffe, N. Souther, A. Dobson, T. Duong, M. Callanan, S. Lick, A. Hamrick, R. Cano and T.R. Klaenhammer. 2005. Complete genome sequence of the probiotic lactic acid bacterium *Lactobacillus acidophilus* NCFM. Proc. Natl. Acad. Sci. USA 102: 3906–12.

Ammor, M.S. and B. Mayo. 2007. Selection criteria for lactic acid bacteria to be used as functional starter cultures in dry sausage production: An update. Meat Sci. 76: 138–46.

An, H., Z. Zhai, S. Yin, Y. Luo, B. Han and Y. Hao. 2011. Coexpression of the superoxide dismutase and the catalase provides remarkable oxidative stress resistance in *Lactobacillus rhamnosus*. J. Agric. Food Chem. 59: 3851–56.

Anal, A.K. and H. Singh. 2007. Recent advances in microencapsulation of probiotics for industrial applications and targeted delivery. Trends Food Sci. Technol. 18: 240–51.

Andrighetto, C., L. Zampese and A. Lombardi. 2001. RAPD-PCR characterization of lactobacilli isolated from artisanal meat plants and traditional fermented sausages of Veneto region (Italy). Lett. Appl. Microbiol. 33: 26–30.

Angelis, M.D., R. Di Cagno, C. Huet, C. Crecchio, P.F. Fox and M. Gobbetti. 2004. Heat Shock Response in *Lactobacillus plantarum*. Appl. Environ. Microbiol. 70: 1336–46.

Arena, M.E., D. Fiocco, M.C. Manca de Nadra, I. Pardo and G. Spano. 2007. Characterization of a *Lactobacillus plantarum* strain able to produce tyramine and partial cloning of a putative tyrosine decarboxylase gene. Curr. Microbiol. 55: 205–10.

Azcarate-Peril, M.A., E. Altermann, R.L. Hoover-Fitzula, R.J. Cano and T.R. Klaenhammer. 2004. Identification and inactivation of genetic loci involved with *Lactobacillus acidophilus* acid tolerance. Appl. Environ. Microbiol. 70: 5315–22.

Beal, C., F. Fonseca and G. Corrieu. 2001. Resistance to freezing and frozen storage of *Streptococcus thermophilus* is related to membrane fatty acid composition. J. Dairy Sci. 84: 2347–56.

Beales, N. 2004. Adaptation of microorganisms to cold temperatures, weak acid preservatives, low pH, and osmotic stress: A review. Compr. Rev. Food Sci. Food Saf. 3: 1–20.

Bernardeau, M., J.P. Vernoux, S. Henri-Dubernet and M. Guéguen. 2008. Safety assessment of dairy microorganisms: The *Lactobacillus* genus. Int. J. Food Microbiol. 126: 278–85.

Bokhorst-van de Veen, H. van, H. Xie, E. Esveld, T. Abee, H. Mastwijk and M.N. Groot. 2015. Inactivation of chemical and heat-resistant spores of *Bacillus* and *Geobacillus* by nitrogen cold atmospheric plasma evokes distinct changes in morphology and integrity of spores. Food Microbiol. 45: 26–33.

Booth, I.R. 2002. Stress and the single cell: Intrapopulation diversity is a mechanism to ensure survival upon exposure to stress. Int. J. of Food Microbiol., 18th International Symposium of the International Committee on Food Microbiology and Hygeine, August 18–23, 2002, Lillehammer Norway. Necessary and Unwanted Bacteria in Food—Microbial Adaption to changing Environments 78: 19–30.

Broadbent, J.R., R.L. Larsen, V. Deibel and J.L. Steele. 2010. Physiological and transcriptional response of *Lactobacillus casei* ATCC 334 to acid stress. J. Bacteriol. 192: 2445–58.

Burgain, J., C. Gaiani, M. Linder and J. Scher. 2011. Encapsulation of probiotic living cells: From laboratory scale to industrial applications. J. Food Eng. 104: 467–83.

Caggianiello, G., M. Kleerebezem and G. Spano. 2016. Exopolysaccharides produced by lactic acid bacteria: From health-promoting benefits to stress tolerance mechanisms. Appl. Microbiol. Biotechnol. 100: 3877–86.

Capozzi, V., D. Fiocco, M.L. Amodio, A. Gallone and G. Spano. 2009. Bacterial stressors in minimally processed food. Int. J. Mol. Sci. 10: 3076–3105.

Capozzi, V., P. Russo, M.T. Dueñas, P. López and G. Spano. 2012. Lactic acid bacteria producing B-group vitamins: Agreat potential for functional cereals products. Appl. Microbiol. Biotechnol. 96: 1383–94.

Carr, F.J., D. Chill and N. Maida. 2002. The lactic acid bacteria: A literature survey. Crit. Rev. Microbiol. 28: 281–370.

Carvalho, A.S., J. Silva, P. Ho, P. Teixeira, F.X. Malcata and P. Gibbs. 2004. Relevant factors for the preparation of freeze-dried lactic acid bacteria. Int. Dairy J. 14: 835–47.

Chaillou, S., M.C. Champomier-Vergès, M. Cornet, A.M. Crutz-Le Coq, A.M. Dudez, V. Martin, S. Beaufilset, E. Darbon-Rongère, R. Bossy, V. Loux and M. Zagorec. 2005. The complete genome sequence of the meat-borne lactic acid bacterium *Lactobacillus sakei* 23K. Nat. Biotechnol. 23: 1527–33.

Champomier Vergès, M.C., M. Zuñiga, F. Morel-Deville, G. Pérez-Martínez, M. Zagorec and S.D. Ehrlich. 1999. Relationships between arginine degradation, pH and survival in *Lactobacillus sakei*. FEMS Microbiol. Lett. 180: 297–304.

Charalampopoulos, D., R. Wang, S.S. Pandiella and C. Webb. 2002. Application of cereals and cereal components in functional foods: A review. Int. J. Food Microbiol., Notermans Special Issue 79: 131–41.

Charalampopoulos, D., S.S. Pandiella and C. Webb. 2003. Evaluation of the effect of malt, wheat and barley extracts on the viability of potentially probiotic lactic acid bacteria under acidic conditions. Int. J. Food Microbiol. 82: 133–41.

Chen, H.C., S.Y. Wang and M.J. Chen. 2008. Microbiological study of lactic acid bacteria in kefir grains by culture-dependent and culture-independent methods. Food Microbiol. 25: 492–501.

Chu-Ky, S., T.K. Bui, T.L. Nguyen and P.H. Ho. 2014. Acid adaptation to improve viability and X-prolyl dipeptidyl aminopeptidase activity of the probiotic bacterium *Lactobacillus fermentum* HA6 exposed to simulated gastrointestinal tract conditions. Int. J. Food Sci. Technol. 49: 565–70.

Cinquin, C., G. Le Blay, I. Fliss and C. Lacroix. 2006. Comparative effects of exopolysaccharides from lactic acid bacteria and fructo-oligosaccharides on infant gut microbiota tested in an *in vitro* colonic model with immobilized cells. FEMS Microbiol. Ecol. 57: 226–38.

Cocolin, L., L.F. Bisson and D.A. Mills. 2000. Direct profiling of the yeast dynamics in wine fermentations. FEMS Microbiol. Lett. 189: 81–87.

Coda, R., C.G. Rizzello, A. Trani and M. Gobbetti. 2011. Manufacture and characterization of functional emmer beverages fermented by selected lactic acid bacteria. Food Microbiol. 28: 526–36.

Collado, M.C. and Y. Sanz. 2007. Induction of acid resistance in *Bifidobacterium*: Amechanism for improving desirable traits of potentially probiotic strains. J. Appl. Microbiol. 103: 1147–57.

Cook, M.T., G. Tzortzis, D. Charalampopoulos and V.V. Khutoryanskiy. 2012. Microencapsulation of probiotics for gastrointestinal delivery. J. Control Release 162: 56–67.

Corcoran, B.M., C. Stanton, G.F. Fitzgerald and R.P. Ross. 2007. Growth of probiotic lactobacilli in the presence of oleic acid enhances subsequent survival in gastric juice. Microbiology 153: 291–99.

Corcoran, B.M., C. Stanton, G. Fitzgerald and R.P. Ross. 2008. Life under stress: The probiotic stress response and how it may be manipulated. Curr. Pharm. Des. 14: 1382–99.

Cotter, P.D. and C. Hill. 2003. Surviving the acid test: Responses of gram-positive bacteria to low pH. Microbiol. Mol. Biol. Rev. 67: 429–53.

De Prisco, A. and G. Mauriello. 2016. Probiotication of foods: A focus on microencapsulation tool. Trends Food Sci. Technol. 48: 27–39.

Derzelle, S., B. Hallet, K. P. Francis, T. Ferain, J. Delcour and P. Hols. 2000. Changes in *cspL*, *cspP*, and *cspC* mRNA abundance as a function of Cold Shock and growth phase in *Lactobacillus plantarum*. J. Bacteriol. 182: 5105–13.

Feehily, C. and K.A.G. Karatzas. 2013. Role of glutamate metabolism in bacterial responses towards acid and other stresses. J. Appl. Microbiol. 114 : 11–24.

Fiocco, D., V. Capozzi, P. Goffin, P. Hols and G. Spano. 2007. Improved adaptation to heat, cold, and solvent tolerance in *Lactobacillus plantarum*. Appl. Microbiol. Biotechnol. 77: 909–15.

Fiocco, D., M. Collins, L. Muscariello, P. Hols, M. Kleerebezem, T. Msadek and G. Spano. 2009. The *Lactobacillus plantarum fts H* gene is a novel member of the CtsR stress response regulon. J.Bacteriol. 191: 1688–94.

Fiocco, D., V. Capozzi, M. Collins, A. Gallone, P. Hols, J. Guzzo, S. Weidmann, A. Rieu, T. Msadek and G. Spano. 2010. Characterization of the CtsR stress response regulon in *Lactobacillus plantarum*. J. Bacteriol. 192: 896–900.

Fonseca, F., C. Béal and G. Corrieu. 2001. Operating conditions that affect the resistance of lactic acid bacteria to freezing and frozen storage. Cryobiology 43: 189–98.

Foster, P.L. 2007. Stress-induced mutagenesis in bacteria. Crit. Rev. Biochem. Mol. Biol. 42: 373–97.

Frees, D., F.K. Vogensen and H. Ingmer. 2003. Identification of proteins induced at low pH in *Lactococcus lactis*. Int. J. Food Microbiol. 87: 293–300.

Furet, J.P., P. Quénée and P. Tailliez. 2004. Molecular quantification of lactic acid bacteria in fermented milk products using real-time quantitative PCR. Int. J. Food Microbiol. 97: 197–207.

Furumoto, H., T. Nanthirudjanar, T. Kume, Y. Izumi, S.B. Park, N. Kitamura, S. Kishino, J. Ogawa, T. Hirata and T. Sugawara. 2016. 10-Oxo-trans-11-octadecenoic acid generated from linoleic acid by a gut lactic acid bacterium *Lactobacillus plantarum* is cytoprotective against oxidative stress. Toxicol. Appl. Pharmacol. 296: 1–9.

Galhardo, R.S., P.J. Hastings and S.M. Rosenberg. 2007. Mutation as a stress response and the regulation of evolvability. Crit. Rev. Biochem. Mol. Biol. 42: 399–435.

Gandhi, A. and N. P. Shah. 2016. Effect of salt stress on morphology and membrane composition of *Lactobacillus acidophilus*, *Lactobacillus casei*, and *Bifidobacterium bifidum*, and their adhesion to human intestinal epithelial-like Caco-2 cells. J. Dairy Sci. 99: 2594–2605.

Garrido, C., S. Gurbuxani, L.Ravagnan and G. Kroemer. 2001. Heat Shock Proteins: endogenous modulators of apoptotic cell death. Biochem. Biophys. Res. Commun. 286: 433–42.

Gerez, C.L., M.I. Torino, G. Rollán and G. Font de Valdez. 2009. Prevention of bread mould spoilage by using lactic acid bacteria with antifungal properties. Food Control 20: 144–48.

Giraffa, G. 2012. Selection and design of lactic acid bacteria probiotic cultures. Eng. Life Sci. 12: 391–98.

Gobbetti, M., M. De Angelis, A. Corsetti and R. Di Cagno. 2005. Biochemistry and physiology of sourdough lactic acid bacteria. Trends Food Sci.Technol., Second International Symposium on Sourdough—From Fundamentals to Applications, 16: 57–69.

Guchte, M. van de, P. Serror, C. Chervaux, T. Smokvina, S. D. Ehrlich and E. Maguin. 2002. Stress responses in lactic acid bacteria. pp. 187–216. *In*: R.J. Siezen, J. Kok, T. Abee and G. Schafsma (eds.). Lactic Acid Bacteria: Genetics, Metabolism and Applications. Springer, Dordrecht, Netherlands.

Guchte, M. van de, S. Penaud, C. Grimaldi, V. Barbe, K. Bryson, P. Nicolas, C. Robert, S. Oztas, S. Mangenot, A. Couloux, V. Loux, R. Dervyn, R. Bossy, A. Bolotin, J.-M. Batto, T. Walunas, J.-F. Gibrat, P. Bessières, J. Weissenbach, S.D. Ehrlich and E. Maguin. 2006. The complete genome sequence of *Lactobacillus bulgaricus* reveals extensive and ongoing reductive evolution. Proc. Natl. Acad. Sci. USA 103: 9274–79.

Gueimonde, M. and B. Sánchez. 2012. Enhancing probiotic stability in industrial processes. Microb. Ecol. Health Dis. 23: 18562.

Guerzoni, M.E., R. Lanciotti and P.S. Cocconcelli. 2001. Alteration in cellular fatty acid composition as a response to salt, acid, oxidative and thermal stresses in *Lactobacillus helveticus*. Microbiology 147: 2255–64.

Guillot, A., D. Obis and M.Y. Mistou. 2000. Fatty acid membrane composition and activation of glycine-betaine transport in *Lactococcus lactis* subjected to osmotic stress. Int. J. Food Microbiol. 55: 47–51.

Halliwell, B. 2009. The wanderings of a free radical. Free Radic. Biol. Med. 46: 531–42.

Hao, P., H. Zheng, Y. Yu, G. Ding, W. Gu, S. Chen, Z. Yu, S. Ren, M. Oda, T. Konno, S. Wang, X. Li, Z.S. Ji and G. Zhao. 2011. Complete sequencing and pan-genomic analysis of *Lactobacillus delbrueckii* subsp. *bulgaricus* reveal its genetic basis for industrial yogurt production. PLOS ONE 6: e15964.

Higuchi, M., Y. Yamamoto and Y. Kamio. 2000. Molecular biology of oxygen tolerance in lactic acid bacteria: functions of NADH oxidases and Dpr in oxidative stress. J. Biosci. Bioeng. 90: 484–93.

Hoe, C.H., C. A. Raabe, T. S. Rozhdestvensky and T. H. Tang. 2013. Bacterial sRNAs: regulation in stress. Int. J. Med. Microbiol. 303: 217–29.

Huq, T., A. Khan, R. A. Khan, B. Riedl and M. Lacroix. 2013. Encapsulation of probiotic bacteria in biopolymeric system. Crit. Rev. Food Sci. Nutr. 53: 909–16.

Kajfasz, J. K. and R. G. Quivey. 2011. Responses of lactic acid bacteria to acid stress. pp. 23-53. *In:* E. Tsakalidou and K. Papadimitriou (eds.). Stress Responses of Lactic Acid Bacteria. Food Microbiology and Food Safety. Springer, Boston, MA.

Kleerebezem, M., J. Boekhorst, R. van Kranenburg, D. Molenaar, O.P. Kuipers, R. Leer, R. Tarchini, S.A. Peters, H.M. Sandbrink, M.W.E.J. Fiers, W. Stiekema, R.M.K. Lankhorst, P.A. Bron, S.M. Hoffer, M.N.N. Groot, R. Kerkhoven, M. de Vries, B. Ursing, W.M. de Vos and R.J. Siezen. 2003. Complete genome sequence of *Lactobacillus plantarum* WCFS1. Proc. Natl. Acad. Sci. USA 100: 1990–95.

Konings, W.N. 2002. The cell membrane and the struggle for life of lactic acid bacteria. pp. 3–27. *In:* R.J. Siezen, J. Kok, T. Abee and G. Schaafsma (eds.). Lactic Acid Bacteria: Genetics, Metabolism and Applications. Springer, Dordrecht, Netherlands.

Kulkarni, S., S.F. Haq, S. Samant and S. Sukumaran. 2017. Adaptation of *Lactobacillus acidophilus* to thermal stress yields a thermotolerant variant which also exhibits improved survival at pH 2. Probiotics Antimicrob. Proteins. doi:10.1007/s12602-017-9321-7.

Kullen, M.J. and T.R. Klaenhammer. 1999. Identification of the pH-inducible, proton-translocating F_1F_0-ATPase (AtpBEFHAGDC) operon of *Lactobacillus acidophilus* by differential display: gene structure, cloning and characterization. Mol. Microbiol. 33: 1152–61.

Leroy, F. and L. De Vuyst. 2004. Lactic acid bacteria as functional starter cultures for the food fermentation industry. Trends Food Sci. Technol. 15: 67–78.

Ljungh, Å. and T. Wadström. 2009. Lactobacillus Molecular Biology: From Genomics to Probiotics. Horizon Scientific Press.

Lund, P., A. Tramonti and D. De Biase. 2014. Coping with low pH: Molecular strategies in neutralophilic bacteria. FEMS Microbiol. Rev. 38: 1091–1125.

Malmo, C., A. La Storia and G. Mauriello. 2013. Microencapsulation of *Lactobacillus reuteri* DSM 17938 cells coated in alginate beads with chitosan by spray drying to use as a probiotic cell in a chocolate soufflé. Food Bioprocess Technol. 6: 795–805.

Martín, M.J., F.L. Villoslada, M.A. Ruiz and M.E. Morales. 2015. Microencapsulation of bacteria: A review of different technologies and their impact on the probiotic effects. Innov. Food Sci. Emerg. Technol. 27: 15–25.

Mills, D.A., H. Rawsthorne, C. Parker, D. Tamir and K. Makarova. 2005. Genomic analysis of *Oenococcus oeni* PSU-1 and its relevance to wine making. FEMS Microbiol. Rev. 29: 465–75.

Mills, S., C. Stanton, G.F. Fitzgerald and R.P. Ross. 2011. Enhancing the stress responses of probiotics for a lifestyle from gut to product and back again. Microb. Cell. Fact. 10: S19.

Mortazavian, A., S.H. Razavi, M.R.E. and S. Sohrabvandi. 2007. Principles and methods of microencapsulation of probiotic microorganisms. Iran. J. Biotechnol. 5: 1–18.

Mozzi, F., R.R. Raya and G.M. Vignolo. 2010. Biotechnology of Lactic Acid Bacteria: Novel Applications. John Wiley & Sons.

Muller, J.A., R.P. Ross, W.F.H. Sybesma, G.F. Fitzgerald and C. Stanton. 2011. Modification of the technical properties of *Lactobacillus johnsonii* NCC 533 by supplementing the growth medium with unsaturated fatty acids. Appl.Environ.Microbiol. 77: 6889–98.

Nezhad, M.H., M.A. Hussain and M.L. Britz. 2015. Stress responses in probiotic *Lactobacillus casei*. Crit. Rev. Food Sci. Nutr. 55: 740–49.

Papadimitriou, K., Á. Alegría, P.A. Bron, M. de Angelis, M. Gobbetti, M. Kleerebezem, J.A. Lemos, D.M. Linares, P. Ross, C. Stanton, F. Turroni, D. van Sinderen, P. Varmanen, M. Ventura, M. Zúñiga, E. Tsakalidou and J. Kok. 2016. Stress physiology of lactic acid bacteria. Microbiol. Mol. Biol. Rev. 80: 837–90.

Petersohn, A., M. Brigulla, S. Haas, J.D. Hoheisel, U. Völker and M. Hecker. 2001. Global analysis of the general stress response of *Bacillus subtilis*. J. Bacteriol. 183: 5617–31.

Phadtare, S.and K. Severinov. 2010. RNA remodeling and gene regulation by cold shock proteins. RNA Biol. 7: 788–95.

Piuri, M., C. Sanchez-Rivas and S.M. Ruzal. 2003. Adaptation to high salt in *Lactobacillus:* Role of peptides and proteolytic enzymes. J. Appl. Microbiol. 95: 372–79.

Piuri, M., C. Sanchez-Rivas and S.M. Ruzal. 2005. Cell wall modifications during osmotic stress in *Lactobacillus casei*. J. Appl. Microbiol. 98: 84–95.

Pophaly, S.D., R. Singh, S.D. Pophaly, J.K. Kaushik and S.K. Tomar. 2012. Current status and emerging role of glutathione in food grade lactic acid bacteria. Microb. Cell Fact. 11: 114.

Prasad, J., P.l McJarrow and P. Gopal. 2003. Heat and osmotic stress responses of probiotic *Lactobacillus rhamnosus* HN001 (DR20) in relation to viability after drying. Appl. Environ. Microbiol. 69: 917–25.

Renault, P. 2002. Genetically modified lactic acid bacteria: Applications to food or health and risk assessment. Biochimie 84: 1073–87.

Ruiz, L., B. Sánchez, P. Ruas-Madiedo, D.L. Reyes-Gavilán, C.G and A. Margolles. 2007. Cell envelope changes in *Bifidobacterium animalis* ssp. *lactis* as a response to bile. FEMS Microbiol. Lett. 274: 316–22.

Russo, P., P. López, V. Capozzi, P.F. de Palencia, M.T. Dueñas, G. Spano and D. Fiocco. 2012. Beta-glucans improve growth, viability and colonization of probiotic microorganisms. Int. J. Mol. Sci. 13: 6026–39.

Salminen, S. and A. von Wright. 2004. Lactic Acid Bacteria: Microbiological and Functional Aspects, Third Edition. CRC Press.

Saraoui, T., J. Cornet, E. Guillouet, M.F. Pilet, F. Chevalier, J.J. Joffraud and F. Leroi. 2017. Improving simultaneously the quality and safety of cooked and peeled shrimp using a cocktail of bioprotective lactic acid bacteria. Int. J. Food Microbiol. 241: 69–77.

Sarkar, S. 2010. Approaches for enhancing the viability of probiotics: A review. Br. Food J. 112: 329–49.

Savini, M., C. Cecchini, M.C. Verdenelli, S. Silvi, C. Orpianesi and A. Cresci. 2010. Pilot-scale production and viability analysis of freeze-dried probiotic bacteria using different protective agents. Nutrients 2: 330–39.

Schroeter, J. and T. Klaenhammer. 2009. Genomics of lactic acid bacteria. FEMS Microbiol. Lett. 292: 1–6.

Scott, C., H. Rawsthorne, M. Upadhyay, C.A. Shearman, M.J. Gasson, J.R. Guest and J. Green. 2000. Zinc uptake, oxidative stress and the FNR-like proteins of *Lactococcus lactis*. FEMS Microbiol. Lett. 192: 85–89.

Silhavy, T.J., D. Kahne and S. Walker. 2010. The bacterial cell envelope. Cold Spring Harb. Perspect. Biol. 2: a000414.

Simova, E., D. Beshkova, A. Angelov, T.S. Hristozova, G. Frengova and Z. Spasov. 2002. Lactic acid bacteria and yeasts in kefir grains and kefir made from them. J. Ind. Microbiol. Biotechnol. 28: 1–6.

Sleator, R.D. and C. Hill. 2002. Bacterial osmoadaptation: The role of osmolytes in bacterial stress and virulence. FEMS Microbiol. Rev. 26: 49–71.

Smit, G., B.A. Smit and W.J.M. Engels. 2005. Flavour formation by lactic acid bacteria and biochemical flavour profiling of cheese products. FEMS Microbiol. Rev. 29: 591–610.

Song, S., D.W. Bae, K. Lim, M.W. Griffiths and S. Oh. 2014. Cold stress improves the ability of *Lactobacillus plantarum* L67 to survive freezing. Int. J. Food Microbiol. 191: 135–43.

Storz, G. and J.A. Imlayt. 1999. Oxidative stress. Curr. Opin. Microbiol. 2: 188–94.

Streit, F., J. Delettre, G. Corrieu and C. Béal. 2008. Acid adaptation of *Lactobacillus delbrueckii* subsp. *bulgaricus* induces physiological responses at membrane and cytosolic levels that improves cryotolerance. J. Appl. Microbiol. 105: 1071–80.

Streit, F., J. Delettre, G. Corrieu and C. Béal. 2008. Acid adaptation of *Lactobacillus delbrueckii* subsp. *bulgaricus* induces physiological responses at membrane and cytosolic levels that improves cryotolerance. J. Appl. Microbiol. 105: 1071–80.

Talwalkar, A.and K. Kailasapathy. 2004. The role of oxygen in the viability of probiotic bacteria with reference to *Lactobacillus acidophilus* and *Bifidobacterium* spp. Curr. Issues Intest. Microbiol. 5: 1–8.

Tamang, J.P., B. Tamang, U. Schillinger, C.M.A.P. Franz, M. Gores and W.H. Holzapfel. 2005. Identification of predominant lactic acid bacteria isolated from traditionally fermented vegetable products of the Eastern Himalayas. Int. J. Food Microbiol. 105: 347–56.

Taranto, M.P., M.L. F.Murga, G. Lorca, e G.F. de Valdez. 2003. Bile salts and cholesterol induce changes in the lipid cell membrane of *Lactobacillus reuteri*. J. Appl. Microbiol. 95: 86–91.

Taranto, M.P., G. Perez-Martinez and G.F. de Valdez. 2006. Effect of bile acid on the cell membrane functionality of lactic acid bacteria for oral administration. Res. Microbiol. 157: 720–25.

Tonon, T. and A. Lonvaud-Funel. 2000. Metabolism of arginine and its positive effect on growth and revival of *Oenococcus oeni*. J. Appl. Microbiol. 89: 526–31.

Trias, R., L. Bañeras, E. Montesinos and E. Badosa. 2008. Lactic acid bacteria from fresh fruit and vegetables as biocontrol agents of phytopathogenic bacteria and fungi. Int. Microbiol. 11: 231–236.

Tsakalidou, E. and K. Papadimitriou. 2011. Stress Responses of Lactic Acid Bacteria. Springer Science & Business Media.

Vermeiren, L., F. Devlieghere and J. Debevere. 2004. Evaluation of meat born lactic acid bacteria as protective cultures for the biopreservation of cooked meat products. Int. J. Food Microbiol. 96: 149–64.

Vos, P. de, M.M. Faas, M. Spasojevic and J. Sikkema. 2010. Encapsulation for preservation of functionality and targeted delivery of bioactive food components. Int. Dairy J., US/Ireland Functional Foods Conference 20: 292–302.

Wang, C., Y. Cui and X. Qu. 2018. Mechanisms and improvement of acid resistance in lactic acid bacteria. Arch. Microbiol. 200: 195–201.

Wen-ying, Z. and K. Zhen-kui. 2013. Advanced progress on adaptive stress response of *Oenococcus oeni*. J. of Northeast Agric.Univ. (English Edition) 20: 91–96.

Wu, C., J. Zhang, W. Chen, M. Wang, G. Du and J. Chen. 2012. A combined physiological and proteomic approach to reveal lactic-acid-induced alterations in *Lactobacillus casei* Zhang and its mutant with enhanced lactic acid tolerance. Appl. Microbiol. Biotechnol. 93: 707–22.

Wu, C., J. Zhang, G. Du and J. Chen. 2013. Aspartate protects *Lactobacillus casei* against acid stress. Appl. Microbiol. Biotechnol. 97: 4083–93.

Wu, C., J. Huang and R. Zhou. 2014. Progress in engineering acid stress resistance of lactic acid bacteria. Appl. Microbiol. Biotechnol. 98: 1055–1063.

Wu, R., W. Zhang, T. Sun, J. Wu, X. Yue, H. Meng and H. Zhang. 2011. Proteomic analysis of responses of a new probiotic bacterium *Lactobacillus casei* Zhang to low acid stress. Int. J. Food Microbiol. 147: 181–87.

Xiong, T., Q. Guan, S. Song, M. Hao and M. Xie. 2012. Dynamic changes of lactic acid bacteria flora during Chinese sauerkraut fermentation. Food Control 26: 178–81.

Yost, C.K. and F.M. Nattress. 2000. The use of multiplex PCR reactions to characterize populations of lactic acid bacteria associated with meat spoilage. Lett. Appl. Microbiol. 31: 129–33.

Zavaglia, A.G., E.A. Disalvo and G.L. De Antoni. 2000. Fatty acid composition and freeze–thaw resistance in lactobacilli. J. Dairy Res. 67: 241–47.

Zhang, J., C. Wu, G. Du and J. Chen. 2012. Enhanced acid tolerance in *Lactobacillus casei* by adaptive evolution and compared stress response during acid stress. Biotechnol. Bioprocess Eng. 17: 283–89.

Zhu, Y., Y. Zhang and Y. Li. 2009. Understanding the industrial application potential of lactic acid bacteria through genomics. Appl. Microbiol. Biotechnol. 83: 597–610.

10

Stress Response in Yeasts Used for Food Production

Helena Orozco,[2] Emilia Matallana[1] and Agustín Aranda[1,]*

1. Introduction

The ability to sense adverse conditions, and rearrange their metabolism and gene expression profile accordingly, are key factors for the success of yeasts as biotechnological tools. Fast adaptation to a changing environment enables yeast to proliferate under a range of harsh industrial growth conditions without incurring in a high negative impact on their function which, in the field of fermented foods, consists of metabolizing substrates, mainly sugars, into products, mainly ethanol and carbon dioxide. Stress response systems are interwoven with nutrient sensing pathways, as expected for a single-cell organism with a free way of life. In this chapter, we describe the molecular mechanisms involved in stress response in the most used yeast in the food industry, baker's yeast *Saccharomyces cerevisiae*. As this is also a model organism in molecular biology, information on the mechanisms that control stress response is plentiful and, as we will see, is key in understanding yeast performance under biotechnological conditions. An increasing number of global analyses are being done at the transcriptomic, proteomic and metabolomic levels during industrial processes. These analyses are providing a clearer picture of the molecular stress response during biotechnological uses of yeast. We now go on to analyze winemaking as our model case of an industrial process using yeast. We describe the stress conditions that wine yeasts face, from biomass production to grape juice fermentation, in order to then analyze the mechanisms described for stress response under these different conditions and how they can be biotechnologically improved. Finally, we describe the role of

[1] Institute for Integrative Systems Biology (UV-CSIC). Paterna, Spain.
[2] Institute of Agrochemistry and Food Technology (IATA-CSIC). Paterna, Spain.
Postal address: I2SysBio, Parc Cientific Universitat de València. Av. Agustín Escardino 9 Paterna 46980 Spain.
Email: helena.orozco@uv.es; emilia.matallana@uv.es
* Corresponding author: agustin.aranda@csic.es

yeasts in beer, bread and sake manufacturing by remarking on the factors that are unique to each of these processes and that could be improved.

2. Yeast Stress Response Mechanisms

The natural environment in which the yeast *S. cerevisiae* typically lives is usually adverse and changing, thus, survival requires a strong ability for stress response and adaptation. The yeast stress response is a multilayered process that relies on multiple pathways, which have been extensively studied in laboratory strains (Hohmann and Mager 2003). These pathways regulate gene expression, which acts at many levels, including the regulation of gene transcription, mRNA processing and translation, and allows the reorganization of protein synthesis. Therefore, a connection of nuclear events, such as transcriptional initiation and elongation and mRNA capping, splicing and export, to cytoplasmic events, such as protein translation, subcellular localization and degradation, becomes increasingly apparent (Das et al. 2017).

A crosstalk exists between different pathways, and also between the core stress response and nutrient sensing pathways (Conrad et al. 2014, Rodkaer and Faergeman 2014). This is mainly due to the fact that stress response is incompatible with growth, so nutrient signaling pathways are also sensitive to specific stress. The fine regulation of stress response and nutrient scarcity is critical in ensuring cell survival. When nutrients are present, the protein kinase A (PKA) and TORC1 pathways are active and promote both cell growth and division. PKA senses abundance of sugars and TORC1 is mainly devoted to signaling when nitrogen sources are abundant. When nutrients are scarce, survival is an essential feature and these and other pathways,such as AMPK kinase Snf1p, that promote the use of non-fermenting carbon sources, and the General Amino Acid Control kinase Gcn2p, rearrange the exit of yeast cells from the mitotic cell cycle to enter the stationary phase in order to ensure longevity (Winderickx et al. 2003, De Virgilio 2012).

Yeast cells can attain distinct quiescent states, and each one allows survival under the condition of a particular nutrient deprivation. The impact of starvation on stress response by signaling pathways is channeled mainly through kinase Rim15p, the main intermediate component in PKA and TORC1, which activates general stress Msn2/4p transcription factors that stimulate postdiauxic growth and stress-response gene expression. These factors bind to a DNA sequence element, STRE (Stress Response Element), which is present in most of the gene promoters induced by stress. PKA also regulates Msn2/4p directly by controlling their nuclear localization. Heat shock factor Hsf1p is an essential protein that plays a central role in the transcriptional response to stress, including oxidative and nutrient starvation, and is similarly regulated in order to control the expression of heat shock proteins (De Virgilio 2012). The major effector of TORC1 is GATA transcription factors Gln3p and Gat1p, which regulate a large number of genes devoted to the assimilation of non-preferred nitrogen sources. During starvation, transcription factor Gcn4p, devoted to amino acid biosynthesis, is also active through the activity of protein kinase Gcn2p, and is also required in order to regulate the expression of oxidative stress response genes (Mascarenhas et al. 2008). Another important nutrient signaling kinase that regulates stress response is the effector of TORC1, protein kinase Sch9p, which is dephosphorylated and activated in response to the starvation of carbon, nitrogen, phosphate or specific amino acid, which thus promotes activation of stress response via cytosolic localization by the phosphorylation of Rim15p and the subsequent activation of Msn2/4p. PKA and TORC1 regulation converge on Sch9p activation in order to control autophagy, a cellular recycling process by which cytosolic components are targeted for degradation under stress conditions in order to maintain cellular homeostasis (Klionsky et al. 2016).

Nutrient signaling pathways are, therefore, lifespan regulators and starvation signals are integrated into stress response by the signals transmitted by the negative regulators of these pathways, such as Snf1p, Rim15p and Sch9p (Zhang and Cao 2017). There are two paradigms in *S. cerevisiae* longevity: Chronological lifespan (CLS), which determines the mean and maximum survival of quiescence cells, and replicative lifespan (RLS), which measures the potential of mother cells to produce daughter cells (Longo et al. 2012). The starvation signals transmitted by the negative regulators of TOR and PKA are integrated so as to enable metabolic reprogramming and acquisition of stress resistance, which support the notion that metabolic reprogramming flexibility and the balance between stress resistance and energetics are key determinants of lifespan extension. However, the coordination between metabolic adaptation and stress response remains poorly understood.

Although many genes change expression following environmental stress, only a subset of these genes is regulated by only one transcriptional program. Regulation of gene expression also takes place by regulating mRNA processing, and export from the nucleus, translation and modulating pools of cytoplasmatic RNAs (Bond 2006). Many transcripts accumulate specifically during stress and quiescence in P-bodies, which contain the mRNA decay factors, or stress granules (SG) that contain translation initiation components. A transition between these aggregates and polysomes exists and, depending on the transcripts, they are degraded or activated to translation. Inhibition of translation by stress leads to the output of mRNAs of polysomes to P-bodies. The mRNAs that form complexes with mRNA binding proteins and translation initiation factors, and which accumulate into stress granules, could be re-routed to polysomes so that they may be translated (Decker and Parker 2012). Translation, decay and storage in the cytoplasmic foci of transcripts is, however, a quick and efficient way to regulate gene expression to exit from quiescence. The inhibition of both global translation and gene reprogramming, which involve the translation of specific proteins, plays an important role in adaptation to environmental changes (Simpson and Ashe 2012). The translation initiation process is mediated by conserved initiation factors (eIFs) and involves protein interactions, activities and stabilities. During the translational adaptation to amino acid changes, General Amino Acid Control (GAAC) is particularly important. It responds to starvation and leads to a global inhibition of translation initiation. However, Gcn4p transcription factor translation is up-regulated, which leads to the transcriptional induction of amino acid biosynthesis. An increase in uncharged tRNAs following amino acid depletion activates protein kinase Gcn2p, which phosphorylates eIF2α and inhibits global protein synthesis. Glucose depletion also causes the rapid inhibition of translation initiation and decreased polysomes, but the mechanism is independent of eIF2α phosphorylation, and the primary cause of this dramatic inhibition remains largely unknown.

Specific stress responses sometimes rely on specific signaling pathways but, as indicated above, mostly regulators are intertwined. Besides starvation, the fundamental stress experienced by yeast cells is temperature. Optimal *S. cerevisiae* growth lies between 25°C and 30°C, and temperatures that exceed 36°C trigger a protective transcriptional program called heat shock response (HSR), in which the major players are transcription factors Hst1p and Msn2/4p (Morano et al. 2012). Furthermore, relevant post-trancriptional regulation takes place by blocking mRNA transport from the nucleus and redirecting cytoplasmic mRNAs of non-heat shock proteins away from ribosomes into SGs and P-bodies. Remarkably, it has been found that TORC1 is sequestered into SGs during heat stress, when TORC1 and Sch9 remain inactive (Takahara and Maeda 2012). During HSR, the abundance of heat shock proteins (HSPs) increases because many of them function as molecular chaperones, which are required for survival after thermal stress. HSPs protect the cell by participating in protein translation and nascent chain folding and are required in order to repair and recover damaged proteins, or to protect them from aggregation after thermal stress (Albanese et al. 2006).

Oxidative stress is another important consequence of cell metabolism. All organisms are exposed to reactive oxygen species (ROS) during normal aerobic metabolism or from environmental insults. These ROS cause wide-ranging damage to macromolecules and eventually lead to cell death. To protect against oxidant damage, yeast cells contain effective defense mechanisms, including antioxidant enzymes and free radical scavengers. Catalases degrade H_2O_2, superoxide dismutases transform the superoxide anion, and various peroxiredoxins degrade a variety of peroxides (Ayer et al. 2014, Herrero et al. 2008). Apart from eliminating oxidative insult, oxidative stress response systems control the redox status of cells by preventing and repairing damage of molecules prone to oxidation, such as the thiol residues of proteins. There are two basic redox-controlling systems: One is based on the tripeptide glutathione, and the other on small proteins called thioredoxins (Ayer et al. 2014). The final reducing power for these systems comes from NAPDH, obtained through metabolic activity, mainly via the pentose phosphate pathway. The transcription of all these genes is controlled byspecific transcription factors Yap1p and Skn7p, which are prone to the redox changes that control their activity. From the post transcriptional regulation perspective, exposure of yeast cells to hydrogen peroxide results in the rapid reversible inhibition of protein synthesis, and this inhibition is mediated by the Gcn2p protein kinase which phosphorylates eIF2α, blocking initiation of translation (Shenton et al. 2006). Inhibition of translation elongation or termination also takes place in response to oxidative stress, independently of Gcn2p, but the mechanisms remain unknown. The effect of oxidative stress on yeast longevity has been extensively studied (Nystrom et al. 2012). Glucose starvation causes robust resistance to oxidative stress and the main cytosolic peroxiredoxin Tsa1p and sulfiredoxin Srx1p, which reduces hyperoxidized Tsa1p, are key players in this resistance. Both proteins are necessary to elicit the life span extension caused by caloric restriction (Molin et al. 2011). Caloric restriction counteracts the hyperoxidation of Tsa1p during aging by increasing PKA and the Gcn2p-dependent translation of the *SRX1* transcript. A new role in protein aggregation of peroxiredoxin Tsa1p has also been described (Hanzen et al. 2016). Tsa1p facilitates the binding of chaperones Hsp70/Hsp104 to damaged proteins formed upon aging through a peroxide redox switch of this peroxiredoxin, and the formed aggregates require the reduction of hyperoxidized Tsa1p by Srx1p.

Osmotic stress is also typically faced by yeast during its lifetime due to variations in environmental water activity. *S. cerevisiae* cells have developed mechanisms that allow them to adapt to high external osmolarity by a process which involves sensing osmotic changes and triggering appropriate cellular responses that aim to maintain cellular activity. Accumulation of a chemically inert osmolyte, such as glycerol, plays a central role in this osmoadaptation (Hohmann 2015). The main osmotic stress signaling is controlled by the HOG Mitogen-Activated Protein Kinase (MAPK) pathway. Hog1 kinase activates the Hot1p transcription factor which, in turn, activates glycerol synthesis from glycolytic intermediate dihydroxyacetone phosphate by up-regulating glycerol-3-phosphate dehydrogenase (*GPD1/2*) and glycerol phosphatase (*GPP1/2*) genes. It also induces a pump called Slt1p, which introduces glycerol from the environment. Hog1p is a master regulator of the massive transcriptional reprogramming that occurs upon osmotic stress, and regulates multiple levels of gene expression (de Nadal and Posas 2015). Hog1p regulates the initiation of transcription by a direct phosphorylation of specific osmostress transcription factors, and by the recruitment of RNA polymerase II and coactivators to osmoresponsive promoters. Postranscriptional modulation of osmogenes by regulating mRNA stability, nuclear export and translation is also important.

All these pathways work in a coordinated fashion and stress response genes are usually regulated by more than one of the above-mentioned transcription factors. This is the molecular cause of cross protection: Previous exposure to a type of stress protects against another type of environmental insult. The cores of stress response genes are labeled as Environmental Stress Responses (ESR), and are a set of around 300 genes whose expression is induced by heat shock, oxidative or reductive stress, osmotic shock, nutrient starvation, DNA damage and extreme pHs (Causton et al. 2001, Gasch et al. 2000). Conversely, around 600 genes are repressed by stress, mainly those devoted to growth and protein synthesis.

3. Winemaking Procedures

The alcoholic fermentation of grape juice sugars into ethanol and CO_2 is the defining process of winemaking (Ribéreau-Gayon et al. 2006). Maturation of grapes produces a juice that contains large amounts of free monosaccharides, a mixture of glucose and fructose at an equimolar ratio, with very low levels of sucrose or other sugars. Alcoholic fermentation is carried out by yeasts that can transform the glycolytic pyruvate into acetaldehyde and CO_2. Strong alcohol dehydrogenases produce ethanol as a final product (Carrascosa Santiago et al. 2011). Some yeasts, such as the genus *Saccharomyces*, are able to perform alcoholic fermentation, even in the presence of oxygen. These Crabree-positive yeasts (Piskur et al. 2006) are the best fermenters as they use a make-accumulate-consume metabolism: Due to their fast glycolytic metabolism, they produce large amounts of ethanol, which accumulates in the medium by whipping off bacteria and competing yeasts. When sugars are finished, they can consume ethanol by respiration as they are facultative anaerobic organisms. Given their specialization to the fermentative niche, *Saccharomyces* species are not found among the epiphytic yeasts present on the surface of grapes, where it is common to find aerobic yeasts, such as species of genera *Hanseniaspora* (*Kloeckera*), *Candida*, *Pichia* and *Hansenula* (Jackson 2000). *Saccharomyces* is easily found in winery equipment, which seems to be the source of inoculation from one vintage to the next. Therefore at the very beginning of fermentation, many yeasts are present in grape juice, and it is the particular ability of *Saccharomyces*, mainly the species *S. cerevisiae*, to adapt to the harsh wine fermentation environment which allows this yeast to outcompete others and to be basically the only one present in central and final fermentation stages by carrying out the bulk of sugar transformation and producing most of the ethanol in the final product. This is why the common practice today involves inoculating juice with a commercial starter of a well-characterized strain with good fermentative behavior and reliable organoleptic properties, generally in the form of active dry yeast (ADY) (Pérez-Torrado et al. 2015). Therefore, the ability of yeasts to sense, react and adapt their physiology to the sequential stress conditions that yeast cells encounter during their biotechnological uses is a key factor for their industrial fitness and success (Fig. 1). Hence increasing stress tolerance is a suitable way of improving yeast industrial performance, which has been extensively studied in *S. cerevisiae* (Pretorius 2000). However, all the yeasts present in grape juice leave their metabolic imprint on the final product, and as the species of yeast present in grapes and their ratios vary from vintage to vintage, and at each geographical location, they contribute to the particular organoleptic character of each wine (Fleet 1993, Jolly et al. 2014).

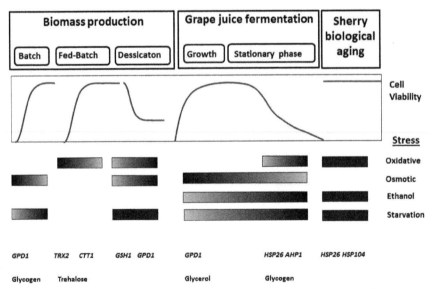

Fig. 1. Stress conditions and stress markers during winemaking. Winemaking steps, from biomass production, grape juice fermentation and Sherry aging, are indicated. The growth and death profiles during the different steps are shown, and the different stress conditions are indicated with bars of different intensities. The genes relevant to stress tolerance that can be used as stress markers and the main protective metabolites are shown. ADY: Active Dry Yeast.

4. Stress Conditions During Biomass Propagation, Drying and Grape Juice Fermentation

This section describes the stress conditions that yeasts face in each step of their industrial uses in chronological order as a guide to understand yeast uses in the winemaking industry (Fig. 1). As mentioned before, modern winemaking relies on the use of commercial starters. As winemaking is a seasonal event, the optimal way to prepare them is in the form of Active Dry Yeast (ADY). Yeasts are propagated first in batch, and then during scaled-up fed-batch growth in fermenters by using diluted cane or beet molasses as a growth substrate (Pérez-Torrado et al. 2015). During growth in batches, the main harmful condition is the hyperosmotic stress caused by high sucrose concentrations in molasses as temperature and pH are controlled in the process. As a respiratory metabolism is later imposed in the fed-batch stage, oxidative stress arises, mainly during the transition from fermentative to oxidative growth (Perez-Torrado et al. 2005). Regarding nutritional stresses, although sugars are plenty, nitrogen and vitamins are scarce, so supplementation with a nitrogen source is usually required. When the fed-batch stage ends, a period without feeding drives cells to enter the stationary phase due to carbon starvation, and to accumulate reserve carbohydrates, trehalose and glycogen, which help yeasts to face upcoming dehydration.

Thermal stress arises as the drying procedure is performed on a hot air bed, but does not exceed 41°C. The sudden loss of water to 8% of the initial moisture in the final product causes strong hyperosmotic shock and also oxidative damage (Pérez-Torrado et al. 2015). The increase in free oxygen radicals caused by water loss maybe due to mitochondrial membrane damage, which affects electron traffic. Yeast starter rehydration for must inoculation causes the opposite effect in osmolarity terms, a hypoosmotic shock caused by the sudden influx of water with just mild heat stress as rehydration is carried out at 37°C (Novo et al. 2003, Rossignol et al. 2006).

The stress conditions that yeast cells face during grape juice fermentation vary throughout the process, together with the chemical transformation of the substrate. Rehydrated yeast cells in inoculated wine fermentations, or natural yeasts from the winery environment in spontaneous fermentations, face a sudden rise in osmotic pressure due to the high sugar concentration in grape must, usually 200–250 g/L (Fleet 1993). pH is relatively low (3.0–3.5), but *S. cerevisiae* is well adapted to this range. However, sulfur dioxide toxicity increases at a low pH, and this chemical is usually added in order to prevent oxidation and microbial spoilage during winemaking. All these conditions play heavily against bacteria and filamentous fungi that fest on grapes. Thus, from the very beginning, fermentation is limited to a succession of yeasts. In modern winemaking, temperature is controlled so the heat produced by fermentation does not pose a huge threat to yeast viability. However in some wines, fermentation temperatures below 15°C are used and a cold response may slow down, or even stop, fermentation. When fermentation proceeds, dissolved oxygen is consumed by microorganisms or reacts against the different chemicals of grape by creating a low oxygen condition that prevents the growth of the more aerobic, apiculate yeasts present on the surface of grapes. Cells stop dividing when sugars are plenty and enter a quiescent state, similarly to the stationary phase observed under laboratory conditions (De Virgilio 2012), but with an active sugars fermentative metabolism, and without both a diauxic shift and respiratory growth on glycogen reserves. The limiting factor for yeast growth in winemaking is nitrogen availability as grape musts are poor in nitrogen, which may lead to stuck and sluggish fermentations (Bisson 1999). To prevent this situation, an inorganic source of nitrogen is often added, usually in the form of diammonium phosphate. Thus, phosphate is also added to improve the glycolysis rate (Ribéreau-Gayon et al. 2006). *Saccharomyces* yeasts prevail during ethanol production. As a Crabtree-positive yeast, ethanol accumulates, and is not consumed, in the presence of sugar, and its toxicity is the most relevant stress at the end of fermentation, together with nutrient starvation. Under these conditions, cells start aging and dying. As previously mentioned, measuring longevity in a non-dividing state is called chronological aging (Longo et al. 2012). Two carbon metabolites accumulated by fermentation (i.e., ethanol, acetic acid and acetaldehyde) promote aging (Orozco et al. 2012a), as does oxidative stress (Orozco et al. 2012b), but this is no obvious reason when mostly considering a fermentative environment.

As traditional fermentations rely on the yeasts present in the cellar to be inoculated by chance the following year (Jackson 2000), this implies that some yeasts cells which enter the quiescent state when nutrients are exhausted must remain in a dormant, but viable, state for a whole year, and must tolerate seasonal temperature changes. Alternatively, they can be kept in some reservoirs in the wild, such as the gut of certain insects, which may explain the transfer from one cellar to another. The *S. cerevisiae* strains used in winemaking have been found in social wasp guts, and these fungal cells are passed to the progeny (Stefanini et al. 2012). Obviously, understanding the physiology of the cells left out in the cellar or in the digestive tract is difficult, but studies with laboratory strains in the stationary phase could allow us to gain some insights (Herman 2002, De Virgilio 2012). As humidity may be lost from the cellar when the vintage is over, it is not far-fetched to assume that yeast cells must be dry at some cellar locations. Hence, this ability to resist drying and rehydration cycles could have been selected in wine yeast to, thus, enable them to be used as ADY.

Some wines undergo post-fermentative processes, where specialized yeasts face specific stress conditions, which are usually very harsh. The process of preparing sparkling wine by the champanoise method involves a second fermentation inside the bottle after adding sucrose. This fermentation in a confined space causes CO_2 pressure to increase, a further rise in ethanol levels, and all this in a nutrient-poor environment. Under these conditions, cell lysis contributes to the final organoleptic properties of wine (Jackson 2000, Cebollero and Gonzalez 2007). The preparation of Sherry wines of the *fino* variety involves biological aging (Angeles Pozo-Bayon and Victoria Moreno-Arribas 2011). After regular white wine fermentation, ethanol is added up to 15–15.5% and a biofilm (vellum) of a specific kind of yeasts (flor yeasts) forms on the surface (Legras et al. 2016). The floating ability of flor yeast allows it to reach the air and to adopt a respiratory metabolism in order to obtain energy through ethanol consumption (Alexandre 2013). This metabolic activity leads to the accumulation of acetaldehyde, a specific stressor of this particular class of wines (Aranda et al. 2002). Therefore, the characteristic stress marks in the biological aging of Sherry wines are high ethanol and acetaldehyde concentrations, together with poor nutrients and high oxygen concentrations.

Many of the described stress conditions, which occur during industrial uses of wine yeasts, are related to or caused by their metabolic activity. Therefore, they can be considered metabolic stress conditions. Excess sugars cause hyperosmotic shock, which makes the main carbon source for yeast growth a stressor. Ethanol and acetaldehyde are toxic metabolic products caused, respectively, by sugar fermentation and ethanol respiration which, at low doses, are harmless metabolites, but become stressors. Reactive oxygen species (ROS) are produced mainly in mitochondria during normal respiratory metabolism, and also due to the imbalances that take place during metabolic transitions from fermentation to respiration, which cause oxidative stress. Therefore, as yeast is a single-celled organism which lives while remaining in contact with the environment, stress response is not easily distinguishable from metabolic adaptation to nutrients, and the pathways that deal with stress are tightly controlled by nutrient availability, as already indicated.

5. Mechanisms of Stress Response During Wine Making: Tools to Improve it

Stress tolerance is a positive trait when selecting wine yeasts from the cellar to be used as biotechnological tools as their success during propagation and fermentation relies on their ability to cope with harsh conditions. It has been demonstrated that high tolerance to several stress conditions, relevant for winemaking (e.g., high osmotic pressure and ethanol), and good fermentative capacity correlate (Ivorra et al. 1999). Therefore, stress tolerance is a good criterion for selecting new yeast strains that may prove to be enologically useful (Zuzuarregui and del Olmo 2004). When competition is set between different strains, even from the same species, the most tolerant yeasts adapt better to their environment and thrive in it. For instance, for sampling during the biological aging that gives Sherry wines, the more prevalent genotypes in the different stages ended up being the strains that better tolerate ethanol and acetaldehyde, which are the most defining stress conditions of this process (Aranda et al. 2002).

The following sections describe the specific stress response in wine strains of the yeast *Saccharomyces cerevisiae* mainly, as this is the most common yeast species in winemaking, although knowledge on other minority species is growing. The first section describes stress pathways that have been reported in laboratory strains of *S. cerevisiae* grown in laboratory media. However, commercial strains display some noteworthy genetic differences when compared to laboratory strains. Laboratory strains have been genetically modified to be heterotallic, so products of sporulation do not mate, while wild yeasts are homotallic (Carrascosa Santiago et al. 2011). However, wine yeasts have low-frequency sporulation, even though they are roughly diploid. In some cases this is due to the presence of aneuploidies. Industrial strains obviously do not have artificially produced auxotrophies, so they are prototrophic. Commercial wine strains are genetically very similar, but genome sequencing reveals differences when compared to reference strains. These differences affect discrete genomic regions and rearrangements that contain genes which confer biotechnological advantages, but conserve basic genetic information (Borneman et al. 2016, Novo et al. 2009). As described below, molecular studies of wine yeast under biotechnological conditions have proven to be in accordance mostly with the expected behavior observed in laboratory strains, so basics systems apply. Nevertheless, some minor genetic variations may explain different responses to specific stresses, e.g., the differences found in the response to weak acids between wine yeast strains (Brion et al. 2013). Next, we focus on the sequential events that a wine yeast must go through.

a) Biomass Propagation and Drying

Although ADY production is a well-established industry, yeast behavior during the process is much less characterized at the molecular level than during alcoholic fermentation. The first molecular studies on stress response during biomass propagation from molasses in fermenters, in both the batch and fed-batch stages, analyzed selected genes that could serve as markers of specific stress insults (Perez-Torrado et al. 2005). It was hardly surprising that the typical osmogene *GDP1*, which codes for enzyme glycerol-3-P dehydrogenase, the first step in glycerol synthesis, was induced at the beginning of the process, when sugars are plentiful. However, the induction of cytosolic thioredoxin *TRX2*, a typical oxidative stress gene, that took place also in the batch stage, suggested unexpected oxidative stress at the point of transition from fermentation to respiration, during the diauxic shift. A detailed transcriptomic analysis during batch and fed-batch growth in molasses has confirmed the induction of many oxidative stress genes (thioredoxins, peroxiredoxins, glutaredoxins, etc.), at this point (Gomez-Pastor et al. 2010a). A similar pattern of induction of the proteins involved in dealing with oxidative insult has been observed in a proteomic analysis (Gomez-Pastor et al. 2010a), and indicated that regulation probably happens at the transcriptional initiation level. Therefore, oxidative stress seems to be the main negative condition that operates against the yeast cells growing in industrial biomass propagation. To reinforce this, the overexpression of the aforementioned *TRX2* gene leads to a wine strain with increased biomass production (Gomez-Pastor et al. 2010b) as it protects many proteins involved in alcoholic fermentation from oxidation, for instance, the main alcohol dehydrogenase Adh1p (Gomez-Pastor et al. 2012).

Oxidative stress has been related to dehydration in many biological systems, including *S. cerevisiae* (Franca et al. 2007). Markers of oxidative damage, e.g., lipid peroxidation and glutathione levels, increase while wine yeasts are drying, and induction of the genes that perform an antioxidant function, such as thioredoxin reductase *TRR1* and glutaredoxin *GRX5*, is observed (Garre et al. 2010). Good performance during wine yeast drying has been linked to high levels of protective disaccharide trehalose, and to strong catalase and glutathione reductase activities (Gamero-Sandemetrio et al. 2014). As oxidative stress is relevant for ADY production, ROS reduction by using antioxidants, or a natural source of them (e.g., argan oil), has proved useful in improving biomass growth and in enhancing its fermentative capacity after drying by reducing markers of oxidative damage (Gamero-Sandemetrio et al. 2015). Further evidence that increased oxidative stress tolerance may improve dehydration comes from specific screenings for gene functions which, in overexpression, can improve dehydration tolerance where genes such as glutaredoxin *GRX5* (Rodriguez-Porrata et al. 2012) and hydrophilin *SIP18* (Lopez-Martinez et al. 2013) have been found. As oxidative stress is relevant for biomass production, it is worth looking at the main source of ROS, this being mitochondrial respiration. Indeed, some variation in this organelle exists because while making hybrids between two wine strains of genera *Saccharomyves, S. cerevisiae* and *S. uvarum*, the mitochondria from *S. cerevisiae* confer better stress tolerance and provide better protection during dehydration (Picazo et al. 2015a).

As previously mentioned, yeast species other than *Saccharomyces* contribute to the full organoleptic properties of wine, hence, the clearly commercial interest in designing mixed starters by combining a yeast with good fermentation capacity, such as *S. cerevisiae*, with other yeasts that contribute some positive metabolites (Jolly et al. 2014). Unfortunately, these yeasts are usually more stress-sensitive and do not perform well under industrial conditions to produce ADY, which have been designed to match the requirements of *S. cerevisiae* (Pereira Ede et al. 2003, Pérez-Torrado et al. 2015). Their biomass yield after growth in molasses is lower, and, even more importantly, they show a vast loss of viability and fermentative capacity when dehydrated. Even though their genomes could be unknown, several biochemical parameters, such as antioxidant enzymes that detoxify ROS, like catalase and superoxide dismutase, can be measured in them all, and have been shown to be good markers of stress tolerance that can be used as good selection criteria (Gamero-Sandemetrio et al. 2013).

b) Rehydration

Rehydration is a sudden event that triggers a marked change in gene expression, as transcriptomic analyses show (Rossignol et al. 2006, Novo et al. 2007). The physiology of yeasts changes when warm water is added, and the cell undergoes a quick re-programming of its transcription and mRNA stability. When taking the ADY as a reference, the transcripts from general stress response genes (e.g., those that code for heat shock proteins), and also specific oxidative stress genes, are down-regulated very quickly after rehydration (Rossignol et al. 2006). These data reinforce the notion that dehydration triggers a very strong stress response due to the very damaging effects of water loss, which are much more stressful than rehydration conditions. However, some specific genes are induced after rehydration, such as the genes that respond to acid stress, like pump *PDR12* (which deals with transporting organic acids), as well as the genes involved in proton transport, like *PMA1* (a H^+ pump). This indicates that acid imbalances are the main stress condition to occur during quick rehydration, probably due to damage caused to membranes, which requires restoration of the ion balance by pumping out any excess acid. The down-regulation of the general stress response may be necessary to resume growth, and may be regulated by the pathways that control general stress response transcription factors Msn2/4p by PKA. Rehydration may also be performed by adding glucose to the rehydration medium in order to help induce the genes involved in hyperosmotic stress (see below) (Jimenez-Marti et al. 2011a).

c) Grape Juice Fermentation

The most relevant molecular mechanism to deal with the hyperosmotic shock that yeasts face as they are inside grape juice, is glycerol synthesis, as it is the main compatible osmolyte. The gene that encodes enzyme glycerol-3P dehydrogenase *GPD1* is a well-characterized osmogene. It is activated within the first hour after inoculation from a preculture in the stationary phase (Perez-Torrado et al. 2002a). Regarding the general stress response, induction of the disaggregase *HSP104* gene takes about 7 hours and LEA-like plasma gene *HSP12* expression does not change. If gene expression is analyzed from an ADY inoculum, *GPD1* is not induced at the beginning of fermentation (Rossignol et al. 2006), and the same happens for most general stress response genes. This indicates that stress conditions during drying and rehydration are severe, and that the whole stress response machinery is activated, which likely ensures the success of inoculated strains. A proteomic analysis carried out shortly after inoculation matched this profile, and showed that Gpd1p, together with other general stress proteins (Hsp12p again, but also Hsp60p and Tps1p, the enzyme involved in trehalose synthesis) and oxidative stress protectors (peroxiredoxin Tsa1p and superoxide dismutase Sod1p) were repressed (Salvado et al. 2008). These results also indicate that the physiological status of the starter is a determinant in stress response during vinification. The HOG pathway is activated by osmotic shock caused by high sugar levels, which occurs in winemaking (Jimenez-Marti et al. 2011b). The HOG pathway has been described as being active during grape juice fermentation and to control *GPD1* expression, but only partially (Remize et al. 2003). This suggests that there are additional pathways which control Gpd1p, and perhaps at more than one level, such as post-transcriptional mechanisms, e.g., the translation of Gpd1p, which also happens during winemaking through the action of mRNA binding protein Pub1p (Orozco et al. 2016).

Taking a broader look at the whole fermentation by transcriptomic analyses, it is seen that most stress response genes (the bulk of the ESR ones) are induced in later fermentation stages, when cells enter the stationary phase (Rossignol et al. 2003, Varela et al. 2005). Oxidative stress response is surprisingly high throughout fermentation, which indicates that this might be a relevant stress, even under low respiration and low oxygen conditions. The original definition of ESR genes is response to transient stress (Gasch et al. 2000). Yet stress conditions are sustained throughout fermentation, so the situation is not the same as for laboratory conditions, although ESR is induced in fermentation. A novel set of 223 additional genes is defined as a more specific fermentation stress response (FSR) (Marks et al. 2008). This new set contains canonical ESR genes, but 62% of them have not been related to stress response before. The genes involved in respiratory metabolism and gluconeogenesis are expressed during fermentation, despite high glucose concentrations. According to the authors, entry into the stationary phase is linked to ethanol stress, and not to nutrient depletion. However, nutrient starvation should be a factor in controlling the genes that belong to ESR as most of these genes are regulated by general stress transcription factors Msn2/4p, whose action is relieved when nutrients are scarce by stopping the PKA and TOR pathways (Rodkaer and Faergeman 2014). As a matter of fact, Msn2p overexpression in wine yeast improves stress response and increases fermentative rates (Cardona et al. 2007), which reinforce this point. When comparing the transcriptomes of different wine yeast strains, differences in gene expression can be linked to specific phenotypic differences, although no clear stress response pathways are differentially regulated (Rossouw and Bauer 2009).

Proteomic analyses have indicated that most heat shock proteins are repressed in later fermentation stages, which suggests a post-transcriptional regulatory mechanism in the stationary phase, maybe due to protein degradation or to lack of translation (Rossignol et al. 2009, Varela et al. 2005). However, some stress proteins, such as Hsp26p and peroxiredoxin Ahp1p, are induced at the protein level as they are transcriptionally up-regulated. Hsp26p, together with Hsp12p, is also induced at the protein level when low-temperature wine fermentation is carried out (Salvado et al.

2012). Therefore, chaperones like Hsp26p are good molecular markers that mark stress under many biotechnological conditions, both at the mRNA or protein level. Sulfur and glutathione metabolism at the proteome level have proven relevant to better adapt to low-temperature fermentations (Garcia-Rios et al. 2014).

Cells die when fermentation ends. Aging under these conditions is not a well-described event. Chronological longevity reflects the viability of a yeast culture in the quiescent state (Longo et al. 2012), as is most grape juice fermentation, which takes place without cell division. In some fermentations, cells start dying when fermentation is incomplete (Ribéreau-Gayon et al. 2006), but in other cases yeast only starts losing viability when sugars are depleted (Boulton 1996). Such aging is, therefore, biotechnologically relevant as it may alter fermentation rates. A good stress response delays the cellular aging process, while environmental insults, such as heat, low pH, and two carbon metabolites produced by fermentation (mainly ethanol, but also acetaldehyde and acetic acid), shorten the wine yeast life span (Orozco et al. 2012c). Comparing the transcriptomics of long- and short-lived commercial *S. cerevisiae* strains has indicated that oxidative stress response is necessary in order to achieve full longevity (Orozco et al. 2012b). Protection against ROS during high sugar fermentation relies on detoxifying enzymes, such as superoxide dismutase and protective disaccharide trehalose (Landolfo et al. 2008).

Starvation is an important stress condition that occurs at the end of fermentation, but is not just another abiotic stress in aging terms because it regulates nutrient-sensing pathways. Reducing nutrient intake without malnutrition is known as dietary restriction, and it extends life span, mainly by lowering the activity of nutrient-sensing pathways, such as TOR, and by increasing stress response (Longo et al. 2012). This is also the case during wine fermentation, where low nitrogen and TOR inhibition extend chronological life span (Orozco et al. 2012a, Picazo et al. 2015b). High PKA activity also shortens life spans in wine yeasts, probably due to its control of the general stress response (Orozco et al. 2012b). Nitrogen starvation triggers autophagy in order to recycle cellular components, which usually benefits longevity, but induction of autophagy is deleterious under winemaking conditions as a mutant with an autophagy block has extended chronological longevity (Orozco et al. 2012c).

An exogenous stress condition that takes place during wine fermentation is imposed by adding preservatives to the grape juice, such as sulfite, in order to prevent spoilage and oxidation (Ribéreau-Gayon et al. 2006). The evolutive adaptation of wine yeasts to this particular harmful agent is not based directly on gene expression regulation, but on chromosomal rearrangement, which creates a novel allele of sulfite efflux pump *SSU1* with a constitutively higher expression that is present only in *S. cerevisiae* wine strains (Perez-Ortin et al. 2002). Its expression in other wine yeasts, such as *S. uvarum*, increases their tolerance to this compound (Liu et al. 2017). Sulfur and adenine metabolism are also affected in response to this additive under winemaking conditions (Aranda et al. 2006) as these pathways may contribute to its assimilation.

d) Singular Wines

During the biological aging of fino sherry wines, yeast cells have to deal with high doses of ethanol and acetaldehyde, so the flor yeasts devoted to this process are adapted and respond with an increased expression of heat shock protein (*HSP*) genes, particularly *HSP26* and *HSP104*, whose expression correlates with ethanol tolerance in flor yeasts (Aranda et al. 2002). The strains with higher HSP induction are the most abundant ones in cellars. Acetaldehyde also brings about the induction of a similar set of *HSP*s, together with higher levels of pumps that export polyamines, such as TPO2, 3 and 4, and the activation of sulfur metabolism as acetaldehyde reacts with sulfur-containing products to render inert adducts (Aranda and del Olmo 2004). Overexpression of stress response genes, such as superoxide dismutases *SOD1* and *SOD2*, and of *HSP12*, improves vellum formation and cell viability in flor yeasts (Fierro-Risco et al. 2013). Proteomic analyses have detected an increase in heat shock proteins, like Hsp82p and Hsp104p, in flor yeasts under biofilm conditions (Moreno-García et al. 2017). This indicates that oxidative stress is highly relevant during this mainly aerobic process.

During sparkling wine second fermentations, nutritional stress conditions activate autophagy, which eventually leads to yeast cell death and lysis (Cebollero and Gonzalez 2006). Global transcriptomic analyses have indicated oxidative stress and cold stress response as the main gene expression responses to this particular environment (Penacho et al. 2012).

6. Genetic Improvement of Stress Tolerance in Wine Yeast

Given the impact of environmental challenges on technological yeast performance, improving stress tolerance has been an objective to optimize industrial yeasts (Pretorius 2000, Pretorius and Bauer 2002). The genetic improvement of wine yeast can be achieved by many different approaches: Clonal selection of variants, mutagenesis, directed evolution, hybridization, rare-mating, spheroplast fusion and genetic manipulation. Gene manipulation is the most specific one as it targets genes of known functions. For instance, the obvious way to increase osmotolerance is to increase the compatible osmolyte. To do so, it is clear that the best approach would be to increase the activity of glycerol 3-phosphate dehydrogenase, the first and limiting step in glycerol biosynthesis. Overexpression of *GPD1* produces larger amounts of glycerol, but also of acetic acid, which is not a desirable metabolite in wine (Michnick et al. 1997). Therefore, manipulations that affect stress tolerance may have metabolic consequences that need to be carefully monitored. Another obvious target to improve stress tolerance is to increase reserve and protective carbohydrates, trehalose and glycogen. Trehalose has proven fundamental for drying. A mild overexpression of its biosynthetic gene *TPS1*, which codes for trehalose-6-phosphate synthase, reduces ethanol. This indicates that the carbon flux is being redirected (Rossouw et al. 2013). Alternatively, trehalose accumulation may be increased by deleting trehalase genes, such as neutral trehalase gene *NTH1* (Perez-Torrado and Matallana 2015). Its deletion improves fermentative capacity after biomass production. Glycogen is the main reserve carbohydrate in *S. cerevisiae*, and its overproduction by up-regulating glycogen synthase enzyme *GSY2* improves biomass production (Perez-Torrado and Matallana 2015) and improves viability during sugar starvation under winemaking conditions (Perez-Torrado et al. 2002b).

A more direct approach involves the manipulation of stress response mechanisms. For instance, it is possible to improve ESR genes by increasing the amount of general stress response transcription factor *MSN2* (Cardona et al. 2007). However, its levels have to be carefully monitored as strong overexpression blocks growth with disappointing results. Nevertheless, regulated *MSN2* overexpression by a stationary phase-induced promoter, like *SPI1*, improves yeast performance during winemaking by increasing sugar consumption. Therefore, precise targeting in both the space and time of the stress response is essential for such manipulations to be successful. As metabolism and stress response are so intertwined, the mutations that produce less acetic acid (anundesirable flavor) happen to be located in oxidative stress response transcription factor *YAP1* (Cordente et al. 2013). Overexpression of cytosolic thioredoxin *TRX2* during yeast industrial propagation increases biomass yield by protecting the enzymes involved in fermentation, such as alcohol dehydrogenase, against oxidation (Gomez-Pastor et al. 2012).

7. Stress Tolerance and Longevity in Brewing Yeasts

Beer fermentation shares many points with wine fermentation as both are batch fermentations on a natural substrate, in this case wort, an aqueous extract of malted barley that contains hops, which is fermented by *Saccharomyces* to produce beer. The usual challenges met by brewing yeasts are changes in oxygen concentration, osmotic potential, pH, ethanol concentration, nutrient availability and temperature (Briggs 2004). These do not differ much from those faced during winemaking, but some differences are worth noting. The use of dry yeast is less common, and cells are usually propagated in vessels with wort under aerobic conditions through the inoculation of sterile air or oxygen. The cells prepared in this way are transferred to fermentation vessels, where fermentation under anaerobic conditions is carried out. As wort is less rich in sugars that grape juice, there is a shortage of carbon and other nutrients at the end of fermentation (Gibson et al. 2007). This triggers cells to enter a quiescent state, which is much more similar to a canonical stationary phase, as described under laboratory conditions, since fermentable sugars are low. When fermentation finishes, cells are stored cold (3–4°C) in beer before being used again (repitched). To do so, they have to be acid-washed at pH 2.2 in order to prevent bacterial spoilage. So these low temperatures and acid stress are trademarks of this fermentation process. Plasma membrane H⁺ATPase has been described as being essential for survival during acid washing, and trehalose, glycerol and fatty acid desaturation contribute to cold stress tolerance. Besides, as the biomass is reused, the replicative aging of cells may be a relevant issue as the number of generations that a particular strain performs under industrial conditions is higher than the single-batch fermentation that defines grape juice winemaking. This parameter varies among different brewery polyploid strains (Maskell et al. 2003), which makes it a factor that is worth taking into account when a strain is selected. As cells are decanted from the bottom of fermenters, a fraction of virgin cells is lost while collecting the biomass (Powell et al. 2000). Thus, after serial fermentations, the population tends to be old and performs worse. However, different studies claim that serial repitching has no impact on population age as cells with a large numbers of scars do not increase (Buhligen et al. 2014). This would indicate that rejuvenation processes take place, and these may depend on different brewing practices which, in turn, depend on wort aeration.

Oxidative stress is important through propagation and the first fermentation stages due to exposure to oxygen. The activities of catalases and superoxide dismutases

Plasma membrane H^+ATPase has been described as being essential for survival during acid washing

Sod1p and Sod2p of brewing yeasts change with oxygen exposure during fermentation (Clarkson et al. 1991), so yeasts are able to react to this stimulus. Extracellular thioredoxin is also higher when the amount of dissolved oxygen rises. However when oxidative parameters are carefully measured during propagation and fermentation, it is clear that antioxidants activities, such as catalase, and the transcription of oxidative response genes, peak at the beginning of the stationary phase when oxygen is low, which suggests a general stress response to growth-limiting conditions, even in the absence of oxygen (Gibson et al. 2008b). Yet when comparing the trasncriptomic analysis of serial fermentations, variation in oxygenation influences the oxidative stress gene expression, particularly those for mitochondrial functions (Bühligen et al. 2013). Good tolerance to oxidative stress has been rendered necessary in order to avoid respiratory incompetent petite cells from appearing when cells are exposed to oxygen (Gibson et al. 2008a). The petite ratio rises during chronological aging, but only when oxygen is present. The fate of antioxidant proteins in final fermentation stages, when lysis takes place, is more confusing. A proteomic analysis had indicated that superoxide dismutases decreases, but peroxiredoxin Tsa1p increases (Xu et al. 2014). In any case, the level of ROS lowers during autolysis, which means that oxidative damage does not cause this process.

8. Stress in Sake Fermentation

Sake is made from rice by parallel fermentations of the fungus *Aspergillus oryzae,* which produce saccharification enzymes, and the yeast *Saccharomyces cerevisiae,* which ferments them. The most relevant stress conditions during sake fermentation are similar to those observed in grape juice fermentation, e.g., anaerobic conditions, high ethanol concentrations ($\sim 20\%$) and low temperatures ($\sim 15°C$). As the levels of ethanol reached in this kind of fermentation exceed other beverages, ethanol tolerance has, therefore, been extensively studied in sake-producing strains. Amino acid recruitment and storage have proven positive for ethanol tolerance. For instance, engineering sake strains to accumulate proline by increasing its synthesis and blocking its degradation leads to better tolerance to ethanol (Takagi et al. 2005). A comparative transcriptomic analysis of sake and laboratory strains under ethanol stress has proven useful in identifying tryptophan biosynthesis as a pathway whose up-regulation leads to improved performance with ethanol challenge (Hirasawa et al. 2007).

However, a similar comparative transcriptomic analysis during sake fermentation has led to the identification of a defective *MSN4* gene in modern sake strains (Watanabe et al. 2011), and to defective ESR during fermentation which, in turn, favors the enhanced ethanol fermentation rate that is the trademark of these strains. However, the trade-off of this enhanced fermentative performance is a deficiency when entering the stationary phase after growth stops, compared to laboratory strains (Urbanczyk et al. 2011). This defect has been linked to a truncated version of protein kinase Rim15p in modern sake strains (Watanabe et al. 2012). This kinase orchestrates the changes that lead to entry in the quiescent state, including ethanol fermentation cessation. Its regulation under a gluconeogenic promoter is a way to enhance fermentative performance in non-sake yeasts (Watanabe et al. 2017)

Gene expression controlled by HSE is also impaired in sake strains due to the hyperphosphorylation of Heat Shock Factor Hsf1p (Noguchi et al. 2012). Sake and related rice wine yeast strains have also shown a high polymorphism of HOG pathway osmosensors Snl1p and Msb2p, which may explain the increased osmotic tolerance of the aforementioned strains (Li et al. 2013).

9. Baking Yeast and its Challenges

The process of making bread or leavening other kind of doughs imposes a very different environment on yeast compared to the aforementioned liquid fermentations (Shima and Takagi 2009) as they are performed on a moist solid substrate. Obviously, this exposes yeast cells to hyperosmotic stress (Randez-Gil et al. 2013). The production of a yeast starter as ADY or as instant dry yeast also implies osmotic stress, as seen for wine starters. In addition to oxidative stress and mitochondrial malfunction, which are common to all dehydration procedures, vacuolar acidification is also observed. Hence, the vacuolar H+-ATPase has been identified as a key factor in air-drying tolerance for baker's yeast (Shima et al. 2008). Small molecules, such as proline and trehalose, prevent damage after drying (Shima and Takagi 2009), similarly to the way that they protect against other stressful conditions (see above). Transcriptomic analyses of solid state fermentations have indicated that the transcription of bread strains changes at lower levels than wine and bioethanol strains when facing this new environment, which indicates better adaptation to this environment (Aslankoohi et al. 2013). The global analysis suggests the activation of HOG-regulated genes at the beginning of fermentation as a key adaptation feature. Increasing glycerol levels improves dough fermentation rates (Aslankoohi et al. 2015), despite the fact that it lowers CO_2 production, which indicates that stress protection can offer an advantage to the cells grown under these conditions. Overexpression of salt-dependent Crz1p transcription factor produces increased tolerance in sweet doughs, where osmotic stress is very high (Randez-Gil et al. 2013). The impact of oxygen on the bread-making procedure and its potential to generate harmful ROS have also been studied in detail (Decamps et al. 2016), although the solid matrix is not permeable to gases and retains large amounts of CO_2, which may also cause a particular stress condition.

The current trend in bakery is to use frozen doughs. Therefore, tolerance to freezing/thawing is an interesting trait that needs to be explored and improved (Randez-Gil et al. 2013, Loveday et al. 2012). As water is frozen, osmolarity increases and accumulation of protective molecules against drying, such as trehalose and proline, has been found to be useful to increase cryotolerance (Sasano et al. 2012a). Regarding the involved molecular mechanisms, PKA is detrimental to freezing as overexpressing *PDE2*, a phosphodiesterase that lowers cAMP, increases viability during freezing (Nakagawa et al. 2017). This canbe explained by the up-regulation of the general stress response; in fact, the overexpression of transcription factor Msn2p leads to increased tolerance to freezing (Sasano et al. 2012b). Another transcription factor that improves the fermentation of frozen doughs is Crz1p (Panadero et al. 2007), a transcription factor that was first described for its functions in saline stress defense. Once again, the machinery that deals with freezing is closely related to that involved in tolerance to high osmolarity. During thawing, oxidative stress seems more prevalent as ROS accumulate inside cells (Loveday et al. 2012).

Another strategy when using frozen doughs is to use alternative cryotolerant non-*Saccharomyces* yeasts, such as the strains of *Torulaspora delbrueckii* (Oda and

Tonomura 1993) and *Kluyveromyces thermotolerans* (Hino et al. 1990). In both cases, accumulation of trehalose seems to be the molecular cause behind their ability to resist freezing. This yeast metabolizes trehalose at low rates, possesses low invertase activity and has the ability to respond rapidly to osmotic stress, which are all key factors that explain its performance (Hernandez-Lopez et al. 2003). Although many yeast species can conferpositive aroma complexity to bread, it seems clear that *S. cerevisiae* and *T. delbrueckii* are the best adapted to these conditions and give a strong fermentation profile (Aslankoohi et al. 2016).

Concluding Remarks

It is a very exciting time for molecular studies of yeast stress response during yeast biotechnological processes. The amount of data produced by global analyses of transcriptomic, proteomic and metabolomics analyses of the industrial strains of yeast *S. cerevisiae* under biotechnological conditions is geometrically growing. The new genome sequencing and RNAseq methods are providing genetic information on an increasing number of strains and, what is more important, on different non-*Saccharomyces* species that can be useful for improving specific products that require specific interventions. Even the new metagenomics approach enables us to study the microbiota of complex systems, like grape juice fermentation, and is able to identify and quantify not only the different yeasts involved, but also the lactic acid bacteria that perform malolactic fermentation (Spano and Torriani 2016).

All this information is already being processed in an integrative manner using Systems Biology tools to provide information of organisms and processes as a whole. Meanwhile, Synthetic Biology uses new approaches in order to improve and better understand yeasts (Jagtap et al. 2017), like DNA editing by CRISPR-Cas, and genomic engineering, which is on its way to create a completely synthetic *S. cerevisiae* (Richardson et al. 2017). Recently a mushroom edited by CRISPR-Cas was approved by the FDA as a non-GMO since it contains no exogenous DNA (Waltz 2016). Hence, this is a tool that is potentially ready to be used in yeasts of biotechnological interest.

References

Albanese, V., A.Y. Yam, J. Baughman, C. Parnot and J. Frydman. 2006. Systems analyses reveal two chaperone networks with distinct functions in eukaryotic cells. Cell 124 (1): 75–88.

Alexandre, H. 2013. Flor yeasts of *Saccharomyces cerevisiae*: Their ecology, genetics and metabolism. Int. J. Food Microbiol. 167(2): 269–275.

Angeles Pozo-Bayon, M. and M. Victoria Moreno-Arribas. 2011. Sherry wines. Adv. Food Nutr. Res.63:17–40.

Aranda, A., A. Querol and M. del Olmo. 2002. Correlation between acetaldehyde and ethanol resistance and expression of HSP genes in yeast strains isolated during the biological aging of sherry wines. Arch. Microbiol. 177(4): 304–312.

Aranda, A. and M.L. del Olmo. 2004. Exposure of *Saccharomyces cerevisiae* to acetaldehyde induces sulfur amino acid metabolism and polyamine transporter genes, which depend on Met4p and Haa1p transcription factors, respectively. Appl. Environ. Microbiol. 70(4): 1913–1922.

Aranda, A., E. Jimenez-Marti, H. Orozco, E. Matallana and M. Del Olmo. 2006. Sulfur and adenine metabolisms are linked, and both modulate sulfite resistance in wine yeast. J. Agric. Food Chem. 54(16): 5839–5846.

Aslankoohi, E., B. Herrera-Malaver, M.N. Rezaei, J. Steensels, C.M. Courtin and K.J. Verstrepen. 2016. Non-conventional yeast strains increase the aroma complexity of bread. PLoS One 11(10): e0165126.

Aslankoohi, E., M.N. Rezaei, Y. Vervoort, C.M. Courtin and K.J. Verstrepen. 2015. Glycerol production by fermenting yeast cells is essential for optimal bread dough fermentation. PLoS One 10(3): e0119364.

Aslankoohi, E., B. Zhu, M.N. Rezaei, K. Voordeckers, D. De Maeyer, K. Marchal, E. Dornez, C.M. Courtin and K.J. Verstrepen. 2013. Dynamics of the *Saccharomyces cerevisiae* transcriptome during bread dough fermentation. Appl. Environ. Microbiol. 79(23): 7325–7333.

Ayer, A., C.W. Gourlay and I.W. Dawes. 2014. Cellular redox homeostasis, reactive oxygen species and replicative ageing in *Saccharomyces cerevisiae*. FEMS Yeast Res. 14(1): 60–72.

Bisson, L.F. 1999. Stuck and sluggish fermentations. Am. J. Enol. Vitic. 50: 107–119.

Bond, U. 2006. Stressed out! Effects of environmental stress on mRNA metabolism. FEMS Yeast Res. 6(2): 160–170.

Borneman, A.R., A.H. Forgan, R. Kolouchova, J.A. Fraser and S.A. Schmidt. 2016. Whole genome comparison reveals high levels of inbreeding and strain redundancy across the spectrum of commercial wine strains of *Saccharomyces cerevisiae*. G3 (Bethesda) 6(4): 957–971.

Boulton, Roger, B. 1996. Principles and practices of winemaking. Chapman & Hall, New York.

Briggs, D.E. 2004. Brewing: Science and practice.Woodhead Publishing Series in Food Science, Technology and Nutrition. Elsevier, Boca Raton, FL.

Brion, C., C. Ambroset, I. Sanchez, J.L. Legras and B. Blondin. 2013. Differential adaptation to multi-stressed conditions of wine fermentation revealed by variations in yeast regulatory networks. BMC Genomics 14: 681.

Buhligen, F., P. Lindner, I. Fetzer, F. Stahl, T. Scheper, H. Harms and S. Muller. 2014. Analysis of aging in lager brewing yeast during serial repitching. J. Biotechnol. 187: 60–70.

Bühligen, F., P. Rüdinger, I. Fetzer, F. Stahl, T. Scheper, H. Harms and S. Muller. 2013. Sustainability of industrial yeast serial repitching practice studied by gene expression and correlation analysis. J. Biotechnol. 168(4): 718–728.

Cardona, F., P. Carrasco, J.E. Perez-Ortin, M. del Olmo and A. Aranda. 2007. A novel approach for the improvement of stress resistance in wine yeasts. Int. J. Food Microbiol. 114(1): 83–91.

Carrascosa Santiago, A.V., R. Muñoz and R. González Garcia. 2011. Molecular wine microbiology. 1st. ed. Academic Press, Boston..

Causton, H.C., B. Ren, S.S. Koh, C.T. Harbison, E. Kanin, E.G. Jennings, T. I. Lee, H.L. True, E.S. Lander and R.A. Young. 2001. Remodeling of yeast genome expression in response to environmental changes. Mol. Biol. Cell. 12(2): 323–337.

Cebollero, E. and R. Gonzalez. 2006. Induction of autophagy by second-fermentation yeasts during elaboration of sparkling wines. Appl. Environ. Microbiol. 72(6): 4121–4127.

Cebollero, E. and R. Gonzalez. 2007. Autophagy: From basic research to its application in food biotechnology. Biotechnol. Adv. 25(4): 396–409.

Clarkson, S.P., P.J. Large, C.A. Boulton and C.W. Bamforth. 1991. Synthesis of superoxide dismutase, catalase and other enzymes and oxygen and superoxide toxicity during changes in oxygen concentration in cultures of brewing yeast. Yeast 7(2): 91–103.

Conrad, M., J. Schothorst, H.N. Kankipati, G. Van Zeebroeck, M. Rubio-Texeira and J.M. Thevelein. 2014. Nutrient sensing and signaling in the yeast *Saccharomyces cerevisiae*. FEMS Microbiol. Rev. 38(2): 254–299.

Cordente, A.G., G. Cordero-Bueso, I.S. Pretorius and C.D. Curtin. 2013. Novel wine yeast with mutations in YAP1 that produce less acetic acid during fermentation. FEMS Yeast Res. 13(1): 62–73.

Das, S., D. Sarkar and B. Das. 2017. The interplay between transcription and mRNA degradation in *Saccharomyces cerevisiae*. Microb. Cell 4(7): 212–228.

de Nadal, E. and F. Posas. 2015. Osmostress-induced gene expression: A model to understand how stress-activated protein kinases (SAPKs) regulate transcription. FEBS J. 282(17): 3275–3285.

De Virgilio, C. 2012. The essence of yeast quiescence. FEMS Microbiol. Rev. 36(2): 306–339.

Decamps, K., I.J. Joye, D.E. De Vos, C.M. Courtin and J.A. Delcour. 2016. Molecular oxygen and reactive oxygen species in bread-making processes: Scarce, but nevertheless important. Crit. Rev. Food Sci. Nutr. 56(5): 722–736.

Decker, C.J. and R. Parker. 2012. P-bodies and stress granules: Possible roles in the control of translation and mRNA degradation. Cold Spring Harb. Perspect. Biol. 4(9): a012286.

Fierro-Risco, J., A.M. Rincon, T. Benitez and A.C. Codon. 2013. Overexpression of stress-related genes enhances cell viability and velum formation in Sherry wine yeasts. Appl. Microbiol. Biotechnol. 97(15): 6867–6881.

Fleet, G.H. 1993. Wine microbiology and biotechnology. CRC Science, Boca Raton, FL. USA.

Franca, M.B., A.D. Panek and E. Eleutherio. 2007. Oxidative stress and its effects during dehydration. Comp. Biochem. Physiol. A Mol. Integr. Physiol. 146: 621–631.

Gamero-Sandemetrio, E., R. Gomez-Pastor and E. Matallana. 2013. Zymogram profiling of superoxide dismutase and catalase activities allows *Saccharomyces* and non-*Saccharomyces* species differentiation and correlates to their fermentation performance. Appl. Microbiol. Biotechnol. 97(10): 4563–4576.

Gamero-Sandemetrio, E., R. Gomez-Pastor and E. Matallana. 2014. Antioxidant defense parameters as predictive biomarkers for fermentative capacity of active dried wine yeast. Biotechnol. J. 9(8): 1055–1064.

Gamero-Sandemetrio, E., M. Torrellas, M.T. Rabena, R. Gomez-Pastor, A. Aranda and E. Matallana. 2015. Food-grade argan oil supplementation in molasses enhances fermentative performance and antioxidant defenses of active dry wine yeast. AMB Express 5(1): 75.

Garcia-Rios, E., M. Lopez-Malo and J.M. Guillamon. 2014. Global phenotypic and genomic comparison of two *Saccharomyces cerevisiae* wine strains reveals a novel role of the sulfur assimilation pathway in adaptation at low temperature fermentations. BMC Genomics 15: 1059.

Garre, E., F. Raginel, A. Palacios, A. Julien and E. Matallana. 2010. Oxidative stress responses and lipid peroxidation damage are induced during dehydration in the production of dry active wine yeasts. Int. J. Food Microbiol. 136(3): 295–303.

Gasch, A.P., P.T. Spellman, C.M. Kao, O. Carmel-Harel, M.B. Eisen, G. Storz, D. Botstein and P.O. Brown. 2000. Genomic expression programs in the response of yeast cells to environmental changes. Mol. Biol. Cell 11(12): 4241–4257.

Gibson, B.R., S.J. Lawrence, J.P. Leclaire, C.D. Powell and K.A. Smart. 2007. Yeast responses to stresses associated with industrial brewery handling. FEMS Microbiol. Rev. 31(5): 535–569.

Gibson, B.R., K.A. Prescott and K.A. Smart. 2008a. Petite mutation in aged and oxidatively stressed ale and lager brewing yeast. Lett. Appl. Microbiol. 46(6): 636–642.

Gibson, B.R., S.J. Lawrence, C.A. Boulton, W.G. Box, N.S. Graham, R.S. Linforth and K.A. Smart. 2008b. The oxidative stress response of a lager brewing yeast strain during industrial propagation and fermentation. FEMS Yeast Res. 8(4): 574–585.

Gomez-Pastor, R., R. Perez-Torrado, E. Cabiscol and E. Matallana. 2010a. Transcriptomic and proteomic insights of the wine yeast biomass propagation process. FEMS Yeast Res. 10(7): 870–884.

Gomez-Pastor, R., R. Perez-Torrado, E. Cabiscol, J. Ros and E. Matallana. 2010b. Reduction of oxidative cellular damage by overexpression of the thioredoxin TRX2 gene improves yield and quality of wine yeast dry active biomass. Microb. Cell Fact. 9: 9.

Gomez-Pastor, R., R. Perez-Torrado, E. Cabiscol, J. Ros and E. Matallana. 2012. Engineered Trx2p industrial yeast strain protects glycolysis and fermentation proteins from oxidative carbonylation during biomass propagation. Microb. Cell Fact. 11: 4.

Hanzen, S., K. Vielfort, J. Yang, F. Roger, V. Andersson, S. Zamarbide-Fores, R. Andersson, L. Malm, G. Palais, B. Biteau, B. Liu, M. B. Toledano, M. Molin and T. Nystrom. 2016. Lifespan control by redox-dependent recruitment of chaperones to misfolded proteins. Cell 166(1): 140–151.

Herman, P.K. 2002. Stationary phase in yeast. Curr. Opin. Microbiol. 5(6): 602–607.

Hernandez-Lopez, M.J., J.A. Prieto and F. Randez-Gil. 2003. Osmotolerance and leavening ability in sweet and frozen sweet dough. Comparative analysis between *Torulaspora delbrueckii* and *Saccharomyces cerevisiae* baker's yeast strains. Antonie van Leeuwenhoek 84(2): 125–134.

Herrero, E., J. Ros, G. Belli and E. Cabiscol. 2008. Redox control and oxidative stress in yeast cells. Biochim. Biophys. Acta 1780(11): 1217–1235.

Hino, A., K. Mihara, K. Nakashima and H. Takano. 1990. Trehalose levels and survival ratio of freeze-tolerant versus freeze-sensitive yeasts. Appl. Environ. Microbiol. 56(5): 1386–1391.

Hirasawa, T., K. Yoshikawa, Y. Nakakura, K. Nagahisa, C. Furusawa, Y. Katakura, H. Shimizu and S. Shioya. 2007. Identification of target genes conferring ethanol stress tolerance to *Saccharomyces cerevisiae* based on DNA microarray data analysis. J. Biotechnol. 131(1): 34–44.

Hohmann, S. and W.H. Mager. 2003. Introduction. pp. 1–9. *In*: S. Hohmann and W.H. Mager (eds.). Yeast Stress Responses. Springer, Heidelberg.

Hohmann, S. 2015. An integrated view on a eukaryotic osmoregulation system. Curr. Genet. 61(3): 373–382.

Ivorra, C., J.E. Perez-Ortin and M. del Olmo. 1999. An inverse correlation between stress resistance and stuck fermentations in wine yeasts. A molecular study. Biotechnol. Bioeng. 64(6): 698–708.

Jackson, R. S. 2000. Wine science: Principles, practice, perception. 2nd ed, Food Science and Technology International Series. Academic Press, San Diego

Jagtap, U.B., J.P. Jadhav, V.A. Bapat and I.S. Pretorius. 2017. Synthetic biology stretching the realms of possibility in wine yeast research. Int. J. Food Microbiol. 252: 24–34.

Jimenez-Marti, E., M. Gomar-Alba, A. Palacios, A. Ortiz-Julien and M.L. del Olmo. 2011a. Towards an understanding of the adaptation of wine yeasts to must: Relevance of the osmotic stress response. Appl. Microbiol. Biotechnol. 89(5): 1551–1561.

Jimenez-Marti, E., A. Zuzuarregui, M. Gomar-Alba, D. Gutierrez, C. Gil and M. del Olmo. 2011b. Molecular response of *Saccharomyces cerevisiae* wine and laboratory strains to high sugar stress conditions. Int. J. Food Microbiol. 145(1): 211–220.

Jolly, N.P., C. Varela and I.S. Pretorius. 2014. Not your ordinary yeast: Non-*Saccharomyces* yeasts in wine production uncovered. FEMS Yeast Res. 14(2): 215–237.

Klionsky, D.J., K. Abdelmohsen, A. Abe et al. 2016. Guidelines for the use and interpretation of assays for monitoring autophagy (3rd edition). Autophagy 12(1): 1–222.

Landolfo, S., H. Politi, D. Angelozzi and I. Mannazzu. 2008. ROS accumulation and oxidative damage to cell structures in *Saccharomyces cerevisiae* wine strains during fermentation of high-sugar-containing medium. Biochim. Biophys. Acta 1780(6): 892–898.

Legras, J.L., J. Moreno-Garcia, S. Zara, G. Zara, T. Garcia-Martinez, J.C. Mauricio, I. Mannazzu, A.L. Coi, M.B. Zeidan, S. Dequin, J. Moreno and M. Budroni. 2016. Flor yeast: New perspectives beyond wine aging. Front. Microbiol. 7: 503.

Li, Y., W. Chen, Y. Shi and X. Liang. 2013. Molecular cloning and evolutionary analysis of the HOG-signaling pathway genes from *Saccharomyces cerevisiae* rice wine isolates. Biochem. Genet. 51(3-4): 296–305.

Liu, X.Z., M. Sang, X.A. Zhang, T.K. Zhang, H.Y. Zhang, X. He, S.X. Li, X.D. Sun and Z.M. Zhang. 2017. Enhancing expression of SSU1 genes in *Saccharomyces uvarum* leads to an increase in sulfite tolerance and transcriptome profiles change. FEMS Yeast Res. 17(3): fox023.

Longo, V.D., G.S. Shadel, M. Kaeberlein and B. Kennedy. 2012. Replicative and chronological aging in *Saccharomyces cerevisiae*. Cell Metab. 16(1): 18–31.

Lopez-Martinez, G., R. Pietrafesa, P. Romano, R. Cordero-Otero and A. Capece. 2013. Genetic improvement of *Saccharomyces cerevisiae* wine strains for enhancing cell viability after desiccation stress. Yeast 30 (8): 319–330.

Loveday, S.M., V.T. Huang, D.S. Reid and R.J. Winger. 2012. Water dynamics in fresh and frozen yeasted dough. Crit. Rev. Food Sci. Nutr. 52(5): 390–409.

Marks, V.D., S.J. Ho Sui, D. Erasmus, G. K. van der Merwe, J. Brumm, W.W. Wasserman, J. Bryan and H.J. van Vuuren. 2008. Dynamics of the yeast transcriptome during wine fermentation reveals a novel fermentation stress response. FEMS Yeast Res. 8(1): 35–52.

Mascarenhas, C., L.C. Edwards-Ingram, L. Zeef, D. Shenton, M.P. Ashe and C.M. Grant. 2008. Gcn4 is required for the response to peroxide stress in the yeast *Saccharomyces cerevisiae*. Mol. Biol. Cell 19(7): 2995–3007.

Maskell, D.L., A.I. Kennedy, J.A. Hodgson and K.A. Smart. 2003. Chronological and replicative lifespan of polyploid *Saccharomyces cerevisiae* (syn. *S. pastorianus*). FEMS Yeast Res. 3(2): 201–209.

Michnick, S., J.L. Roustan, F. Remize, P. Barre and S. Dequin. 1997. Modulation of glycerol and ethanol yields during alcoholic fermentation in *Saccharomyces cerevisiae* strains overexpressed or disrupted for GPD1 encoding glycerol 3-phosphate dehydrogenase. Yeast 13(9): 783–793.

Molin, M., J. Yang, S. Hanzen, M.B. Toledano, J. Labarre and T. Nystrom. 2011. Life span extension and H(2)O(2) resistance elicited by caloric restriction require the peroxiredoxin Tsa1 in *Saccharomyces cerevisiae*. Mol. Cell 43(5): 823–833.

Morano, K.A., C.M. Grant and W.S. Moye-Rowley. 2012. The response to heat shock and oxidative stress in *Saccharomyces cerevisiae*. Genetics 190(4): 1157–1195.

Moreno-García, J., J. Mauricio, J. Moreno and T. García-Martínez. 2017. Differential proteome analysis of a flor yeast strain under biofilm formation. Int. J. Mol. Sci. 18(4): 720.

Nakagawa, Y., H. Ogihara, C. Mochizuki, H. Yamamura, Y. Iimura and M. Hayakawa. 2017. Development of intra-strain self-cloning procedure for breeding baker's yeast strains. J. Biosci. Bioeng. 123(3): 319–326.

Noguchi, C., D. Watanabe, Y. Zhou, T. Akao and H. Shimoi. 2012. Association of constitutive hyperphosphorylation of Hsf1p with a defective ethanol stress response in *Saccharomyces cerevisiae* sake yeast strains. Appl. Environ. Microbiol. 78(2): 385–392.

Novo, M., G. Beltran, N. Rozes, J. M. Guillamon, S. Sokol, V. Leberre, J. Francois and A. Mas. 2007. Early transcriptional response of wine yeast after rehydration: Osmotic shock and metabolic activation. FEMS Yeast Res. 7(2): 304–316.

Novo, M., F. Bigey, E. Beyne, V. Galeote, F. Gavory, S. Mallet, B. Cambon, J.-L. Legras, P. Wincker, S. Casaregola and S. Dequin. 2009. Eukaryote-to-eukaryote gene transfer events revealed by the genome sequence of the wine yeast *Saccharomyces cerevisiae* EC1118. Proc. Natl. Acad. Sci. USA 106(38): 16333–16338.

Novo, M.T., G. Beltran, M.J. Torija, M. Poblet, N. Rozes, J.M. Guillamon and A. Mas. 2003. Changes in wine yeast storage carbohydrate levels during preadaptation, rehydration and low temperature fermentations. Int. J. Food Microbiol. 86(1-2): 153–161.

Nystrom, T., J. Yang and M. Molin. 2012. Peroxiredoxins, gerontogenes linking aging to genome instability and cancer. Genes Dev. 26(18): 2001–2008.

Oda, Y. and K. Tonomura. 1993. Selection of a novel baking strain from the *Torulaspora* yeasts. Biosci. Biotechnol. Biochem. 57(8): 1320–1322.

Orozco, H., E. Matallana and A. Aranda. 2012a. Wine yeast sirtuins and Gcn5p control aging and metabolism in a natural growth medium. Mech. Ageing Dev. 133(5): 348–358.

Orozco, H., E. Matallana and A. Aranda. 2012b. Oxidative stress tolerance, adenylate cyclase and autophagy are key players in the chronological life span of *Saccharomyces cerevisiae* during winemaking. Appl. Environ. Microbiol. 78(8): 2748–2757.

Orozco, H., E. Matallana and A. Aranda. 2012c. Two-carbon metabolites, polyphenols and vitamins influence yeast chronological life span in winemaking conditions. Microb. Cell Fact. 11(1): 104.

Orozco, H., A. Sepulveda, C. Picazo, E. Matallana and A. Aranda. 2016. RNA binding protein Pub1p regulates glycerol production and stress tolerance by controlling Gpd1p activity during winemaking. Appl. Microbiol. Biotechnol. 100(11): 5017–5027.

Panadero, J., M.J. Hernandez-Lopez, J.A. Prieto and F. Randez-Gil. 2007. Overexpression of the calcineurin target CRZ1 provides freeze tolerance and enhances the fermentative capacity of baker's yeast. Appl. Environ. Microbiol. 73(15): 4824–4831.

Penacho, V., E. Valero and R. Gonzalez. 2012. Transcription profiling of sparkling wine second fermentation. Int. J. Food Microbiol. 153(1-2): 176–182.

Pereira Ede, J., A.D. Panek and E.C. Eleutherio. 2003. Protection against oxidation during dehydration of yeast. Cell Stress Chaperones 8(2): 120–124.

Perez-Ortin, J.E., A. Querol, S. Puig and E. Barrio. 2002. Molecular characterization of a chromosomal rearrangement involved in the adaptive evolution of yeast strains. Genome Res. 12(10): 1533–1539.

Perez-Torrado, R., P. Carrasco, A. Aranda, J. Gimeno-Alcaniz, J.E. Perez-Ortin, E. Matallana and M. L. del Olmo. 2002a. Study of the first hours of microvinification by the use of osmotic stress-response genes as probes. Syst. Appl. Microbiol. 25(1): 153–161.

Perez-Torrado, R., J.V. Gimeno-Alcaniz and E. Matallana. 2002b. Wine yeast strains engineered for glycogen overproduction display enhanced viability under glucose deprivation conditions. Appl. Environ. Microbiol. 68(7): 3339–3344.

Perez-Torrado, R., J.M. Bruno-Barcena and E. Matallana. 2005. Monitoring stress-related genes during the process of biomass propagation of *Saccharomyces cerevisiae* strains used for wine making. Appl. Environ. Microbiol. 71(11): 6831–6837.

Perez-Torrado, R. and E. Matallana. 2015. Enhanced fermentative capacity of yeasts engineered in storage carbohydrate metabolism. Biotechnol. Prog. 31(1): 20–24.

Pérez-Torrado, R., E. Gamero, R. Gómez-Pastor, E. Garre, A. Aranda and E. Matallana. 2015. Yeast biomass, an optimised product with myriad applications in the food industry. Trends Food Sci. Technol. 46(2): 167–175.

Picazo, C., E. Gamero-Sandemetrio, H. Orozco, W. Albertin, P. Marullo, E. Matallana and A. Aranda. 2015a. Mitochondria inheritance is a key factor for tolerance to dehydration in wine yeast production. Lett. Appl. Microbiol. 60(3): 217–222.

Picazo, C., H. Orozco, E. Matallana and A. Aranda. 2015b. Interplay among Gcn5, Sch9 and mitochondria during chronological aging of wine yeast is dependent on growth conditions. PLoS One 10(2): e0117267.

Piskur, J., E. Rozpedowska, S. Polakova, A. Merico and C. Compagno. 2006. How did *Saccharomyces* evolve to become a good brewer? Trends Genet. 22(4): 183–186.

Powell, C.D., S.M. Van Zandycke, D.E. Quain and K.A. Smart. 2000. Replicative ageing and senescence in *Saccharomyces cerevisiae* and the impact on brewing fermentations. Microbiology 146(Pt 5): 1023–1034.

Pretorius, I.S. 2000. Tailoring wine yeast for the new millennium: Novel approaches to the ancient art of winemaking. Yeast 16(8): 675–729.

Pretorius, I.S. and F.F. Bauer. 2002. Meeting the consumer challenge through genetically customized wine-yeast strains. Trends Biotechnol. 20(10): 426–432.

Randez-Gil, F., I. Corcoles-Saez and J.A. Prieto. 2013. Genetic and phenotypic characteristics of baker's yeast: Relevance to baking. Annu. Rev. Food Sci. Technol. 4: 191–214.

Remize, F., B. Cambon, L. Barnavon and S. Dequin. 2003. Glycerol formation during wine fermentation is mainly linked to Gpd1p and is only partially controlled by the HOG pathway. Yeast 20(15): 1243–1253.

Ribéreau-Gayon, P., D. Dubourdieu and B. Donèche. 2006. Handbook of enology. 2nd ed. 2 vols. John Wiley & Sons, Chichester, West Sussex, England.

Richardson, S.M., L.A. Mitchell, G. Stracquadanio, K. Yang, J. S. Dymond, J. E. DiCarlo, D. Lee, C. L. V. Huang, S. Chandrasegaran, Y. Cai, J.D. Boeke and J.S. Bader. 2017. Design of a synthetic yeast genome. Science 355(6329): 1040–1044.

Rodkaer, S.V. and N.J. Faergeman. 2014. Glucose- and nitrogen sensing and regulatory mechanisms in *Saccharomyces cerevisiae*. FEMS Yeast Res. 14(5): 683–696.

Rodriguez-Porrata, B., D. Carmona-Gutierrez, A. Reisenbichler, M. Bauer, G. Lopez, X. Escote, A. Mas, F. Madeo and R. Cordero-Otero. 2012. Sip18 hydrophilin prevents yeast cell death during desiccation stress. J. Appl. Microbiol. 112(3): 512–525.

Rossignol, T., L. Dulau, A. Julien and B. Blondin. 2003. Genome-wide monitoring of wine yeast gene expression during alcoholic fermentation. Yeast 20(16): 1369–1385.

Rossignol, T., O. Postaire, J. Storai and B. Blondin. 2006. Analysis of the genomic response of a wine yeast to rehydration and inoculation. Appl. Microbiol. Biotechnol. 71(5): 699–712.

Rossignol, T., D. Kobi, L. Jacquet-Gutfreund and B. Blondin. 2009. The proteome of a wine yeast strain during fermentation, correlation with the transcriptome. J. Appl. Microbiol. 107(1): 47–55.

Rossouw, D. and F.F. Bauer. 2009. Comparing the transcriptomes of wine yeast strains: Toward understanding the interaction between environment and transcriptome during fermentation. Appl. Microbiol. Biotechnol. 84(5): 937–954.

Rossouw, D., E.H. Heyns, M.E. Setati, S. Bosch and F.F. Bauer. 2013. Adjustment of trehalose metabolism in wine *Saccharomyces cerevisiae* strains to modify ethanol yields. Appl. Environ. Microbiol. 79(17): 5197–5207.

Salvado, Z., R. Chiva, S. Rodriguez-Vargas, F. Randez-Gil, A. Mas and J.M. Guillamon. 2008. Proteomic evolution of a wine yeast during the first hours of fermentation. FEMS Yeast Res. 8(7): 1137–1146.

Salvado, Z., R. Chiva, N. Rozes, R. Cordero-Otero and J.M. Guillamon. 2012. Functional analysis to identify genes in wine yeast adaptation to low-temperature fermentation. J. Appl. Microbiol. 113(1): 76–88.

Sasano, Y., Y. Haitani, K. Hashida, I. Ohtsu, J. Shima and H. Takagi. 2012a. Simultaneous accumulation of proline and trehalose in industrial baker's yeast enhances fermentation ability in frozen dough. J. Biosci. Bioeng. 113(5): 592–595.

Sasano, Y., Y. Haitani, K. Hashida, I. Ohtsu, J. Shima and H. Takagi. 2012b. Overexpression of the transcription activator Msn2 enhances the fermentation ability of industrial baker's yeast in frozen dough. Biosci. Biotechnol. Biochem. 76(3): 624–627.

Shenton, D., J.B. Smirnova, J.N. Selley, K. Carroll, S.J. Hubbard, G.D. Pavitt, M.P. Ashe and C.M. Grant. 2006. Global translational responses to oxidative stress impact upon multiple levels of protein synthesis. J. Biol. Chem. 281(39): 29011–29021.

Shima, J., A. Ando and H. Takagi. 2008. Possible roles of vacuolar H+-ATPase and mitochondrial function in tolerance to air-drying stress revealed by genome-wide screening of *Saccharomyces cerevisiae* deletion strains. Yeast 25(3): 179–190.

Shima, J. and H. Takagi. 2009. Stress-tolerance of baker's-yeast (*Saccharomyces cerevisiae*) cells: Stress-protective molecules and genes involved in stress tolerance. Biotechnol. Appl. Biochem. 53(Pt 3): 155–164.

Simpson, C.E. and M.P. Ashe. 2012. Adaptation to stress in yeast: To translate or not? Biochem. Soc. Trans. 40(4): 794–799.

Spano, G. and S. Torriani. 2016. Editorial: Microbiota of Grapes: Positive and Negative Role on Wine Quality. Front. Microbiol. 7: 2036.

Stefanini, I., L. Dapporto, J.L. Legras, A. Calabretta, M. Di Paola, C. De Filippo, R. Viola, P. Capretti, M. Polsinelli, S. Turillazzi and D. Cavalieri. 2012. Role of social wasps in *Saccharomyces cerevisiae* ecology and evolution. Proc. Natl. Acad. Sci. USA 109(33): 13398–13403.

Takagi, H., M. Takaoka, A. Kawaguchi and Y. Kubo. 2005. Effect of L-proline on sake brewing and ethanol stress in *Saccharomyces cerevisiae*. Appl. Environ. Microbiol. 71(12): 8656–8662.

Takahara, T. and T. Maeda. 2012. Transient sequestration of TORC1 into stress granules during heat stress. Mol.Cell 47(2): 242–252.

Urbanczyk, H., C. Noguchi, H. Wu, D. Watanabe, T. Akao, H. Takaqi and H. Shimoi. 2011. Sake yeast strains have difficulty in entering a quiescent state after cell growth cessation. J. Biosci. Bioeng. 112(1): 44–48.

Varela, C., J. Cardenas, F. Melo and E. Agosin. 2005. Quantitative analysis of wine yeast gene expression profiles under winemaking conditions. Yeast 22(5): 369–383.

Waltz, E. 2016. Gene-edited CRISPR mushroom escapes US regulation. Nature 532(7599): 293.

Watanabe, D., H. Wu, C. Noguchi, Y. Zhou, T. Akao and H. Shimoi. 2011. Enhancement of the initial rate of ethanol fermentation due to dysfunction of yeast stress response components Msn2p and/ or Msn4p. Appl. Environ. Microbiol. 77(3): 934–941.

Watanabe, D., Y. Araki, Y. Zhou, N. Maeya, T. Akao and H. Shimoi. 2012. A loss-of-function mutation in the PAS kinase Rim15p is related to defective quiescence entry and high fermentation rates of *Saccharomyces cerevisiae* sake yeast strains. Appl. Environ. Microbiol. 78(11): 4008–4016.

Watanabe, D., A. Kaneko, Y. Sugimoto, S. Ohnuki, H. Takagi and Y. Ohya. 2017. Promoter engineering of the *Saccharomyces cerevisiae* RIM15 gene for improvement of alcoholic fermentation rates under stress conditions. J. Biosci. Bioeng. 123(2): 183–189.

Winderickx, J., I. Holsbeeks, O. Lagatie, F. Giots, J. Thevelein and H. Winde. 2003. From feast to famine; adaptation to nutrient availability in yeast. pp. 305–386. *In*: S. Hohmann and W.H. Mager (eds.). Yeast Stress Responses. Springer, Heidelberg.

Xu, W., J. Wang and Q. Li. 2014. Comparative proteome and transcriptome analysis of lager brewer's yeast in the autolysis process. FEMS Yeast Res. 14(8): 1273–1285.

Zhang, N. and L. Cao. 2017. Starvation signals in yeast are integrated to coordinate metabolic reprogramming and stress response to ensure longevity. Curr. Genet. 63(5): 839–843.

Zuzuarregui, A. and M. del Olmo. 2004. Analyses of stress resistance under laboratory conditions constitute a suitable criterion for wine yeast selection. Antonie Van Leeuwenhoek 85(4): 271–280.

11

Genomic Insights into Gram-Negative Food Spoilers

Spiros Paramithiotis and *Eleftherios H. Drosinos**

1. Introduction

The term 'spoilage' is meaningful only in the context of a certain food commodity. In other words, spoilage may be considered as any deviation from the desired properties of a given product. Such deviations may occur due to physical damage, chemical reactions, microbial growth or their combination. In the majority of the cases, deviations from the anticipated quality characteristics of a product result from microbial growth during storage and is attributed to the production of biomass, enzymes and metabolites.

Bacteria, yeasts and molds may drive food spoilage depending upon the intrinsic physicochemical characteristics of the product (e.g., a_w, pH, concentration of carbon and nitrogen sources, presence of antimicrobial compounds, etc.), as well as the storage conditions (e.g., temperature, relative humidity, atmosphere composition, etc.). These were first introduced in the literature by the classic publication of Mossel and Ingram that formulated the concept of spoilage (Mossel and Ingram 1955). The potential of bacteria, yeasts and molds as food spoilers has been extensively studied and the conditions that allow their development and production of the spoilage phenotype have been identified and adequately reviewed (Petruzzi et al. 2017).

Gram-negative bacteria include a wide variety of human, animal and plant pathogenic species, as well as species that may proliferate in the conditions used for storage of certain food commodities. Especially regarding the latter, *Pseudomonas* spp., *Shewanella* spp. and several species belonging to the *Enterobacteriaceae* family have been at the epicenter of intensive study, mostly due to their ability to grow at low storage temperatures (i.e., around 4ºC) and produce pigments and enzymes as well as metabolites that result in off-odors.

Laboratory of Food Quality Control and Hygiene, Department of Food Science and Human Nutrition, Agricultural University of Athens, Iera Odos 75, 118 55 Athens, Greece.
* Corresponding author: ehd@aua.gr

The molecular microbiology approach has been extensively considered, primarily regarding the taxonomic assignment of these food spoilers and secondarily to obtain insights into their genomic determinants that trigger and coordinate their metabolic activities. In the first case, the inability of 16S-rRNA gene sequence to accurately depict their exact phylogenetic affiliation has been adequately highlighted. For that purpose, the use of other genes, such as *gyrB*, *rpoB*, *rpoD* or their concatenated sequences, following a multi-locus sequence analysis (MLSA) strategy, has been proposed. As far as the genomic and transcriptomic organization of spoilage-related genes and their responses to various environmental stimuli is concerned, research has been focused on deciphering the role of quorum sensing in the regulation of attributes that account for the spoilage potential, such as production of enzymes, metabolites and biofilm formation. The quorum sensing of Gram-negative bacteria has recently been reviewed by Banerjee and Ray (2016). In brief, quorum sensing involves the production and extracellular release of molecules, termed autoinducers (AIs). These molecules are accumulated in the growth environment and, upon reaching a certain threshold, these AIs are sensed by membrane-bound or cytoplasm receptors and regulate gene expression. Since this occurs at microecosystem level, it is perceived as a coordinated action that may involve production of biofilm, enzymes, metabolites, sporulation, virulence factors, etc. (Ng and Bassler 2009, Rutherford and Bassler 2012). Four classes of AIs have been described: (i) AI-1 class consists of N-acyl homoserine lactones (AHLs), which are produced by Gram-negative bacteria, (ii) AI-2 class consists of a furanosyl borate diester the production of which is S-adenosylmethionine dependent and occurs in both Gram-positive and -negative bacteria, (iii) AI-3, which are produced by human intestinal biota and to which, along with epinephrine and norepinephrine, EHEC responds by activating transcription of the genes involved in AE lesion formation and (iv) autoinducing peptides (AIPs) that are only produced by Gram-positive bacteria (Winzer et al. 2003, Lyon and Novick 2004, De Keersmaecker et al. 2006, Smith et al. 2006, Reading et al. 2007). Regarding Gram-negative bacteria, the most common pathway consists of *luxI* (*lasI* in *P. aeruginosa*) that encodes an acyl homoserine lactone synthase, which produces 3-oxo-C12-homoserine lactone that is transported outside the cell. When the necessary concentration threshold is reached, it activates *luxR* (*lasR* in *P. aeruginosa*) that act as a transcription factor for a series of genes. In another similar procedure, *rhlI*, which encodes for another acyl homoserine lactone synthase, produces butanoyl homoserine lactone which, in turn, activates *rhlR*, with similar function as *luxR*. The third system that has been described, involves the production of 2-Heptyl-4(1H)-quinolone and 2-Heptyl-3-hydroxy-quinolone by *phnA*, *phnB* and *pqsABCDEH* that activate *pqsR*, a LysR family transcriptional regulator that is involved in the regulation of the transcription of several genes (Pesci et al. 1999, Gallagher et al. 2002, Deziel et al. 2004, Schafhauser et al. 2014).

The importance of quorum sensing in spoilage of food commodities has been adequately highlighted. Indeed, AIs have been detected during spoilage of foodstuff of both plant and animal origin (Riedel et al. 2001, Bruhn et al. 2004, Lu et al. 2004, Rasch et al. 2005, Liu et al. 2006, Medina-Martinez et al. 2006a,b, van Houdt et al. 2006, Silagyi et al. 2009, Ammor et al. 2008, Blana and Nychas 2014).

In the present chapter, current knowledge regarding the applications of the molecular microbiology approach in taxonomy, genomic and transcriptomic organization of spoilage associated species belonging to the *Pseudomonas* and *Shewanella* genera, as well as to the *Enterobacteriaceae* family, is critically reviewed and presented.

2. *Pseudomonas* spp.

The genus *Pseudomonas* includes more than 200 species and subspecies; among which, there are several plant pathogenic species, such as *P. syringae* and *P. savastanoi* (Hofte and de Vos 2007), as well as animal and human pathogenic species such as *P. aeruginosa* (Valentini et al. 2018). Based on the wide variety of niches from which they have been isolated, this genus may be characterized as ubiquitous.

Initially, sequencing of the 16S-rRNA gene was used to assess the phylogenetic relationships between the *Pseudomonas* spp. and delineate their taxonomic position. In Fig. 1, the evolutionary history of the major species belonging to the *Pseudomonas* genus is inferred by the maximum likelihood method based on the Hasegawa-Kishino-Yano model (Hasegawa et al. 1985) that was identified as the DNA substitution pattern that best describes the dataset used, namely 16S-rRNA sequences. Three major groups are separated: Group I, or the *P. aeruginosa* group, consists of *P. aeruginosa*, *P. putida*, *P. oryzihabitans*, *P. oleovorans*, *P. straminea* and *P. stutzeri*, group II consists only of *P. pertucinogena* and group III, or the *P. fluorescens* group consists of *P. fluorescens*, *P. syringae*, *P. lutea* and *P. anguilliseptica*. However, this gene is highly conserved and, therefore, accurate assignment to the genus level is not accompanied by a respective at species level, especially when the species are closely related. This was also the case of *Pseudomonas* spp. Thus, the use of other genes, such as *atpD*, *gyrB*, *recA*, *rpoB* and *rpoD*, was also considered (Gomila et al. 2015). The most effective differentiation was achieved with the use of a multi-locus sequence analysis (MLSA) strategy. More accurately, the concatenated sequences of 16S rDNA, *gyrB*, *rpoB* and *rpoD* provided with a novel insight into the phylogenetic relationships of *Pseudomonas* spp. and effectively differentiated closely related species (Mullet et al. 2010). Most recently, Peix et al. (2018) after having comprehensively reviewed all aspects related to *Pseudomonas* spp. taxonomy, suggested specific improvements to the scheme proposed by Mullet et al. (2010). More accurately, they proposed that former *P. pertucinogena*-group could be regarded as a third lineage, together with the previously described

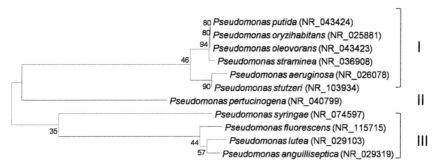

Fig. 1. The evolutionary history of *Pseudomonas* spp. as inferred by using the Maximum Likelihood method based on the Hasegawa-Kishino-Yano model (Hasegawa et al. 1985) of the 16S-rRNA gene sequences. Bootstrap values at the node were calculated from 1000 samplings. GenBank accession numbers are given in parenthesis. Latin numbers (I, II, III) designate the major taxonomic groups.

P. fluorescens and *P. aeruginosa* ones. In addition, *P. oryzihabitans*-group was transferred from the *P. aeruginosa* lineage to the *P. fluorescens* one, and *P. anguilliseptica*- and *P. straminea*-groups were transferred from the *P. fluorescens* lineage to the *P. aeruginosa* one. Taking into consideration that the same sequences were employed for both analyses, these differences were explained by the variances in the concatenation placement and the methodology used to infer phylogenetic relationship (Peix et al. 2018). However, these differences may also indicate the inability of the selected genetic markers to accurately depict the phylogenetic affiliation of these groups and possibly reveal the need for the addition or substitution of a genetic marker in the MLSA approach.

From a food technology perspective, *P. fluorescens*, *P. lundensis*, *P. fragi* and *P. putida* are the most interesting species, since they are the most frequently isolated ones during food spoilage. The latter has been attributed to their ability to grow at 4°C, produce pectinolytic, proteolytic and lipolytic enzymes, exopolyssaccharides and pigments, as well as a series of metabolites that result in the production of off-odors. More accurately, spoilage of animal-derived products is very often correlated to lipolytic and proteolytic strains of all four species, whereas spoilage of fresh produce during storage through induction of soft rot is very often attributed to pectinolytic strains of the first species (Liao 2006). Apart from these species, occurrence of several others, including *P. azotoformans*, *P. brennerii*, *P. chlororaphis*, *P. gessardii*, *P. grimontii*, *P. jessenii*, *P. lurida*, *P. marginalis*, *P. moraviensis*, *P. poae*, *P. proteolytica*, *P. psychrophila*, *P. rhodesiae*, *P. salomonii*, *P. stutzeri*, *P. syringae*, *P. tolaasii* and *P. veronii*, has also been reported (von Neubeck et al. 2015, Caldera et al. 2016, Vithanage et al. 2016).

The genetic and transcriptional organization primarily of *P. aeruginosa* and secondarily of *P. syringae* pathovars have been the epicenter of significant research over the last few decades due to their significance as human and plant pathogens, respectively. On the contrary, gene regulation of food-associated *Pseudomonas* spp. has not been studied to the same extent. *P. fragi*, besides being an important milk and meat spoiler (Wiedmann et al. 2000, Ercolini et al 2007), is also one of the most interesting species from a biotechnological perspective. Especially regarding the latter, a wide range of applications have been described, including plant growth promotion through solubilization of zinc and phosphate (Selvakumar et al. 2009, Kamran et al. 2017) or a range of plant-probiotic activities (Agaras et al. 2018), production of substances such as trehalose and biosurfactants (Ben Belgacem et al. 2015, Mei et al. 2016), degradation of toxic compounds (Adelowo et al. 2006, Yanzhen et al. 2016, Paikhomba Singha et al. 2017), protection from corrosion (Jayaraman et al. 1998), etc. Food spoilage potential has been extensively assessed, mainly through the production of lipolytic and proteolytic enzymes (von Neubeck et al. 2015, Meng et al. 2017, Pinto Junior et al. 2017, Xin et al. 2017). Alternatively, PCR detection of the *aprX* gene, which encodes for an extracellular caseinolytic metalloprotease, has also been employed (Dufour et al. 2008, Marchand et al. 2009, Ribeiro Junior et al. 2018). However, detection of proteolytic activity is not always accompanied by the detection of this gene, most probably due to the presence of other proteolytic enzymes (Ercolini et al. 2010). An insight into the genetic determinants that account for this spoilage potential was offered by Stanborough et al. (2018). In this study, the genome of 12 *P. fragi* and 7 *P. lundensis*, mostly isolated from meat products, was sequenced. Analysis of the sequences revealed a large diversity that may indicate the need for classification reassessment within this genus. In addition, the large number of regulatory genes detected suggest a great flexibility regarding transcriptional organization.

The alternative sigma factor σ^s, encoded by *rpoS*, has been recognized as a key factor for survival of stationary phase bacterial cells to stressful environments (Dodd and Aldsworth 2002, Alvarez-Ordonez et al. 2015, Jaishankar and Srivastava 2017). Indeed, this has been exhibited for several foodborne bacteria, including *Salmonella enterica* (Spector and Kenyon 2012), *E. coli* (Bae and Lee 2017), *S. aureus* (Cebrian et al. 2015) and *L. monocytogenes* (Ait-Ouazzou et al. 2012). However, this role is often variable among bacteria and it should, therefore, be verified in every case. In the case of *P. fluorescens*, this role was recently assessed by Liu et al. (2018). It was reported that RpoS had no effect on resistance to NaCl, positively affected resistance to crystal violet, ethanol, heat and H_2O_2 and negatively regulated resistance to acetic acid (Liu et al. 2018). A more integrated insight into *P. putida* gene regulation was provided by Mohareb et al. (2015). In that study, a whole genome microarray was employed in order to assess gene expression of *P. putida* strain KT2440 during growth in Luria Bertani broth supplemented with glucose at 10 and 30°C. A total of 430 and 568 genes were overexpressed, while the expression of 309 and 701 genes was repressed, at 30 and 10°C, respectively, at decreasing glucose concentration. However, the only trend that was reported were the repression of genes involved in the production of malodorous end-products at 10°C. Regarding growth at 30°C, the majority of the differentially expressed genes were involved in fundamental cellular activities, such as transcription regulation.

Control of bacterial metabolism by quorum sensing is a field of active research over the last decade and the direct control over a wide range of bacterial processes, such as secretion of virulence factors and biofilm production has been reported (for recent reviews see Goo et al. 2015, Banerjee and Ray 2016). The insights gained have allowed the development of alternative strategies to combat bacterial growth and development. Targeting quorum sensing systems seems a promising alternative to traditional bactericidal approaches since it may also reduce the likelihood of resistance development (for a recent review see Coughlan et al. 2016). However, little is known regarding the role of quorum sensing systems during growth of *Pseudomonas* spp. in food. Pinto et al. (2010) reported that the proteolytic activity of *P. fluorescens* 07A was not regulated by quorum sensing since no N-acyl-homoserine lactones (AHLs) were detected in the growth medium of the bacterium and exogenous addition had no effect on its proteolytic activity. Similar results were reported by Martins et al. (2014); biofilm formation, swarming motility and proteolytic activity of *P. fluorescens* strains 07A and 041 were not AHLs-dependent, indicating the possible absence of the respective quorum sensing system. On the contrary, Bai and Vittal (2014) reported the detection of C4-HSL signaling molecule during growth of *P. psychrophila* strain PSPF19 in Luria Bertani broth. In addition, furanone C-30, a synthetic quorum sensing inhibitor, negatively affected production of protease and lipase as well as biofilm formation. More recently, Liu et al. (2018) reported that the AHL quorum sensing system of *P. fluorescens* strain UK4 was directly regulated by RpoS. In addition, RpoS contributed to the overall spoilage potential of this strain by regulating the production of total volatile basic nitrogen and extracellular proteases during growth in sterilized salmon juice.

3. *Shewanella* spp.

The genus *Shewanella* includes more than 60 species, among which are *S. putrefaciens* and *S. algae* that have been characterized as opportunistic human pathogens, as well as several interesting species from a biotechnological perspective, such as *S. oneidensis*, *S. sediminis, S. halifaxensis, S. canadensis,* etc. (Ganesh et al. 1997, Fredrickson et al. 2002, Carpentier et al. 2003, 2005, Zhao et al. 2005, 2006, 2007). From a taxonomic perspective, *Shewanella* spp. are divided into three groups (Satomi 2014): the high GC content group, the psychrotolerant and non-halophilic group, and lastly the psychrotolerant and psychrophilic sodium ion-requiring group. However, this seems to be a rather phenotype-based distinction and, therefore, is not visible in phylogenetic trees based on 16S-rRNA. Indeed, in Fig. 2 the relative phylogenetic affiliation of the *Shewanella* spp., as inferred by the maximum likelihood method based on the Jukes-Cantor model (Jukes and Cantor 1969), is shown. The species are distinguished into three major clusters that, with the exception of cluster II, contain species from all three aforementioned groups. As it happens with phylogeny based on 16S-rRNA gene, closely related species may not be separated. For that purpose, *gyrB* (a gene encoding the b subunit of DNA gyrase and is characterized as rapidly evolving) has been effectively employed (Satomi et al. 2003).

Shewanella spp. have been associated with spoilage of protein-rich foods of high pH value stored at low temperatures, even at or near 0°C. Production of H_2S and reduction of trimethylamine oxide to trimethylamine are considered as characteristic for the species belonging to this genus.

The effect of exogenously added autoinducers, namely 4,5-dihydroxypentane-2,3-dione (DPD), cyclo-(L-Pro-L-Leu) and cyclo-(L-Pro-L-Phe) on the spoilage phenotype, in terms of biofilm formation, production of biogenic amines, total volatile basic nitrogen (TVB-N) and trimethylamine (TMA), along with the expression of *luxR, luxS, torA* and ornithine decarboxylase gene (ODC) of *S. baltica* was studied by Zhu et al. (2016). Biofilm, TMA and putrescine production significantly increased in the presence of cyclo-(L-Pro-L-Leu). In addition, upregulation of *luxR, torA* and ODC was reported; on the contrary, no effect on the transcription levels of *luxS* was observed. Overall, it was indicated that *S. baltica* spoilage may be regulated by diketopiperazines (DKP)-based quorum sensing. The co-regulation of *S. baltica* spoilage potential by DKPs and N-acyl-L-homoserine lactones (AHLs) was reported by Zhao et al. (2016). The latter study verified the stimulatory effect of cyclo-(L-Pro-L-Leu) on biofilm, TMA and biogenic amine production and, on the other hand, reported the inhibitory effect of AHLs.

Fu et al. (2018) sequenced the genome of 2 *S. baltica* strains possessing different spoilage potential (Gu et al. 2013) and reported significant differences in the amino acid sequence of the TMAO reductase system periplasmic protein TorT (encoded by *torT*), cysteine synthase B (encoded by *cysM*) and thioredoxin reductase (encoded by *trxB*). An attempt to gain further insight was made by comparing the amino acid sequences of TorT of an additional fourteen strains of known spoilage potential. Several differences were observed and some correlation was made between them and spoilage potential, which has to be experimentally verified. In the same study, the effect of four DKPs, namely cyclo-(L-Pro-L-Gly), cyclo-(L-Pro-L-Leu), cyclo-(L-Leu-L-Leu) and cyclo-(L-Pro-L-Phe), on the expression of *torT, cysM, trxB* and *torCAD* of the aforementioned two strains, was assessed. In the strain that was characterized

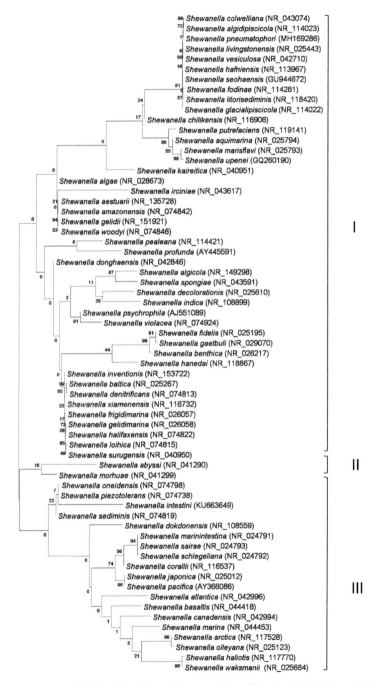

Fig. 2. Phylogenetic affiliation of the *Shewanella* spp. as inferred by using the Maximum Likelihood method based on the Jukes-Cantor model (Jukes and Cantor 1969) of the 16S-rRNA gene sequences. Bootstrap values at the node were calculated from 1000 samplings. GenBank accession numbers are given in parenthesis. Latin numbers (I, II, III) designate the major taxonomic groups.

by higher spoilage potential, the mRNA levels of *torT* and *cysM* were higher and the mRNA levels of *trxB* and *torCAD* were lower compared to the strain with the reduced spoilage potential under control conditions, i.e. without exogenous addition of DKPs. Their addition resulted in upregulation of *torT* and *trxB* and downregulation of *cysM* in both strains. In addition, *torCAD* expression was enhanced by cyclo-(L-Pro-L-Leu) in the strain that was characterized by enhanced spoilage ability and by cyclo-(L-Leu-L-Leu) in the strain with the reduced spoilage ability.

The coordination of biofilm and motility in *S. putrefaciens* strain CN32, and very likely in all *Shewanella* spp., by FlrA, was reported by Cheng et al. (2017). More accurately, FlrA was reported to act as a repressor of the *brfA* operon with c-di-GMP and FlhG inhibiting this action. In addition, in FlrA-deficient mutants, a decrease in the transcription of several flagellar genes, as well as the concomitant cell motility, was observed.

Zhang et al. (2017) assessed the role of AHL in *S. putrefaciens* spoilage potential through a proteomic approach. For that purpose, *S. putrefaciens* strain Z4 was grown in Luria-Bertani broth until early logarithmic phase; then addition of autoinducers, namely C4-HSL, C6-HSL and O-C6-HSL took place, followed by further incubation until stationary phase. Analysis of proteins with pI 4–7 and MW from approx. 6 to 200 kDa revealed that a total of 224 proteins were downregulated and 131 upregulated as a result of autoinducer addition. Twenty-three proteins were characterized as more interesting and identified; they were either associated to growth and the respective regulation, environmental adaptation or *de novo* purine synthesis.

4. Enterobacteriaceae

This family includes the following genera (the number of species within each genus is given in parenthesis): *Escherichia* (8), *Alterococcus* (1), *Arsenophonus* (3), *Atlantibacter* (2), *Biostraticola* (1), *Brenneria* (9), *Buchnera* (1), *Budvicia* (2), *Buttiauxella* (7), *Calymmatobacterium* (1), *Cedecea* (3), *Chania* (1), *Citrobacter* (15), *Cosenzaea* (1), *Cronobacter* (10), *Dickeya* (9), *Edwardsiella* (4), *Enterobacillus* (1), *Enterobacter* (34), *Erwinia* (39), *Franconibacter* (3), *Gibbsiella* (4), *Hafnia* (3), *Izhakiella* (2), *Klebsiella* (17), *Kluyvera* (5), *Kosakonia* (7), *Koserella* (1), *Leclercia* (1), *Lelliottia* (2), *Leminorella* (2), *Levinea* (2), *Lonsdalea* (4), *Mangrovibacter* (3), *Metakosakonia* (1), *Moellerella* (1), *Morganella* (2), *Obesumbacterium* (1), *Pantoea* (25), *Pectobacterium* (11), *Phaseolibacter* (1), *Photorhabdus* (4), *Phytobacter* (1), *Plesiomonas* (1), *Pluralibacter* (2), *Pragia* (1), *Proteus* (10), *Providencia* (10), *Pseudescherichia* (1), *Pseudocitrobacter* (2), *Rahnella* (6), *Raoultella* (4), *Rosenbergella* (4), *Rouxiella* (3), *Saccharobacter* (1), *Salmonella* (2), *Samsonia* (1), *Serratia* (20), *Shigella* (4), *Shimwellia* (2), *Siccibacter* (2), *Sodalis* (2), *Tatumella* (6), *Trabulsiella* (2), *Wigglesworthia* (1), *Xenorhadbus* (26), *Yersinia* (19) and *Yokenella* (1) (www.bacterio.net). From a taxonomic point of view, 16S-rRNA sequences have provided the information regarding the phylogenetic relationships between the above genera and the species within. However, the accuracy of the obtained phylogenetic affiliation is often compromised by the inherent disadvantages that characterize the use of this gene as a marker. This is also the case of the *Enterobacteriaceae* family. In Fig. 3, the phylogenetic position of representative members of the *Enterobacteriaceae* family,as inferred by using the Maximum Likelihood method based on the Tamura-

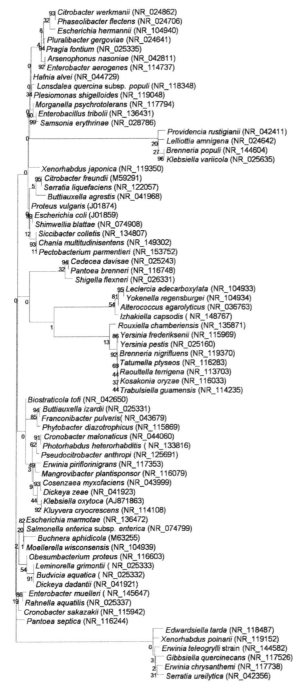

Fig. 3. Phylogenetic position of representative species belonging to the *Enterobacteriaceae* family as inferred by using the Maximum Likelihood method based on the Tamura-Nei model (Tamura and Nei 1993) of the 16S-rRNA gene sequences. Bootstrap values at the node were calculated from 1000 samplings. GenBank accession numbers are given in parenthesis.

Nei model (Tamura and Nei 1993) of the 16S-rRNA gene sequences, is presented. When more than one species from the same genus were included in the analysis (e.g., *Citrobacter* spp., *Escherichia* spp., *Buttiauxella* spp., *Cronobacter* spp., *Klebsiella* spp., etc.), these failed to cluster together, with the exception of *Yersinia frederiksenii* and *Y. pestis*. In order to improve the accuracy and gain additional insights, several other genes, such as *gyrB*, *dnaJ*, *oriC* and *recA*, have also been considered for that purpose (Fukushima et al. 2002, Pham et al. 2007, Roggenkamp 2007, Tailliez et al. 2010). The MLSA strategy has also been effectively applied; *tuf* and *atpD* were used to delineate the phylogenetic relationship between 78 species belonging to 31 genera of the family (Paradis et al. 2005), *atpD*, *carA* and *recA* to distinguish between species of the genera *Brenneria*, *Dickeya*, *Enterobacter*, *Erwinia*, *Pantoea*, *Pectobacterium*, and *Samsonia* (Young and Park 2007), *gyrB*, *rpoB*, *atpD* and *infB* for the identification of *Pantoea* spp. (Brady et al. 2008) and the taxonomic evaluation of the genus *Enterobacter* (Brady et al. 2013), *atpD*, *carA*, *gyrB*, *infB*, *recA*, and *rpoB* were proposed as the most effective combination for taxonomic studies within the *Erwinia*/*Pantoea* phylogeny (Zhang and Qiu 2015), *fusA*, *pyrG*, *rplB*, *rpoB* and *sucA* for the differentiation of *Chania multitudinisentens* from closely related genera (Ee et al. 2016).

Several important plant and human pathogenic bacteria have been taxonomically classified in the *Enterobacteriaceae* family. From a food technology perspective, a few genera have been associated with food spoilage, namely *Citrobacter*, *Enterobacter*, *Hafnia*, *Klebsiella*, *Pantoea*, *Proteus* and *Serratia* (Doulgeraki et al. 2011, Sade et al. 2013, Ntuli et al. 2016, Carrizosa et al. 2017, Silbande et al. 2018).Their spoilage potential of meat and dairy products has been attributed to their simple nutritional requirements, ability to grow at low storage temperatures, favorably at or above 5°C, ability to grow under aerobic or anaerobic conditions and production of heat-stable proteolytic and lipolytic enzymes, amino acid decarboxylases as well as pigments. In addition, soft rot induction in fresh fruits and vegetables has been often attributed to the ability of *Erwinia* spp. to produce pectinolytic enzymes.

The importance of quorum sensing in spoilage caused by members of the *Enterobacteriaceae* family has also been highlighted and a wide range of signal molecules associated with spoilage modulation have been detected (Riedel et al. 2001, Bruhn et al. 2004, Rasch et al. 2005, Liu et al. 2006, van Houdt et al. 2006, Ammor et al. 2008, Blana and Nychas 2014). In addition, their possible effect on biofilm formation (Gerstel and Romling 2001, Chorianopoulos et al. 2010, Blana et al. 2017) has also been evaluated. Furthermore, the efficacy of molecules such as eugenol, coumarin, dihydrocoumarin, petunidin and hexanal to interfere with quorum sensing signals and concomitantly inhibit biofilm production, and attenuate spoilage potential has also been reported (Makhfian et al. 2015, Gopu et al. 2016, Hou et al. 2017, Reen et al. 2018, Zhang et al. 2018). However, no studies are currently available on the transcriptomic organization of the related genes and their responses to environmental stimuli.

Conclusions and Future Perspectives

Taxonomy of Gram negative food spoilers as well as several aspects regarding their genomic and transcriptomic organization have been studied, to some extent, from a molecular microbiology perspective. However, further study is still necessary in order to obtain insights that will allow the design and implementation of appropriate growth-preventive strategies. Such research is expected to take place in the near future. In addition, the transcriptional organization of several metabolic pathways with increased significance in biotechnological applications is also expected to draw specific attention.

References

Adelowo, O.O., S.O. Alagbe and A.A. Ayandele. 2006. Time-dependent stability of used engine oil degradation by cultures of *Pseudomonas fragi* and *Achromobacter aerogenes*. Afr. J. Biotechnol. 5: 2476–2479.

Agaras, B.C., A. Iriarteand C.F. Valverde. 2018. Genomic insights into the broad antifungal activity, plant-probiotic properties, and their regulation, in *Pseudomonas donghuensis* strain SVBP6. PLoS ONE 13(3): e0194088.

Ait-Ouazzou, A., P. Mañas, S. Condón, R. Pagán and D. García-Gonzalo. 2012. Role of general stress-response alternative sigma factors σ^S (RpoS) and σ^B (SigB) in bacterial heat resistance as a function of treatment medium pH. Int. J. Food Microbiol. 153: 358–364.

Alvarez-Ordonez, A., V. Broussolle, P. Colin, C. Nguyen-The and M. Prieto. 2015. The adaptive response of bacterial food-borne pathogens in the environment, host and food: Implications for food safety. Int. J. Food Microbiol. 213: 99–109.

Ammor, M.S., C. Michaelidis and G.-J.E. Nychas. 2008. Insights into the role of quorum sensing in food spoilage. J. Food Prot. 71: 1510–1525.

Bae, Y.-M. and S.-Y. Lee. 2017. Effect of salt addition on acid resistance response of *Escherichia coli* O157:H7 against acetic acid. Food Microbiol. 65: 74–82.

Bai, J.A. and R.R. Vittal. 2014. Quorum sensing regulation and inhibition of exoenzyme production and biofilm formation in the food spoilage bacteria *Pseudomonas psychrophila* PSPF19. Food Biotechnol. 28: 293–308.

Banerjee, G. and A.K. Ray. 2016. The talking language in some major Gram-negative bacteria. Arch. Microbiol. 198: 489–499.

Ben Belgacem, Z., S. Bijttebier, C. Verreth, S. Voorspoels. I. Van de Voorde, G. Aerts, K.A. Willems, H. Jacquemyn, S. Ruytersand B. Lievens. 2015. Biosurfactant production by *Pseudomonas* strains isolated from floral nectar. J. Appl. Microbiol. 118: 1370–1384.

Blana, V., A. Georgomanou and E. Giaouris. 2017. Assessing biofilm formation by *Salmonella enterica* serovar Typhimurium on abiotic substrata in the presence of quorum sensing signals produced by *Hafnia alvei*. Food Control 80: 83–91.

Blana, V.A. and G.-J.E. Nychas. 2014. Presence of quorum sensing signal molecules in minced beef stored under various temperature and packaging conditions. Int. J. Food Microbiol. 173: 1–8.

Brady, C., I. Cleenwerck, S. Venter, T. Coutinho and P. De Vos. 2013. Taxonomic evaluation of the genus *Enterobacter* based on multilocus sequence analysis (MLSA): Proposal to reclassify *E. nimipressuralis* and *E. amnigenus* into *Lelliottia* gen. nov. as *Lelliottia nimipressuralis* comb. nov. and *Lelliottia amnigena* comb. nov., respectively, *E. gergoviae* and *E. pyrinus* into *Pluralibacter* gen. nov. as *Pluralibacter gergoviae* comb. nov. and *Pluralibacter pyrinus* comb. nov., respectively, *E. cowanii*, *E. radicincitans*, *E. oryzae* and *E. arachidis* into *Kosakonia* gen. nov. as *Kosakonia cowanii* comb. nov., *Kosakonia radicincitans* comb. nov., *Kosakonia oryzae* comb. nov. and *Kosakonia arachidis* comb. nov., respectively, and *E. turicensis*, *E. helveticus* and *E. pulveris* into *Cronobacter* as *Cronobacter zurichensis* nom. nov., *Cronobacter helveticus* comb. nov. and *Cronobacter pulveris* comb. nov., respectively, and emended description of the genera *Enterobacter* and *Cronobacter*. Syst. Appl. Microbiol. 36: 309–319.

Brady, C., I. Cleenwerck, S. Venter, M. Vancanneyt, J. Swings and T. Coutinho. 2008. Phylogeny and identification of *Pantoea* species associated with plants, humans and the natural environment based on multilocus sequence analysis (MLSA). Syst. Appl. Microbiol. 31: 447–460.

Bruhn, J.B., A.B. Christensen, L.R. Flodgaard, K.F. Nielsen, T.O. Larsen, M. Givskovand L.Gram. 2004. Presence of acylated homoserine lactones (AHLs) and AHL-producing bacteria in meat and potential role of AHL in spoilage of meat Appl. Environ. Microbiol. 70: 4293–4302.

Caldera, L., L. Franzetti, E. Van Coillie, P. De Vos, P. Stragier, J. De Blockand M. Heyndrickx. 2016. Identification, enzymatic spoilage characterization and proteolytic activity quantification of *Pseudomonas* spp. isolated from different foods. Food Microbiol. 54: 142–153.

Carpentier, W., K. Sandra, I. De Smet, A. Brigé, L. De Smet and J. Van Beeumen. 2003. Microbial reduction and precipitation of vanadium by *Shewanella oneidensis*. Appl. Environ. Microbiol. 69: 3636–3639.

Carpentier, W., L. De Smet, J. Van Beeumen and A. Brigé. 2005. Respiration and growth of *Shewanella oneidensis* MR-1 using vanadate as the sole electron acceptor. J. Bacteriol. 187: 3293–3301.

Carrizosa, E., M.J. Benito, S. Ruiz-Moyano, A. Hernández, M. del Carmen Villalobos, A. Martín and M. de Guía Córdoba. 2017. Bacterial communities of fresh goat meat packaged in modified atmosphere. Food Microbiol. 65: 57–63.

Cebrián, G., C. Arroyo, S. Condón and P. Mañas. 2015. Osmotolerance provided by the alternative sigma factors σ^B and rpoS to *Staphylococcus aureus* and *Escherichia coli* is solute dependent and does not result in an increased growth fitness in NaCl containing media. Int. J. Food Microbiol. 214: 83–90.

Cheng, Y.-Y., C. Wu, J.-Y. Wu, H.-L. Jia, M.-Y. Wang, H.-Y. Wang, S.-M. Zou, R.-R. Sun, R. Jia and Y.-Z. Xiao. 2017. FlrA represses transcription of the biofilm-associated *bpfA* operon in *Shewanella putrefaciens*. Appl. Environ. Microbiol. 83: e02410–16.

Chorianopoulos, N.G., E.D. Giaouris, Y. Kourkoutas and G.J. Nychas. 2010. Inhibition of the early stage of *Salmonella enterica* serovar Enteritidis biofilm development on stainless steel by cell-free supernatant of a *Hafnia alvei* culture. Appl. Environ. Microbiol. 76: 2018–2022.

Coughlan, L.M., P.D. Cotter, C. Hill and A. Alvarez-Ordóñez. 2016. New weapons to fight old enemies: Novel strategies for the (bio)control of bacterial biofilms in the food industry. Front. Microbiol. 7: 1641.

De Keersmaecker, S.C., K. Sonck and J. Vanderleyden. 2006. Let LuxS speak up in AI-2 signaling. Trends Microbiol. 14: 114–119.

Deziel, E., F. Lepine, S. Milot, J. He, M.N. Mindrinos, R.G. Tompkins and L.G. Rahme. 2004. Analysis of *Pseudomonas aeruginosa* 4-hydroxy-2-alkylquinolines (HAQs) reveals a role for 4-hydroxy-2-heptylquinoline in cell-to-cell communication. Proc. Natl. Acad. Sci. USA 101: 1339–1344.

Dodd, C.E.R. and T.G. Aldsworth. 2002. The importance of RpoS in the survival of bacteria through food processing. Int. J. Food Microbiol. 74: 189–194.

Doulgeraki, A.I., S. Paramithiotis and G.J.E. Nychas. 2011. Development of *Enterobacteriaceae* community during storage of minced beef under aerobic or modified atmosphere packaging conditions. Int. J. Food Microbiol. 145: 77–83.

Dufour, D., M. Nicodeme, C. Periin, A. Driou, E. Brusseaux, G. Humbert, J.L. Gaillard and A. Dary. 2008. Molecular typing of industrial strains of *Pseudomonas* spp. isolated from milk and genetical and biochemical characterization of an extracellular protease by one of them. Int. J. Food Microbiol. 125: 188–196.

Ee, R., M. Madhaiyan, L. Ji, Y.L. Lim, N.M. Nor, K.K. Tee, J.W. Chen and W.F. Yin. 2016. *Chania multitudinisentens* gen. nov., sp. nov., an N acyl-homoserine-lactone-producing bacterium in the family *Enterobacteriaceae* isolated from landfill site soil. Int. J. Syst. Evol. Microbiol. 66: 2297–2304.

Ercolini, D., F. Russo, G. Blaiotta, O. Pepe, G. Mauriello and F. Villani. 2007. Simultaneous detection of *Pseudomonas fragi*, *P. lundensis* and *P. putida* from meat by a multiplex PCR assay targeting the *carA* gene. Appl. Environ. Microbiol. 73: 2354–2359.

Ercolini, D., A. Casaburi, A. Nasi, I. Ferrocino, R. Di Monaco, P. Ferranti, G. Mauriello and F. Villani. 2010. Different molecular types of *Pseudomonas fragi* have the same overall behaviour as meat spoilers. Int. J. Food Microbiol. 142: 120–131.

Fredrickson, J.K., J.M. Zachara, D.W. Kennedy, C.X. Liu, M.C. Duff, D.B. Hunter and A. Dohnalkova. 2002. Influence of Mn oxides on the reduction of uranium(VI) by the metal-reducing bacterium *Shewanella putrefaciens*. Geochim. Cosmochim. Acta 66: 3247–3262.

Fu, L., C. Wang, N. Liu, A. Ma and Y. Wang. 2018. Quorum sensing system-regulated genes affect the spoilage potential of *Shewanella baltica*. Food Res. Int. 107: 1–9.

Fukushima, M., K. Kakinuma and R. Kawaguchi. 2002. Phylogenetic analysis of *Salmonella*, *Shigella*, and *Escherichia coli* strains on the basis of the *gyrB* gene sequence. J. Clin. Microbiol. 40: 2779–2785.

Gallagher, L.A., S.L. McKnight, M.S. Kuznetsova, E.C. Pesci and C. Manoil. 2002. Functions required for extracellular quinolone signaling by *Pseudomonas aeruginosa*. J. Bacteriol. 184: 6472–6480.

Ganesh, R., K.G. Robinson, G.D. Reed and G.S. Sayler. 1997. Reduction of hexavalent uranium from organic complexes by sulfate- and iron-reducing bacteria. Appl. Environ. Microbiol. 63: 4385–4391.

Gerstel, U. and U. Römling. 2001. Oxygen tension and nutrient starvation are major signals that regulate *agfD* promoter activity and expression of the multicellular morphotype in *Salmonella* Typhimurium. Environ. Microbiol. 3: 638–648.

Gomila, M., A. Peña, M. Mulet, J. Lalucat and E. García-Valdés. 2015. Phylogenomics and systematics in *Pseudomonas*. Front. Microbiol. 6: 214.

Goo, E., J.H. An, Y. Kang and I. Hwang. 2015. Control of bacterial metabolism by quorum sensing. Trends Microbiol. 23: 567–576.

Gopu, V., C.K. Meena, A. Murali and P.H. Shetty. 2016. Petunidin as a competitive inhibitor of acylated homoserine lactones in *Klebsiella pneumoniae*. RSC Adv. 6: 2592–2601.

Gu, Q., L. Fu, Y. Wang and J. Lin. 2013. Identification and characterization of extracellular cyclic dipeptides as quorum-sensing signal molecules from *Shewanella baltica*, the specific spoilage organism of *Pseudosciaena crocea* during 4°C storage. J. Agric. Food Chem. 61: 11645–11652.

Hasegawa, M., H. Kishino and T. Yano. 1985. Dating the human-ape split by a molecular clock of mitochondrial DNA. J. of Mol. Evol. 22: 160–174.

Hofte, M. and P. de Vos. 2007. Plant pathogenic *Pseudomonas* species. pp. 507–533. *In*: S.S. Gnanamanickam (ed.). Plant-Associated Bacteria. Springer, Heidelberg.

Hou, H.M., F. Jiang, G.L. Zhang, J.Y. Wang, Y.H. Zhu and X.Y. Liu. 2017. Inhibition of *Hafnia alvei* H4 biofilm formation by the food additive dihydrocoumarin. J. Food Prot. 80: 842–847.

Jaishankar, J. and P. Srivastava. 2017. Molecular basis of stationary phase survival and applications. Front. Microbiol. 8: 2000.

Jayaraman, A., A.K. Sunand T.K. Wood. 1998. Characterization of axenic *Pseudomonas fragi* and *Escherichia coli* biofilms that inhibit corrosion of SAE 1018 steel. J. Appl. Microbiol. 84: 485–492.

Jukes, T.H. and C.R. Cantor. 1969. Evolution of protein molecules. pp. 21–132. *In*: H.N. Munro (ed.). Mammalian Protein Metabolism. Academic Press, New York.

Kamran, S., I. Shahid, D.N. Baig, M. Rizwan, K.A. Malik and S. Mehnaz. 2017. Contribution of zinc solubilizing bacteria in growth promotion and zinc content of wheat. Front. Microbiol. 8: 2593.

Liao, C.H. 2006. *Pseudomonas* and related genera. pp. 507–540. *In*: C.W. Blackburn (ed.). Food spoilage microorganisms. Woodhead Publishing Limited, Abington Hall, Abington, Cambridge, England.

Liu, M., J.M. Gray and M.W. Griffiths. 2006. Occurrence of proteolytic activity and N-acyl-homoserine lactone signals in the spoilage of aerobically chill-stored proteinaceous raw foods. J. Food Prot. 69: 2729–2737.

Liu, X., L. Ji, X. Wang, J. Li, J. Zhu and A. Sun. 2018. Role of RpoS in stress resistance, quorum sensing and spoilage potential of *Pseudomonas fluorescens*. Int.J. Food Microbiol. 270: 31–38.

Lu, L., M.E. Hume and S.D. Pillai. 2004. Autoinducer-2-like activity associated with foods and its interaction with food additives. J. Food Prot. 67: 1457–1462.

Lyon, G.J. and R.P. Novick. 2004. Peptide signaling in *Staphylococcus aureus* and other Gram-positive bacteria. Peptides 25: 1389–1403.

Makhfian,M., N. Hassanzadeh, E. Mahmoudi and N. Zandyavari. 2015. Anti-quorum sensing effects of ethanolic crude extract of *Anethum graveolens* L. J. Essent. Oil Bear. Pl. 18: 687–696.

Marchand, S., G. Vandriesche, A. Coorevits, K. Coudijzer, V. De Jonghe, K. Dewettinck, P. De Vos, B. Devreese, M. Heyndrickx and J. De Block. 2009. Heterogeneity of heat resistant proteases from milk *Pseudomonas* species. Int. J. Food Microbiol. 133: 68–77.

Martins, M.L., U.M. Pinto, K. Riedel, M.C.D. Vanetti, H.C. Mantovani and E.F. de Araújo. 2014. Lack of AHL-based quorum sensing in *Pseudomonas fluorescens* isolated from milk. Braz. J. Microbiol. 45: 1039–1046.

Medina-Martinez, M.S., M. Uyttendaele, V. Demolder and J. Debevere. 2006a. Effect of temperature and glucose concentration on the N-butanoyl-Lhomoserine lactone production by *Aeromonas hydrophila*. Food Microbiol. 23: 534–540.

Medina-Martinez, M.S., M. Uyttendaele, V. Demolder and J. Debevere. 2006b. Influence of food system conditions on N-acyl-L-homoserine lactonesproduction by *Aeromonas* spp. Int. J. Food Microbiol. 112: 244–252.

Mei, Y.Z., P.W. Huang,Y. Liu, W. He and W.-W. Fang. 2016. Cold stress promoting a psychrotolerant bacterium *Pseudomonas fragi* P121 producing trehaloase. World J. Microbiol. Biotechnol. (2016)32: 134.

Meng, L., Y. Zhang, H. Liu, S. Zhao, J. Wang and N. Zheng. 2017 Characterization of *Pseudomonas* spp. and associated proteolytic properties in raw milk stored at low temperatures. Front. Microbiol. 8: 2158.

Mohareb, F., M. Iriondo, A.I. Doulgeraki, A. Van Hoek, H. Aarts, M. Cauchiand G.-J.E. Nychas. 2015. Identification of meat spoilage gene biomarkers in *Pseudomonas putida* using gene profiling. Food Control 57: 152–160.

Mossel, D.A.A. and M. Ingram. 1955. The physiology of the microbial spoilage of foods. J. Appl. Bacteriol. 18: 232–268.

Mulet, M., J. Lalucat and E. García-Valdés. 2010. DNA sequence-based analysis of the *Pseudomonas* species. Environ. Microbiol. 12: 1513–1530.

Ng, W.L. and B.L. Bassler. 2009. Bacterial quorum-sensing network architectures. Annu. Rev. Genet. 43: 197–222.

Ntuli, V., P.M.K. Njage and E.M. Buys. 2016. Characterization of *Escherichia coli* and other *Enterobacteriaceae* in producer-distributor bulk milk. J. Dairy Sci. 99: 9534–9549.

Paikhomba Singha, L., R. Kotoky and P. Pandey. 2017. Draft genome sequence of *Pseudomonas fragi* strain DBC, which has the ability to degrade high-molecular-weight polyaromatic hydrocarbons. Genome Announc. 5: e01347–17.

Paradis, S., M. Boissinot, N. Paquette, S.D. Belanger, E.A. Martel, D.K. Boudreau, F.J. Picard, M. Ouellette, P.H. Roy and M.G. Bergeron. 2005. Phylogeny of the *Enterobacteriaceae* based on genes encoding elongation factor Tu and F-ATPase beta-subunit. Int. J. Syst. Evol. Microbiol. 55: 2013–2025.

Peix, A., M.-H. Ramírez-Bahena and E. Velázquez. 2018. The current status on the taxonomy of *Pseudomonas* revisited: An update. Infect. Genet. Evol. 57: 106–116.

Pesci, E.C., J.B. Milbank, J.P. Pearson, S. McKnight, A.S. Kende, E.P. Greenberg and B.H. Iglewski. 1999. Quinolone signaling in the celltocell communication system of *Pseudomonas aeruginosa*. Proc. Natl. Acad. Sci. USA 96: 11229–11234.

Petruzzi, L., M.R. Corbo, M. Sinigaglia and A. Bevilacqua. 2017. Microbial spoilage of foods: Fundamentals. pp. 1–21. *In*: A. Bevilacqua, M.R. Corbo and M. Sinigaglia (eds.). The microbiological quality of food foodborne spoilers. Woodhead Publishing, Duxford, United Kingdom.

Pham, H.N., K. Ohkusu, N. Mishima, M. Noda, M. Monir Shah, X. Sun, M. Hayashi, T. Ezaki, and M.M. Shah. 2007. Phylogeny and species identification of the family *Enterobacteriaceae* based on *dnaJ* sequences. Diagn. Microbiol. Infect. Dis. 58: 153–161.

Pinto Junior, W.R., L.O. Joaquim, P.R. Pereira, M. Cristianini, E.M. Del Aguila and V.M. Flosi Paschoalin. 2017. Effect of high isostatic pressure on the peptidase activity and viability of *Pseudomonas fragi* isolated from a dairy processing plant. Int. Dairy J. 75: 51–55.

Pinto, U.M., E.D. Costa, H.C. Mantovani and M.C.D. Vanetti. 2010. The proteolytic activity of *Pseudomonas fluorescens* 07A isolated from milk is not regulated by quorum sensing signals. Braz. J. Microbiol. 41: 91–96.

Rasch, M., J.B. Andersen, K.F. Nielsen, L.R. Flodgaard, H. Christensen, M. Givskov and L. Gram. 2005. Involvement of bacterial quorum-sensing signals in spoilage of bean sprouts Appl. Environ. Microbiol. 71: 3321–3330.

Reading, N.C., A.G. Torres, M.M. Kendall, D.T. Hughes, K. Yamamoto and V. Sperandio. 2007. A novel two-component signaling system that activates transcription of an enterohemorrhagic *Escherichia coli* effector involved in remodeling of host actin. J. Bacteriol. 189: 2468–2476.

Reen, F.J., J A. Gutiérrez-Barranquero, M.L. Parages and F. O′Gara. 2018. Coumarin: A novel player in microbial quorum sensing and biofilm formation inhibition. Appl. Microbiol. Biotechnol. 102: 2063–2073.

Ribeiro Junior, J.C., A.M. de Oliveira, F. de G. Silva, R. Tamanini, A.L.M. de Oliveira and V. Beloti. 2018. The main spoilage-related psychrotrophic bacteria in refrigerated raw milk. J. Dairy Sci. 101: 75–83.

Riedel, K., T. Ohnesorg, K.A. Krogfelt, T.S. Hansen, K. Omori, M. Givskov and L. Eberl. 2001. N-acyl-L-homoserine lactone-mediated regulation of the lip secretion system in *Serratia liquefaciens* MG1. J. Bacteriol. 183: 1805–1809.

Roggenkamp, A. 2007. Phylogenetic analysis of enteric species of the family *Enterobacteriaceae* using the *oriC*-locus. Syst. Appl. Microbiol. 30: 180–188.

Rutherford, S.T. and B.L. Bassler. 2012. Bacterial quorum sensing: Its role in virulence and possibilities for its control. Cold Spring Harb. Perspect. Med. 2(11): a012427.

Sade, E., A. Murros and J. Björkroth. 2013. Predominant enterobacteria on modified-atmosphere packaged meat and poultry. Food Microbiol. 34: 252–258.

Satomi, M., H. Oikawa and Y. Yano. 2003. *Shewanella marinintestina* sp. nov., *Shewanella schlegeliana* sp. nov. and *Shewanella sairae* sp. nov., novel eicosapentaenoic-acid-producing marine bacteria isolated from sea-animal intestines. Int. J. Syst. Evol. Microbiol. 53: 491–499.

Satomi, M. 2014. The Family *Shewanellaceae*. pp. 597–625. *In*: E. Rosenberg, E.F. De Long, S. Lory, E. Stackebrandt and F. Thompson (eds.). The Prokaryotes, Gammaproteobacteria. Fourth Edition, Springer Heidelberg.

Schafhauser, J., F. Lepine, G. McKay, H.G. Ahlgren, M. Khakimova and D. Nguyen. 2014. The stringent response modulates 4-hydroxy-2-alkylquinoline biosynthesis and quorum-sensing hierarchy in *Pseudomonas aeruginosa*. J. Bacteriol. 196: 1641–1650.

Selvakumar, G., P. Joshi, S. Nazim, P.K. Mishra, J.K. Bisht and H.S. Gupta. 2009. Phosphate solubilization and growth promotion by *Pseudomonas fragi* CS11RH1 (MTCC 8984), a psychrotolerant bacterium isolated from a high altitude Himalayan rhizosphere. Biologia 64: 239–245.

Silagyi, K., S.H. Kim, Y.M. Lo and C.I. Wei. 2009. Production of biofilm and quorum sensing by *Escherichia coli* O157:H7 and its transfer from contact surfaces to meat, poultry, ready-to-eat deli, and produce products. Food Microbiol. 26: 514–519.

Silbande, A., J. Cornet, M. Cardinal, F. Chevalier, K. Rochefort, J. Smith-Ravin, S. Adenet and F. Leroi. 2018. Characterization of the spoilage potential of pure and mixed cultures of bacterial species isolated from tropical yellowfin tuna (*Thunnus albacares*) J. Appl. Microbiol. 124: 559–571.

Smith, D., J.H. Wang, J.E. Swatton, P. Davenport, B. Price, H. Mikkelsen, H. Stickland, K. Nishikawa, N. Gardiol, D.R. Spring and M. Welch. 2006. Variations on a theme: Diverse N-acyl homoserine lactone-mediated quorum sensing mechanisms in Gram-negative bacteria. Sci. Prog. 89: 167–211.

Spector, M.P. and W.J. Kenyon. 2012. Resistance and survival strategies of *Salmonella enterica* to environmental stresses. Food Res. Int. 45: 455–481.

Stanborough, T., N. Fegan, S.M. Powell, T. Singh, M. Tamplin and P.S. Chandry. 2018. Genomic and metabolic characterization of spoilage-associated *Pseudomonas* species. Int. J. Food Microbiol. 268: 61–72.

Tailliez, P., C. Laroui, N. Ginibre, A. Paule, S. Pages and N. Boemare. 2010. Phylogeny of *Photorhabdus* and *Xenorhabdus* based on universally conserved protein-coding sequences and implications for the taxonomy of these two genera. Proposal of new taxa: *X. vietnamensis* sp. nov., *P. luminescens* subsp. *caribbeanensis* subsp. nov., *P. luminescens* subsp. *hainanensis* subsp. nov., *P. temperata* subsp. *khanii* subsp. nov., *P. temperata* subsp. *tasmaniensis* subsp. nov., and the reclassification of *P. luminescens* subsp. *thracensis* as *P. temperata* subsp. *thracensis* comb. nov. Int. J. Syst. Evol. Microbiol. 60: 1921–1937.

Tamura, K. and M. Nei. 1993. Estimation of the number of nucleotide substitutions in the control region of mitochondrial DNA in humans and chimpanzees. Mol. Biol. Evol. 10: 512–526.

Valentini, M., D. Gonzalez, D.A. Mavridou and A. Filloux. 2018. Lifestyle transitions and adaptive pathogenesis of *Pseudomonas aeruginosa*. Curr. Opin. Microbiol. 41: 15–20.

Van Houdt, R., P. Moons, M. HuesoBuj and C.W. Michiels. 2006. N-acyl-L-homoserine lactone quorum sensing controls butanediol fermentation in *Serratia plymuthica* RVH1 and *Serratia marcescens* MG1. J. Bacteriol. 188(12): 4570–4572.

Vithanage, N.R., M. Dissanayake, G. Bolge, E.A. Palombo, T.R. Yeager and N. Datta. 2016. Biodiversity of culturable psychrotrophic microbiota in raw milk attributable to refrigeration conditions, seasonality and their spoilage potential. Int. Dairy J. 57: 80–90.

von Neubeck, M., C. Baur, M. Krewinkel, M. Stoeckel, B. Kranz, T. Stressler, L. Fischer, J. Hinrichs, S. Scherer and M. Wenning. 2015. Biodiversity of refrigerated raw milk microbiota and their enzymatic spoilage potential. Int. J. Food Microbiol. 211: 57–65.

Wiedmann, M., D. Weilmeier, S.S. Dineen, R. Ralyea and K.J. Boor. 2000. Molecular and phenotypic characterization of *Pseudomonas* spp. isolated from milk. Appl. Environ. Microbiol. 66: 2085–2095.

Winzer, K., K.R. Hardie and P. Williams. 2003. LuxS and autoinducer-2: Their contribution to quorum sensing and metabolism in bacteria. Adv. Appl. Microbiol. 53: 291–396.

Xin, L., Z. Meng, L. Zhang, Y. Cui, X. Han and H. Yi. 2017. The diversity and proteolytic properties of psychrotrophic bacteria in raw cows' milk from North China. Int. Dairy J. 66: 34–41.

Yanzhen, M., L. Yang, X. Xiangting and H. Wei. 2016. Complete genome sequence of a bacterium *Pseudomonas fragi* P121, a strain with degradation of toxic compounds. J. Biotechnol. 224: 68–69.

Young, J.M. and D.C. Park. 2007. Relationships of plant pathogenic enterobacteria based on partial *atpD*, *carA*, and *recA* as individual and concatenated nucleotide and peptide sequences. Syst. Appl. Microbiol. 30: 343–354.

Zhang, C., S. Zhu, A.-N. Jatt, Y. Pan and M. Zeng. 2017. Proteomic assessment of the role of N-acyl homoserine lactone in *Shewanella putrefaciens* spoilage. Lett. Appl. Microbiol. 65: 388–394.

Zhang, Y. and S. Qiu. 2015. Examining phylogenetic relationships of *Erwinia* and *Pantoea* species using whole genome sequence data. Antonie Leeuwenhoek 108: 1037–1046.

Zhang, Y., J. Kong, F. Huang, Y. Xie, Y. Guo, Y. Cheng, H. Qian and W. Yao. 2018. Hexanal as a QS inhibitor of extracellular enzyme activity of *Erwinia carotovora* and *Pseudomonas fluorescens* and its application in vegetables. Food Chem. 255: 1–7.

Zhao, A., J. Zhu, X. Ye, Y. Ge and J. Li. 2016. Inhibition of biofilm development and spoilage potential of *Shewanella baltica* by quorum sensing signal in cell-free supernatant from *Pseudomonas fluorescens*. Int. J. Food Microbiol. 230: 73–80.

Zhao, J.S., D. Manno, C. Beaulieu, L. Paquet and J. Hawari. 2005. *Shewanella sediminis* sp. nov., a novel Na+ -requiring and hexahydro-1,3,5-trinitro-1,3,5-triazine degrading bacterium from marine sediment. Int. J. Syst. Evol. Microbiol. 55: 1511–1520.

Zhao, J.S., D. Manno, C. Leggiadro, D. O'Neil and J. Hawari. 2006. *Shewanella halifaxensis* sp. nov., a novel obligately respiratory and denitrifying psychrophile. Int. J. Syst. Evol. Microbiol. 56: 205–212.

Zhao, J.S., D. Manno, S. Thiboutot, G. Ampleman and J. Hawari. 2007. *Shewanella canadensis* sp. nov. and *Shewanella atlantica* sp. nov., manganese dioxide and hexahydro-1,3,5-trinitro-1,3,5-triazine-reducing, psychrophilic marine bacteria. Int. J. Syst. Evol. Microbiol. 57: 2155–2162.

Zhu, J., A. Zhao, L. Feng and H. Gao. 2016. Quorum sensing signals affect spoilage of refrigerated large yellow croaker (*Pseudosciaena crocea*) by *Shewanella baltica*. Int. J. Food Microbiol. 217: 146–155.

12

Applications of Nanotechnology in Food and Agriculture

Rout George Kerry,[1] *Jyoti Ranjan Rout,*[2] *Gitishree Das,*[3]
Leonardo Fernandes Fraceto,[4] *Spiros Paramithiotis*[5] *and*
Jayanta Kumar Patra[3,*]

1. Introduction

Self-assembly or manipulation of individual atoms, molecules, or molecular clusters into structures to create devices and materials with unique dimensions and properties is basically carried out in nanotechnology. The term "nanotechnology" is based on a Greek word "nano", which means "dwarf", and was first introduced by Prof. Norio Taniguchi, impaling 10^{-9} meters, one billionth of something, or roughly the length of three atoms kept adjacent to each other in a row (Bodaiah et al. 2016, Khandel and Shahi 2016). For comparison, this scale can be used to denote the size of a DNA strand which is around 2.5 nm wide, a virus 100 nm, a red blood cell 7000 nm, while a human hair is about 80,000 nm wide (Bhattacharyya et al. 2014). The term nanotechnology is commonly used for the materials ranging within the size of 0.1 to 100 nm. As a result of their nano-dimension, the material displays different properties from bulk. The process of nanoparticle synthesis involves either the top down approach, in which

[1] Department of Biotechnology, Utkal University, Vani Vihar, Bhubaneswar, Odisha, India.
[2] School of Biological Sciences, AIPH University, Bhubaneswar, Odisha, India.
[3] Research Institute of Biotechnology & Medical Converged Science, Dongguk University, Goyang-si, Republic of Korea.
[4] Institute of Science and Technology, São Paulo State University (UNESP), Sorocaba, Brazil.
[5] Department of Food Science and Human Nutrition, Agricultural University of Athens, Athens, Greece.
* Corresponding author: jkpatra@dongguk.edu

the size is reduced to nanoscale range, or the bottom up approach, where individual atoms or molecules are engineered into nanostructures (Kavitha et al. 2013). These approaches could be carried out by the means of physical or chemical processes. Currently, synthesized organic or inorganic nanoparticles, by the means of physical or chemical approaches, results in the production of nanoparticles with the desired characteristics, such as structure, size and quantity. However, the production method is usually a bit expensive, labor-intensive, and gives out certain potential hazardous byproducts that may be toxic to the ecosystem (Thatoi et al. 2016). Thus, an alternative method for the synthesis of nanoparticles is an obvious need of the present century, a method which is at least eco-friendly in nature. This is possible by means of biological process, due to its eco-friendly nature; the process is also known as green synthesis. In this green approach of nanoparticle synthesis, biological agents such as unicellular prokaryotes, unicellular eukaryotes and multi-cellular eukaryotes (plants, algae, fungi) and their secondary metabolites and enzymes are used (Umashankari et al. 2012, Suresh et al. 2014, Thatoi et al. 2016).

The engineered nanoparticles have heterogeneous applications in diversified fields of medical sciences. In particular, their application has been widely evaluated for diagnosis and therapeutics by the means of different forms of biomarkers, cell labelling, contrast agents for biological imaging, antimicrobial agents and controlled drug delivery systems for the treatment of various life-threatening diseases (Bala and Arya 2013, Singh et al. 2015, Hussain et al. 2016). The present chapter is mostly focused on the applications of nanotechnology in the field of food and agriculture as food is one of the basic needs of any living creature, whereas agriculture is the source of economic development.

The estimated FAO (Food and Agriculture Organization) food price index in the year 2000 was 91.1, in 2010 it increased to188.0. However, a stabilizing trend was observed between December 2016 and August 2017, in which it was estimated to be 170.3 and 176.6, respectively (FAO 2017a). According to FAO, cereal production in 2013–14 was 2519.7 million tonnes, while, most recently, in September 2017, it reached 2611.0 million tonnes, creating a difference of 91.3 million tonnes. Likewise for wheat, rice and coarse grains, the differences were 37.4, 8.9 and 45 million tonnes, respectively. Similarly, the world food supply for cereal in 2013–14 was 3052.3 million tonnes, most recently in September 2017 it was 3316.4 million tones with a difference in supply of about 264.1 million tones. Likewise for wheat, rice and coarse grains, the differences were 110.7, 18.4 and 135 million tonnes, respectively (FAO 2017b). In 2017, it was expected that large import volumes and the rising shipping costs might lift up the global food import bill to over USD 1.3 trillion, which is an increment of 10.6% from 2016s global food import bill (FAO 2017c). Presently the world population is 7 billion, the current production and supply should be sufficient to meet the growing need of the present population, however more than 800 million people are chronically hungry, 2 billion are suffering from micronutrient deficiency and around 700 million people living in rural areas are extremely poor. If additional effort to promote pro-poor development is not undertaken and continuity in the similar 'business-as-usual' scenario is allowed, then by 2030, some 653 million people will be still undernourished (FAO 2017d). Moreover, if the predicted 30% population explosion over the next 35 years is near to accurate, then the outcome will be critical. It will further be chronically severe by the year 2050, when the population may reach around 9 to 10 billion, and then 12.3 billion by 2100 (Fita et al. 2015, FAO 2017, Ghosh et al. 2017).

Thus, it is high time to prepare the consequence, more importantly with the present improvement in science and technology, it is best to prevent such scenario from occuring. It is a well-known fact that sustainable agriculture is the basis of food supply, therefore the new challenges rising in the sector of agriculture should be addressed. Some of these challenges include the demand for food (crops, grains, vegetables, etc.), contamination of soil, water, and, concomitantly, the increased risk of diseases to agricultural crop plants, poultry animals as well as threat to the agricultural and poultry production from the improper preservation with respect to changing weather patterns. However, these challenges could be addressed by the revolutionary technology called nanotechnology. Over the past few decades, nanotechnology has continuously amazed the agricultural and food industries with new and sophisticated user-friendly tools for the molecular diagnosis and treatment of diseases, which is rapid and, therefore, time saving, precise and enhances plants' ability to absorb nutrients, etc. Smart sensors and tuned delivery nanosystems in the agricultural industry have been developed for retention and release of agrochemicals, as well as to combat viruses and crop pathogens. In the near future, these nanoparticulated systems will replace the conventional pesticides and herbicides. Nanotechnology will solely protect the ecosystem indirectly through the use of alternative, advanced and sustainable energy supplies and, simultaneously, will be helpful in filtering or catalytically reducing the pollution (Fig. 1). The present chapter provides an overview of these nanotechnological benefits in the field of food and agriculture and the factors that limit these applications and the future prospects.

2. Microbial Mediated Synthesis of Nanoparticles

Nanoparticles can be synthesized by physical, chemical or biological (green synthesis) methods. Physical and chemical methods lead to the production of certain by-products that are toxic, sophisticated, laborious and expensive; therefore, research has focused on the biological alternatives for nanoparticle synthesis. In these biological methods, nanoparticle synthesis takes place withthe use of unicellular prokaryotes (bacteria, cyanobacteria), unicellular eukaryotes (yeast) and multi-cellular eukaryotes (plants, algae, fungus) and their secondary metabolites and enzymes (Bala and Arya 2013, Singh et al. 2015, Hussain et al. 2016, Manivasagan et al. 2016, Guilger et al. 2017). Though plant mediated nanoparticle synthesis was previously evaluated, the utilization of microorganisms and their bioactive compounds, such as secondary metabolites and enzyme mediated nanoparticle synthesis, is the frontier of the current research because of their omnipresence and abundance in every ecosystem, as well as their ability to adapt to extreme conditions (Pantidos and Horsfall 2014). Furthermore, microbes are fast growing, less costly and easy to manipulate. The microbial mediated nanoparticle synthesis offers better control over size through the rate of intracellular particle formation or compartmentalization within the periplasmic space, which could be further manipulated by regulating certain parameters, such as pH, temperature, aeration and substrate concentration (Singh et al. 2015).

As per the literature, a wide variety of microorganisms have been used for nanoparticle synthesis. These include bacteria such as *Lactobacillus acidophilus*,

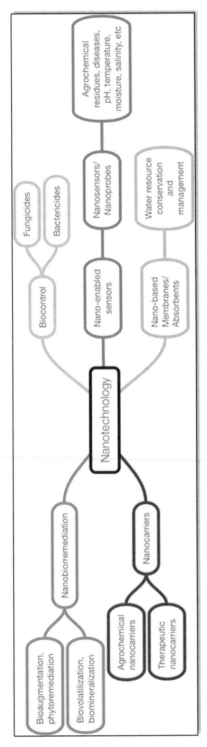

Fig. 1. The diagram depicts the various applications of nanotechnology.

L. fermentum, L. plantarum, L. casei, L. rhamnosus, Magnetotactic sp., *Pseudomonas stutzeri, P. putida, P. fluorescens, Idiomarina baltica, Bacillus subtilis, Shewanella algae, Klebsiella pneumoniae, Stenottrophomonas* sp. and *Marinobacter pelagius* (Garmasheva et al. 2016, Manimaran and Kannabiran 2017), fungi such as *Aspergillus versicolor, A. flavus, A. terreus, A. niger, Fusarium oxysporum, F. acuminatum, Neurospora crassa, Penicillium fellutanum, P. brevicompactum, Thraustochytrium* sp., *Volvariella volvaceae, Trichoderma harzianum* (Netala et al. 2016, Moghaddam et al. 2015, Quester et al. 2016, Guilger et al. 2017) actinomycetes (*Streptomyces hygroscopicus, S. graminofaciens, Streptacidiphilus durhamensis*), yeasts (*Rhodosporidium diobovatum, Pichia capsulata*), algae (*Turbinaria conoides, Colpomenia sinuosa*) and diatoms (*Navicula atomus, Diadesmis gallica*) (Table 1) (Waghmare et al. 2014, Manimaran and Kannabiran 2017).

3. Applications of Nanotechnology in Food and Agriculture

As previously mentioned, the growing population is also associated to the increase in demand for food of about 70–100%. It is also known that the demand for food could be fulfilled through the development of agriculture, which accounts for the largest fraction of energy and water use that is 6–30% which, in turn, results in emission of 20% of all greenhouse gases. The current global water use of 70% is projected to increase up to 83% by 2050 in order to sustain the scarcity of food. Currently 2.5 million tons of pesticides applied per year are lost in run-off and 30–50% is lost in the atmosphere during spray application, which also includes 50–70% loss of nitrogen. Again, between 1.3 and 2 billion tons, which is around 33–50%, of global food produced annually spoils due to microbial contamination, package expiration and is wasted in the supply chain (Rodrigues et al. 2017). All these challenges could be addressed simultaneously with the application of various tools of nanotechnology. Starting from nano-biosensors, which are one of the cutting age technologies applied in the field of agriculture for various purposes such as the determination of moisture content, pH value, salinity, agrochemical residues and disease detection by their high sensitivity, selectivity, fast response and small size (Cheng et al. 2016, Wang et al. 2016). Nanotechnology could also be applicable in biocontrol agents, which includes anti-microbial, anti-fungal or insecticidal, in the form of nanopesticides, nanoherbicides, nanoinceticides, etc. (Chauhan et al. 2016, Garmasheva et al. 2016, Yearla and Padmashree 2016, Zhou et al. 2016). This interdisciplinary technology was also found to be fruitful in the treatment of a wide range of water resources; for example, reuse of agricultural run-off, remediation of polluted water, filtration of harmful water born infections, improvement of conventional energy usage as well as chemical-intensive decontamination processes, such as chlorination and ozonation (Chong et al. 2010, Gehrke et al. 2015). The nanoscale properties are leveraged for the further application of nanotechnology in targeted and controlled delivery of agrochemicals/therapeutic drugs for the improvement of crop yield, nutrition and disease treatment. These controlled delivery vehicles increase the solubility, stability of the bioactive ingredients during storage in the form of nanoemulsions and prevent premature degradation or deterioration, chelation, leaching or volatilization by the means of nanoencapsulation (Nuruzzaman et al. 2016). Finally, nanotechnology also orchestrates a significant role in ameliorating the impacts of previously accumulated

Table 1. Description of implemented microbial mediated nanoparticle synthesis.

Organisms	Nanoparticles	References
Bacterium		
Magnetotactic bacteria	Mg	Xie et al. 2009
Pseudomonas stutzeri	Ag	Klaus et al.1999
Pseudomonas putida	Pd	Gericke and Pinches 2006
Bacillus subtilis	Ag	Vijayaraghavan et al. 2012
Shewanella algae	Ag	Babu et al. 2014
Klebsiella pneumoniae	Au	Malarkodi et al. 2013
Pseudomonas fluorescens	Ag	Prabhawathi et al. 2012
Stenottrophomonas sp	Au	Malhotra et al. 2013
Marinobacter pelagius	Au	Sharma et al. 2012
Fungus		
Aspergillus terreus	Ag	Li et al. 2012
Neurospora crassa	Ag	Quester et al. 2016
Fusarium acuminatum	Ag	Ingle et al. 2008
Penicillium fellutanum	Ag	Priya and Sivakumar 2015
Verticillium sp.	Fe_3O_4	Moghaddam et al. 2015
Thraustochytrium sp.	Ag	Kalidasan et al. 2015
Penicillium brevicompactum	Ag	Saxena et al. 2014
Volvariella volvaceae	Ag-AuF	Moghaddam et al. 2015
Trichoderma harzianum	Ag	Guilger et al. 2017
Actinomycetes		
Streptomyces sp.	Au	Karthik et al. 2013
Thermoactinomyces sp.	Ag	Deepa et al. 2013
Actinomycete	Ag	Narasimha et al. 2013
Yeast		
Rhodosporidium diobovatum	PbS	Seshadri et al. 2011
Pichia capsulata	Ag	Subramanian et al. 2010
Algae		
Turbinaria conoides	Ag, Au	Vijayan et al. 2014
Colpomenia sinuosa	Ag	Kiran and Murugesan 2014
Diatoms		
Navicula atomus	Au	Schrofel et al. 2011
Diadesmis gallica	Au-Silica	

xenobiotics or agrochemical residues which still remained after the runoff water is gone by the means of nano-bioremediation or nano-fertilization (Aghajani and Soleymani 2017, Cecchin et al. 2017). These applications are explained in details in the following sections.

3.1 Nano-Enabled Sensors

Improvement of agricultural productivity in terms of management of water resources, energy and agrochemical input may be achieved through nano-enabled sensors or biosensors and other chromogenic or fluorescent-based probes by supplying extensively dispersed and real-time sensing, which are significant requirements for crops and livestock at the elevated spatial and temporal resolution needed for exactitude agricultural approaches (Rai et al. 2012, Sekhon 2014, Omanovic-Miklicanin and Maksimovic 2016). Nanosensors are devices used for the detection and quantification of a particular substance through rapid and reversible binding, whereas nanoprobesare devices that specifically bind to a particular substance with accuracy and high sensitivity and specificity (Omanovic-Miklicanin and Maksimovic 2016, Kaushal and Wani 2017, Rodrigues et al. 2017).

For agricultural analysis, research and development of nanosensors have shown spectacular accuracy in the measurement of pesticides, fertilizers, nucleic acids and proteins in water and agri-food systems, including DDT, 2,4-dichlorophenoxyacetic acid, carbofuran, triazophos, pyrethroid, fenitrothion, pirimicarb, dichlorvos, paraoxon, methyl parathion, dimethoate or residues of organochlorines, organophosphates and carbamates (Kaushal and Wani 2017, Prasad et al. 2017). Similarly, nanoprobes have also contributed greatly towards the agricultural development by active detection of substances atminimum detectable concentration at femtomolar level. Nanoprobes have also been successfully evaluated for the detection of minute molecules, colorant additives and adulterants in foods, non-invasive gas sensing, for example, to detect oxygen, carbon dioxide, gaseous amines or volatile organics (fruit spoilage) and lastly, foodborne pathogens (Peng et al. 2013, Rodrigues et al. 2017). The advancement and further development of these nanosensors and nanoprobes is a field of active research mostly due to their diversity and their capacity for combination with microfluidic devices and cantilever arrays to build multianalyte array sensors, allowing for the improvement of selectivity and accuracy in the detection of target analytes such as antigens, DNA, urea, glucose and agrochemical residues (Rai et al. 2012, Peng et al. 2013). The efficiency of these nanosensing devices is associated with their increased surface loading of biologically active elements, such as enzymes, antibodies, lectins or DNA/ RNA per mass of material or in the transducers which provide effective selectivity and interface sensitivity (Noah et al. 2012).

Presently, nanosensor mediated sensing is focused mainly in the detection of pathogenic microorganisms or toxins in food and water. For that purpose, gold, silver, cerium oxide, carbon nanotubes, single walled or multi-walled nanotubes, as well as graphene magnetic nanoparticles, are mostly used for sensing. Some of these nanoparticles, such as gold and silver nanoscale structures, were found to provide enhanced Surface-Enhanced Raman Scattering (SERS) signals, thereby broadening the limit of detection and simultaneously speeding up SERS-based detection of chemical contaminants, such as melamine and perchlorate, as well as foodborne pathogens, such as *Salmonella* sp., or viruses (Duncan 2011, Bulbul et al. 2015, Kaushal and Wani 2017). Moreover, the detection of toxins, such as aflatoxin B1, which is a common mycotoxin produced by *A. parasiticus*, *A. flavus* and ochratoxin, a mycotoxin produced by *Aspergillus* sp., could be detected by nano-based electrical sensing devices (Kuswandi et al. 2017). The development of a very interesting nano-based sensing device was reported by Liu et al. (2015). With that device, the detection of *Vibrio parahaemolyticus* in shrimp samples in just about 4.5 hours by an antibody-coated superparamagnetic nanoparticle was achieved. Furthermore, the time required for the detection of *V. parahaemolyticus* in oysters was reduced to 2.5 hours by incorporating enzyme-modified silica nanoparticles (Park and Choi 2017). Development of nanoprobes for effective detection of agrochemical residues and analysts of food contaminants is a mature field of research where companies like AgroMicron Ltd. (Singapore) and Kraft Foods Inc. (United States) are commercializing their products (Rodrigues et al. 2017).

3.2 Nanoparticle Mediated Biocontrol

An estimate of world health organization states that roughly 1 in every 10 people throughout the world experience illness due to consumption of contaminated food and 420,000 die as a result, including 250,000 children under the age of 5, which is approximately 9% of the total population of 2015 (WHO 2015). Furthermore, 33–50% of global food produced annually spoils due to microbial contamination, package expiration and is wasted in the supply chain, which has a wide adverse effect both on the world population and the system which is built to sustain the population (Rodrigues et al. 2017). Thus, in the current era, due to the development of science and technology,new and effective measures of prevention and control have emerged, such as nanotechnological tools, which include nanoparticles with antimicrobial agents.

3.2.1. Antibacterial activity. Nanoparticles possess certain unique attributes, like high surface to volume ratio, which improves efficient penetration, increases ion release, and photocatalytic activity that may concomitantly lead to improved antimicrobial activity. This antimicrobial activity could be further improved by functionalizing the nanoparticle with antibody for selective targeting of specific pathogens. There is a large number of publications in which the antimicrobial (anti-bacterial and/or anti-fungal) activity of nanoparticlesisis evaluated. The following are among the most characteristic ones.

Silver nanoparticles synthesized by the means of *B. methylotrophicus* DC3 were evaluated for their antimicrobial activity against *S. enterica*, *Escherichia coli*, *V. parahaemolyticus* and *Candida albicans*. The results confirmed that enhanced antimicrobial activity was exhibited against *C. albicans* (Wang et al. 2015). Garmasheva et al. (2016) evaluated the nanopartical synthesizing ability and antimicrobial efficacy of 22 different strains of *L. acidophilus* against Gram-positive opportunistic pathogens, such as *Staphylococcus aureus*, *S. epidermidis*, *B. cereus* and some Gram-negative bacteria, like *Pseudomonas aeruginosa*, *Proteus vulgaris*, *E. coli*, *Klebsiella pneumoniae*, *Salmonella enterica*, *Shigella sonnei*, *Sh. flexneri* and a yeast *C. albicans*. It was reported that nanoparticles obtained from *L. acidophilus* 58p were more active against *S. epidermidis*, *E. coli*, *K. pneumoniae*, *Sh. flexneri* and *Sh. sonnei* (Garmasheva et al. 2016). In another study, silver and gold nanoparticles were synthesized by *Sporosarcina koreensis* DC4 and were checked for their antimicrobial activity against *S. enterica*, *E. coli*, *V. parahaemolyticus*, *B. anthracis*, *B. cereus* and *S. aureus*. Moreover, the enhanced antimicrobial activity of silver and gold nanoparticles at a concentration of only 3 µg, with commercial antibiotics, such as vancomycine, rifampicin, oleandomycin, penicillin G, novobiocin and lincomycin, were obtained (Singh et al. 2016). Recently, gold nanoparticles were synthesized using *Brevibacillus formosus* and antimicrobial activity was checked against *S. aureus* and *E. coli*. It was found that the nanoparticles were more active against *S. aureus* and less active against *E. coli* (Srinath et al. 2017). Du et al. (2017) demonstrated the antimicrobial activity of silver nanoparticles synthesized by means of *Novosphingobium* sp. THG-C3 against *S. aureus*, *E. coli*, *V. parahaemolyticus*, *S. enterica*, *B. subtilis*, *B. cereus*,

P. aeruginosa, C. tropicalis, C. albicans, as well as their enhanced antimicrobial activity in combination with commercial antibiotics against *P. aeruginosa, S. enterica, E. coli* and *V. parahaemolyticus* (Table 2).

3.2.2. Antifungal activity. Besides antibacterial activity, nanoparticles have also been effectively explored for their antifungal activity. There are a number of studies exhibiting the antifungal effect of bacteria-mediated synthesized nanoparticles, such as *L. acidophilus* 58p, *B. methylotrophicus* DC3 against *C. albicans* and *Novosphingobium* sp. against both *C. albicans* and *C. tropicalis* (Wang et al. 2015, Garmasheva et al. 2016, Du et al. 2017). Presently, endophytic fungi are of prime concern of agricultural research because of their synergetic function with plants. Besides this, the efficacy of the synthesized nanoparticles as antimicrobial agents further enhances the significance of these endophytic fungi. *A. versicolor* ENT7, an endophytic fungus isolated from an ethnomedicinal plant *Centella asiatica,* was used for synthesizing silver nanoparticles, the antimicrobial activity of which was evaluated against certain pathogenic bacteria and fungi, such as *C. albicans* and *C.* non-*albicans.* It was reported that the nanoparticles synthesized from *A. versicolor* ENT7 possess excellent antifungal properties (Netala et al. 2016). Another study suggested the ability of grape molding fungus *P. citrinum* to synthesize silver nanoparticle that exhibited impressive antifungal activity against aflatoxigenic *A. flavus* var. *columnaris* (Yassin et al. 2017). Despite the efficacy of these nanoparticles as an antifungal agent, their application as fungicides is less explored. Therefore, it can be expected that both agricultural and food science could be further developed if this wing of nanoscience explored a bit more.

3.3 Mechanism of Nanoparticle Mediated Antimicrobial Activity

Currently, nanoparticles are produced in large quantities worldwide because of their wide application in meditational, industrial and agricultural sectors. It is estimated that the production of these nanoparticles reaches 500 tons for silver nanoparticles, 550 tons for zinc oxide nanoparticles, 3,000 tons for titanium oxide nanoparticles and 5,500 tons for silicon dioxide nanoparticles (Hsueh et al. 2015, Calderon-Jimenez et al. 2017). One of the most common applications of these nanoparticles is their use as antimicrobial agents. Indeed, their excellent antimicrobial properties are well documented; however, the exact underlying mechanisms of the nanoparticle mediated microbial toxicity remains the subject of controversy (Hsueh et al. 2015). According to current scientific evidence, nanoparticles containing Ag^+ ions, may lead to microbial cell wall rupture, protein degradation, halt cell respiration and concurrently induce cell mortality. In addition, bacterial contact with the nanoparticles plays also a key role, which depends upon the method of synthesis, size, shape and encapsulation of the nanoparticles (Yamanaka et al. 2005, Fabrega et al. 2009, Kawata et al. 2009).

The effect that the nanoparticles may have on the cellular, physiological and morphological status of a microorganism is concentration-dependent. For example, *B. subtilis,* a symbiotic soil bacterium, is susceptible to silver nanoparticles of concentration as low as 10 ppm. Moreover, at concentrations above 25 ppm, cell growth arrest and chromosomal DNA degradation were evident (Hsueh et al. 2015).

The antimicrobial activity of nanoparticles has been explained by three models: oxidative stress induction, metal ion release and non-oxidative mechanisms (Fig. 2)

Table 2. Application of important nanoparticles in agriculture and food industry.

Nanoparticles	Application	References
Ag	Antibacterial and Antifungal Activity	Shivakrishna et al. 2013
Ag	Antimicrobial Activity	Prabhawathi et al. 2012
Ag	Antifungal Activity	Vijayaraghavan et al. 2012
Ag, Au	Antibiofilm activity	Vijayan et al. 2014
Ag	Anti-diabetic Activity	Kiran and Murugesan 2014
Ag	Antibacterial Activity	Shiny et al. 2013
Ag	Antimicrobial Activity	Kathiresan et al. 2010
Ag	Disinfecting filters and Coating materials	Priya and Sivakumar 2015
Ag	Antimicrobial Activity	Kalidasan et al. 2015
Au	Antimalarial Activity	Karthik et al. 2013
Ag	Antimicrobial Activity	Deepa et al. 2013
Ag	Antimicrobial Activity	Narasimha et al. 2013
Ag	Biocontrol	Lamsal et al. 2011
ZnO	Transparent electronics, ultraviolet light emitters, piezoelectric devices, chemical sensors, spin electronics, enhance crop growth	Sabir et al. 2014
CeO_2	Stress response and tolerance	Zhao et al. 2012
Fe_2O_3	Insecticide	Nhan et al. 2015
Fe_2O_3	Nano-fertilizer	Rui et al. 2016
CeO_2, TiO_2	Nano-fertilizer	Poscic et al. 2016
Nanocomposites	Efficient bioactive packaging of goods	Moreira et al. 2013
Nanocomposites	Nanofertilizers	Giroto et al. 2017
Bionanocomposite	Light energy conversion, biosensing of agrochemical residues	Nagy et al. 2014
Nanobiocomposite	Biosensor	Jamir and Mahato 2016
NP-based sensors	Assessing food safety	Bulbul et al. 2015
Nano-biosensors	Labelling products and automated storage	Ali et al. 2017
Superparamagnetic silica nanoparticles	Water filteration	Gonzalez-Fuenzalida et al. 2014
Quaternized chitosan-capped mesoporous silica nanoparticles	Fungicidal activity	Cao et al. 2016
Hydrous Fe_2O_3 nanoparticles with carboxymethyl cellulose or starch	Detection and removal of arsenate from water	Huo et al. 2017
Electro-conductive nanopolymer	Efficient nanofiltration processes	Mamun et al. 2017, Formoso et al. 2017
Nanostructured lipid carriers	Antimicrobial activity	Cortesi et al. 2017
Optimized diuron nanoformulation	Nanoherbicide	Yearla and Padmashree 2016
Cu $(OH)_2$	Nanopesticides	Zhou et al. 2016
AgNPs/ oxMWCNTs	Insecticide	Hsu et al. 2017
Au nanoparticles-based electrochemical biosensor	Rapid and sensitive detection of plant pathogen DNA	Lau et al. 2017

Table 2 contd....

Table 2 contd....

Nanoparticles	Application	References
Curcumin conjugated diamond nanoparticle	Chicken embryo development	Strojny et al. 2016
MWCNT and cotton CNF	Biosensor of ecotoxicity	Pereira et al. 2014
CNT	Bioremediation (bioaugmentation)	Zhang et al. 2015
Nano zerovalent iron with organic polymer	Promotes the growth and provides greater biodegradation of xenobiotics	Cecchin et al. 2016
Polymers based nanoparticles	Nanofertilizers for plant growth promotion	Nakasato et al. 2016
Chitosan nanoparticles	Plant growth promotion	Pereira et al. 2017a,b
CuO-MWCNTs	Detection of glyphosate in water	Chang et al. 2016

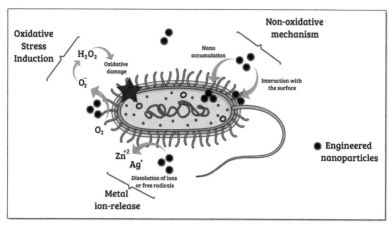

Fig. 2. Mode of antibacterial activity caused by nanoparticles (Concept developed from Wang et al. 2017, Jiang et al. 2010).

(Wang et al. 2017). Based on their surface to volume ratio, nanoparticles or solutes appear as small granular (electron dense) structures, which are either accumulated within the cell or adhered to the cell wall (Tripathi et al. 2017). Attachment and accumulation of these nanoparticles on the bacterial cell wall (composed of lipopolysaccharide, lipotechoic acid, proteins and phospholipid), specifically by silver ions,further leads to the alteration in integrity of the cell by continuous leakage of intracellular potassium which changes the permeability of the membrane and inhibits respiration of the bacterial cell (Vardanyan et al. 2015). The probable mechanism for the invasion of silver ion in the membrane includes its ability to interact with thiol groups found on the surface of proteins or, secondly, the membrane phospholipids (Navarro et al. 2008, Jiang et al. 2010). Similarly, fungi contain chitin as the major cell wall component, which is semipermeable in nature and permits the translocation of small particles while blocking larger ones. Sometimes, when the silver nanoparticles accumulate after the entry, it results in the formation of pits, which later results in the formation of pores the mediate the translocation of larger molecules. This deregulated inflow and outflow of the solute results in the destabilization of the membrane which directly or indirectly leads to respiration inhibition (Tripathi et al. 2017). Further, after entry into the bacterial cell, the nanoparticles interact with basic components of cell, such as DNA, lysosomes, ribosomes and enzymes, etc. This leads to oxidative stress, differential alterations in the membrane permeability, disorder in electrolytic balance, enzyme inhibition, deactivation of proteins and finally diversified changes in gene expression, which may directly or indirectly lead to cell cytotoxicity (Wang et al. 2017).

3.4 Nanotechnology Mediated Water Management

Approximately 71% of earth's surface is covered by water, of which only 3% is freshwater; 69% of freshwater is trapped in glaciers, 30% is underground and less than 1% is distributed in the lakes, rivers and swamps (https://www.e-education. psu.edu/earth103/node/701). Presently, 70% of available freshwater is withdrawn for irrigation, however, according to the estimations, it is concluded that agricultural requirements are met by less than 50% (FAO/IAEA 2017a, Catley-Carlson 2017). An estimation of WWAP report states that 42% of the 70% of water consumed by irrigation is released as agricultural drainages (WWAP 2017).Global requirement of water for agriculture by 2050 is expected to increase by 50% in order to meet the increased food demands of the growing population (FAO/IAEA 2017b). Furthermore, a serious threat for the reuse of water is the relatively high cost and energy consumption of the chemical-mediated treatment that is required. Therefore, to meet the present and future water demand for agriculture, effective measures and advanced technologies should be acknowledged. Nanotechnology is one such technology which could be used as a measure for conservation,treatment or reuse of water in order to meet the future scarcity of water.

Some of the nanotechnology-based achievements for increasing water availability include the solar-driven processes nanoparticle tuned water treatment systems, which are simultaneously less costly and conserve energy in comparison to traditional water treatment systems (Chong et al. 2010, Qu et al. 2013). Nanomaterial-based membranes, such as carbon nanotube (CNT), graphane oxide, dendrimers, polymeric materials (such as polyethersulfone, polyvinylidene difluoride, polyacrylonitrile), nanoporous ceramics, thin layer of nanocomposite, nanofibers, zeolites and nanosponges, are used for higher water recovery rates and lower energy consumption (Kumar et al. 2014, Le and Nunes 2016, Devanathan 2017). Enhanced absorption by superparamagnetic nanomaterials, such as superparamagnetic silica nanoparticles in magnetic in-tube solid phase microextraction coupled to capillary LC, was evaluated for its efficiency to detect analytes like atrazine, terbutylazine and simazine (Gonzalez-Fuenzalida et al. 2014). Hydrous ferric oxide nanoparticles with carboxymethyl cellulose or starch as modifier were synthesized for efficient detection of arsenate and its removal from water (Huo et al. 2017). Nanofiltration membrane fouling is one of the critical obstructions in the path of development of sustainable energy-efficient nanofiltration processes. To counteract or to mitigate fouling, electro-conductive membranes are being developed using electro-conductive polymers, such as polyaniline, polypyrrole and carbon nanotubes, which further enhances separation performances, flux and perm selectivity (Formoso et al. 2017, Mamun et al. 2017). The nano-engineered fabrication process uplifts wettability, surface roughness, electric charge, thickness and provides improved structure of the designed membrane in comparison to conventional fabrication technologies. However, few nano-based products currently using the above technologies, such as nano-adsorbents, nano-enabled membranes, nano-photocatalysts and nano-enabled disinfection systems, are being commercialized for the efficient treatment of water and its reuse (Rodrigues et al. 2017).

3.5 Nanocarriers

Nanocarriers may be defined as "nanomaterial of natural or synthetic origin with the ability to encapsulate or functionalize with single or multiple substances with the intention of delivering/releasing the particular substance at the preferred site or location". Generally, nanocarriers involve different types of inorganic or organic nanomaterials, and the overall sizes along with the load range within 1000 nm (Siafaka et al. 2016). These nanocarriers vary based on their chemical composition and their origin and may either be organic, inorganic or a hybrid. The inorganic nanocarriers include silica dioxide (porous, non-porous, mesoporous), titanium oxide, superparamagnetic and hydrous ferric oxide nanoparticles (Baeza et al. 2017). The organic nanocarriers include carbon-based nanocarriers, such as carbon nanotubes (SWCNT, MWCNT), graphenes, fullerenes, nanodiamonds, as well as polymer-based nanocarriers, such as micelles, emulsions, dendrimers, liposomes, etc. (Baeza et al. 2017, Naz et al. 2017).

3.5.1. Controlled delivery of agrochemicals. The size of these nanoscale-nanoformulated particles maximizes the beneficiary role of nanotechnology in controlled delivery of agrochemicals for enhancing nutrition contents as well as crop yield. The tuned targeted delivery of agrochemicals is also being exploited for the improvement of products during storage through the use of nanoemulsions that increase the stability and solubility of the active ingredients within the products and prohibit their premature degradation. Further, by the means of nanoencapsulation, volatilization/leaching of these active ingredients could be inhibited. Overall, these nanoformulations that are generally a mixture of different agrochemically active compounds and nanocarriers either in the form of nanocapsule or nanosphere are currently being explored for improvement of agricultural productivity (Sekhon et al. 2014, Seabra et al. 2015, Faceto et al. 2016, Grillo et al. 2016). One such nanoformulation is $Cu(OH)_2$ nanopesticide, which was proved to be beneficial for the regulation of nutritional level of *Lactuca sativa*. It was reported that the deposition of copper in the vascular bundle of leaves for foliar application, the levels of which were within recommended guidelines, increased concentration of certain minerals, specifically potassium, and up-regulated many physiologically important proteins, vitamins and phytohormones (Zhao et al. 2016). Optimized diuron nanoformulation is another such nano-product which is a stable nanoherbicide made from diuron (1, 1-dimethyl, 3-(3', 4'-dichlorophenyl) urea) and stem lignin as a matrix for controlled release in *Leucaena leucocephala*. Apart from its function as nanoherbicide, this nanoformulation also offers protection to the active ingredient from microbial degradation and UV damage, and its matrix lignin may also act as a fertilizer, which is helpful tothe farmers from an economic point of view (Yearla and Padmashree 2016).

Another interesting example from the literature is the development of poly-epsilon-caprolactone nanocapsules loaded with the herbicide atrazine (Grillo et al. 2012, Oliveira et al. 2015a,b). In this case, the authors described that the same herbicide activity in the target organism (mustard plants) was achieved through this nanoformulation with ten times reduction in the atrazine concentration in relation to the commercial formulation. Moreover, the authors showed that the nanoatrazine does not produce effect in non-target plants (such as *Zea mays*).

In other studies involving herbicides (Oliveira et al. 2015, Maruyama et al. 2016) and fungicides (Campos et al. 2015), the encapsulation of the active ingredients in polymeric nanoparticles and lipid nanoparticles showed promising results for weed and pest control. It has been shown that the encapsulation changes the release profile of the active ingredient, decreases the toxicity and increases the biological effect, indicating that the association of active ingredients and nanotechnology could be an interesting strategy.

In addition, nanolipid carriers are known for encapsulating natural molecules with antimicrobial activity, such as plumbagin, hydroquinon, eugenol, alpha-asarone and alpha-tocopherol. The efficacy of this system in controlling plants pathogenic microbes by increasing solubility/ rapid penetration and targeted delivery of natural antimicrobial molecules encapsulated within it has been reported by Cortesi et al. (2017). Although promising results have been obtained regarding the application of nanocarriers for the efficient delivery of agrochemicals, there is a wide gap between their laboratory success and their application in the field by the farmers. Therefore, further research is required for their large scale production and widespread use by the agriculturists (Grillo et al. 2016).

3.5.2. Controlled delivery of therapeutics. The role of nanocarriers in the delivery of therapeutics is undisputed. The poultry product's qualities as well as quantity both improved simultaneously with the implementation of nanotechnology for restricting disease prevalence and delivery of nutrient supplements, antioxidants and micronutrients. The delivery of these bioactive compounds improves efficient protein translation and its delivery using micelles, liposomes, nanoemulsions, biopolymeric, solid lipid nanoparticles and dendrimers (Sekhon 2014). Currently, research is mostly focused on the delivery of antineoplastic, antimicrobial, analgesics and inflammatory agents (Nabawy et al. 2014, Mohammadi et al. 2015, Hosoya et al. 2016, Strojny et al. 2016, Stoicea et al. 2017). The efficacy with which these nanocarriers reach the targeted site lies in their ability to accumulate at the site of increased vascular permeability or is facilitated by tuning the surface of the specific targeting moiety (Underwood and van Eps 2012).

Various trials have been made for fertility, purification of bull semen, commercial artificial insemination via sperm-mediated gene-transfer, tagging sperm or loading it with exogenous proteins (Sutovsky and Kennedy 2013). Works were also undertaken, where the antifungal activity of ZnO and Fe_2O_3 nanoparticles against *A. flavus* as feed additives was evaluated (Nabawy et al. 2014). Addition of ZnO nanoparticles to the dry diets in comparison to the wet diet significantly improved carcasses yield and simultaneously increased the relative weight of digestive as well as lymphoid organs of the broilers during the initial periods (Mohammadi et al. 2015). Hosoya et al. (2016) developed a hydrogel-based single nanostructure embedded with tumor targeting, photon-to-heat conversion and triggered drug delivery, which enabled multimodal imaging and controlled release of therapeutic cargo. This multifunctional nanostructure could enable the performance of such at ask, due to the fact that the nanoparticles were densely packed within the nanostructure; their surface plasomon resonance shifted to near-infrared, which enabled cargo release via laser-mediated photothermal mechanism (Hosoya et al. 2016). There area number of FDA-approved nano-analgesics, such as Exparel and DepoDur; the former is a liposome-based bupivacaine topical formulation and the latter another liposome formulation of morphine sulfate, which offers an extended release time and was designed for use in postoperative pain management (Stoicea et al. 2017). Certain natural inflammatory agents, such as curcumin conjugated with diamond nanoparticle (carbon allotropes), were tested in the chicken embryo model. The experiment showed that neither curcumin nor diamond nanoparticle or their bio-complexes affected the development of the embryo (Strojny et al. 2016). These findings indicated that nanocarriers are an effective mode of therapeutic bioactive compound delivery.

3.6 Nano-Bioremediation

Bioremediation is the process of consuming or degrading environmental pollutants with the aid of naturally occurring or deliberately introduced living organisms (plants, bacteria, fungus, algae, etc.), or their products, such as enzymes, in order to clean up the polluted site (Kumar and Gopinath 2016). This ecofriendly process consists of various remediation strategies, such as bioaugmentation mediated by biostimulation, phytoremediation, biovolatilization and biomineralization (Conesa et al. 2012, Zhang et al. 2015). When these strategies of bioremediation are conjugated with nanotechnology, this amalgamating process can be called nano-bioremediation (Kumar and Gopinath 2016).

Bioaugmentation is a natural attenuation process which involves the addition of indigenous or genetically modified microorganisms, stimulated by the addition of biostimulants in the form of nutrients to biodegrade recalcitrant molecules at the site of pollution (Kumar and Gopinath 2016, Nzila et al. 2016). Emerging evidence suggests that this remediation strategy (bioaugmentation) can be greatly improved by functionalizing it with nanotechnology. For example, CNTs have been reported to control the growth of the bacterial species *Arthrobacter* spp. at different concentrations. *Arthrobacter* spp. biodegrades various organic pollutants, like agrochemical pesticides, and wide spread environmental pollutants, like hexavalent chromium, to a reduced form, trivalent chromium, which is less toxic than hexavalent chromium. It has been reported that this bacterium is able to biodegrade atrazine more efficiently in the presence of CNTs, to which the pollutant is initially absorbed (Zhang et al. 2015). The process by which living organisms produce minerals, such as silicates (vertebrates, algae and diatoms), carbonates (invertebrates), calcium and phosphates (vertebrates) and magnetite and/or greigite (magnetotactic bacteria), can be collectively termed as biomineralization (Dhami et al. 2013, Lin et al. 2017, Krajina et al. 2018). Recently, Sakaguchi et al. (2017) reported that efficient biomineralization could also be obtained by certain biomolecules, like silver biomineralization peptide which regulates the morphology of silver nanostructure formation. Microbiota present in the soil are benefited by nano zerovalent iron stabilized with organic polymers, which also promotes growth and provides greater biodegradation of xenobiotics or toxic substances, resulting in the enhancement of the decontamination at the site (Cecchin et al. 2016). There area number of metallic nanoparticles (platinum, rhodium, lead, iridium, silver, copper, cobalt, nickel, gold, aluminum, cadmium, selenium, zinc,etc.), carbon derived nanometerials, such as CNTs (SWCNTs, MWCNTs), graphene and fullerenes, and certain polymeric nanoparticles, such as dendrimers, liposomes, polyaniline and nanowires of polypyrrole, which have being used for ameliorations of containments of soil and water (Rizwan et al. 2014, Zhang et al. 2016, Yadav et al. 2017). Enzyme functionalized nanoparticles for bioremediation processes are currently being acknowledged because of their convenient amalgamation with phytoremediation, which makes their use greener and cheaper. There are several organic compounds, like long-chain hydrocarbons and organochlorines, which cannot be degraded by microorganisms or plants or their enzymes alone; instead, these contaminants could be easily degraded by nanocapsulated enzymes or magnetic iron nanoparticles functionalized with enzymes (Enzyme linked magnetic nanosphere), representing the joint contribution of both biotechnology and nanotechnology (Rizwan et al. 2014, Yadav et al. 2017). Immobilization of enzymes and other biomolecules using these approaches could prove promising in the development and sustainability of these ecofriendly remediation strategies by providing an excellent opportunity to improve the half-life and reusability of enzymes, thus making the processes less costly.

4. Barriers and Overlapping Challenges in the Application of Nanotechnology

Today, nanotechnology is one of the most explored and interactive branches of science and technology. Nanoparticle synthesis and characterization requires the most sensitive, robust instrumentation and sophisticated procedures. The applications have touched almost every aspect that mankind has ever imagined, starting from the basic needs, such as food, clothing and shelter, to the distant stars and galaxies light years away that astrophysicists dream of knowing and understanding (Meador et al. 2010). This extensive application of nano technology has lead us to encounter certain unpredicted barriers and overlapping challenges, such as gaps in the actual utility and field application, fluctuation in the cost-to-value ratio of the innovations made for the conservation of water and energy and the sustainable development of agriculture, and finally, nanoaccumulation resulting in nanotoxicity which raises the public safety concerns regarding the utilization of nanomaterials in food products and water (Rodrigues et al. 2017).

According to the World Economic Outlook, the global grow this projected to be 3.7% in 2018, and by 2020 the global agricultural market is estimated to be worth more than US$3.4 trillion (Duhan et al. 2017, IMF 2017). Presently, there are more than 400 companies throughout the world working on nanotechnological research and development, and in the next 10 years it is expected that this number will increase to 1000 (Duhan et al. 2017). Though almost all the necessary aspects of agricultural development could be bridged by nanotechnological achievements, the complete utilization of the technology is still lagging behind in the present era. This gap may be due to farmers' unawareness of the recent achievements and applicability (Mishra et al. 2017). In addition, the high cost of some nanotechnology-based products, alongwith the related biosafety concerns, fail to convince the consumer to utilize nanotechnology-based products. This leads to decrease in demand of the product which directly affects the production profit margin (Singh and Ikram 2017).

Secondly, uncertainty in the cost-to-value ratio diversified innovations made for the improvement of agricultural and food sectors (Rodrigues et al. 2017).

These innovations include nano-biosensors for the detection of accumulated agrochemicals residues, nutrient deficiency, pH value, salinity and moisture content and soil requirement, as well as nano-biocontrol agents both for disease detection and prevention, nano-filters for the purification and detection, controlled delivery of agrochemicals, therapeutics and nano-remediating agents (Omanovic-Miklicanin and Maksimovic 2016, Le and Nunes 2016, Kaushal and Wani 2017, Srinath et al. 2017, Devanathan 2017, Cortesi et al. 2017, Yadav et al. 2017). The issue is that most of these products will be used by the farmers. However, in developing countries these innovations are inaccessible to the farmer, mostly due to the low profit margins.

Finally, there are a number of public biosafety concerns regarding the utilization of nanomaterials in food products and water because of their bioavailability and estimated bioaccumulation which, over time, results in nanotoxicity (Mishra et al. 2017, Rodrigues et al. 2017). Presently, the direct or indirect hazardous impacts on the environment have raised considerable concern among the public; this could be neutralized to some extent, if obligatory actions are taken at the stage of synthesis of these nanoparticles for the agricultural purpose (Mishra et al. 2017).

Adsorption of nanoparticles to soil matrix and aggregation leads to the reduction in their mobility, mineralization capacity and bioavailability, and also adversely effects the microbiotal diversity (Simonin et al. 2015, Simonin and Richaume 2015, Ruttkay–Nedecky et al. 2017). Similarly, deposition of these metallic nanoparticles also affects certain crop plants, such as *Triticum aestivum*, *Lolium perenne*, *Zea mays*, *Lactuca sativa*, *Allium cepa*, *Linum usitatissimum*, *Lolium perenne*, *Cucurbita pepo*, *Raphanus sativus*, *Brassica napus*, *Phaseolus radiatus*, *Brassica oleracea*, *Daucus carota* and *Lycopersicon esculentum*. The nanoparticles affect these crop plants by reducing germination, decreasing root length/elongation and growth, decreasing shoot length, decreasing mitosis and disturbing metaphase, resulting in sticky chromosome, cell wall disintegration and damage as well as reduction in seedling growth and biomass (Duhan et al. 2017).

If the safety of crop plants is at stake, then the food prepared from these crop plants and, concomitantly, the safety of the consumers is also questionable. The effects caused by the nanotoxicity could hardly be detected at the initial stage because of their minute size-to-volume ratio which was also an advantage at the beginning during the synthesis or the primary motive behind their widespread application (Peters et al. 2014, Omanovic-Miklicanin and Maksimovic 2016). Detection and quantification of such nanoparticles require advanced and sophisticated instrumentation,such as electron microscopy, asymmetric flow field-flow fractionation conjugated with mass spectrometry, single-particle inductively coupled mass spectrometry, etc., as well as highly trained personnel (Peters et al. 2014). Therefore, building a simple, cheaper, more precise and accurate nanoparticle detecting device is also a limitation of the present decade. Therefore, attempts should be made such that,when detection and quantification of nanoparticles during the ingestion of a pizza, for example, is desired, we can accomplish it without difficulty.

5. Future Prospects

Many sectors of global economy are greatly influenced by the promising applications of nanotechnology. Evidence of this can be seen in the double–digit production growth rates of nanomaterials, nanotools and nanodevices. By 2021, the global nanotechnology market, the nanomaterial market and nanodevices market will reach $90.5, $ 77.3 and $195.9 billion, respectively, with a five-year compound annual growth rate of 18.2, 18.9 and 28.2%, respectively (BCC Research 2017). Furthermore, by 2021 the market share of nanomaterials is expected to increase by 85.3% (BCC Research 2017).

Overall, it can be expected that the developments in the field of nanotechnology will be tremendous and will rule the future economic and scientific community, irrespective of its present limitations and barriers (Dhewa 2015). Despite the commercial successes, nanotechnology still requires a great deal of development and improvement in order to actually meet the practical challenges that would emerge in the near future, with respect to the present limitations (Rodrigues et al. 2017). Starting from the most essential application of nanotechnology, which is probing and sensing, complex food matrices remain a challenge for the sensitive detection of pathogens and toxins. Furthermore, the capacity to precisely distinguish between viable and dead pathogens among a variety of other harmful organisms is another tedious task which could be made easier with the use of future nano-enabled sensing devices (Yazgan et al. 2014). Nanoparticles serving as an antimicrobial agent are slowly taking over from conventional antibiotics, however, the actual molecular mechanism is still largely unknown. It could be expected that with the future advancements in understanding of nanomaterials and their impact or the underling role as antimicrobial agents by the help of advanced techniques like cryo-electron microscopy for the high-resolution structure determination (Saibil 2000, Rodrigues et al. 2017). "Prevention is better than cure", therefore, instead of preparing to meet the future challenge of bioaccumulation resulting from the utilization of non-recyclable nanomaterials in food and food packaging, it is possible in the future to develop reusable, recyclable, and biodegradable nanomaterials (Astruc et al. 2005). Most of the nanotechnology-based applications for water treatment are still under laboratory research and very few have progressed as far as field tests;several products are commercially available though, such as nanoadsorbents, nano-enabled membranes, nanophotocatalysts and nano-enabled disinfection systems. However, the lacuna is that there exists no clear speculation about how nano-enabled catalyst or sorbents would perform in the presence of variable organic carbon and biomass (Rodrigues et al. 2017). Future nano-enabled catalyst or sorbents would be conjugated with probes or sensors where the detection would more accurately determined by some advanced branch for femtochemistry. A clear understanding of the mode of action might lead in the development of advanced membranes by tailoring at atomic level with selectivity for specific compounds, which is a challenge of the present scenario.

Similarly, tuned nanocarriers could be developed for the delivery of multiple drugs or agronomically significant compounds, such as fertilizers, pesticides, vaccines, growth regulatory hormones and antimicrobial agents, using nanocapsules or nanospheres (Dhewa 2015, Pereira et al. 2017a,b).

Genetic engineering of plants in order to produce better tolerance towards various biotic and abiotic stresses, such as drought, high temperature, metal accumulation and microbial infections, could be developed with the use of nanotechnology-based gene delivery vehicles (Rai et al. 2015). Efficient bioremediation by means of nanotechnological advancement would greatly influence the future civilizations in solving the problem of waste management and its recycling,one of the major challenges of present civilized society. Finally the sector of energy conservation will be immensely influenced by the application of nanomaterials to solar panels, which would simultaneously be low cost and an effective source of energy that could transmogrify the future (Rodrigues et al. 2017).

Conclusion

From the present chapter, it is concluded that nanotechnology provides a diversified opportunity in the present, as well as in the future, for the sustainable growth of food and for the agricultural sector in general. With the help of this new technology, a great deal of advancement could be achieved that directly benefits the consumer by improving production and management of the agricultural products. Overall, it can be said that nanotechnology-based opportunity could be leveraged in order to boost or promote sustainable agricultural and food systems. Specifically, nano-enabled biosensors and probes for the detection of agrochemical residues, nutrient deficits, estimation of abiotic and biotic parameters influencing plant growth and development. Nano-biocontrol agents protect the plants from pathogenic microorganisms and regulate the number of symbiotic and endophytic microorganisms associated with plants' metabolic or physiologic activity. Nano-filters, nano-adsorbents and nano-enabled membranes can be used for the removal of harmful or hazardous chemical residues from the water and also enable the reuse and disinfection of water, thereby availing or fulfilling the previous deficit of water resources and conserving energy. Controlled delivery of agrochemicals, on the other hand, further boosts agricultural sustainability by mediating regulated modes of nutrient, antimicrobial agent, pesticide, insecticide and herbicide delivery. Finally, nano-bioremediationmay directly or indirectly enhance the durability of mankind by leveraging food and agriculture sustainability via the inhibition of nanotoxicity resulting from bioaccumulation, in the present as well as in the future.

References

Aghajani, A. and A. Soleymani. 2017. Effects of nano-fertilization on growth and yield of bean (*Phaseolus vulgaris* L.) under water deficit conditions. Curr. Nanosci. 13(2): 194–201.

Ali, J., J. Najeeb, M.A. Ali, M.F. Aslam and A. Raza. 2017. Biosensors: Their fundamentals, designs, types and most recent impactful applications: A Review. J. Biosens. Bioelectron. 8: 235.

Astruc, D., F. Lu and J.R. Aranzaes. 2005. Nanoparticles as recyclable catalysts: The frontier between homogeneous and heterogeneous catalysis. Angew. Chem. Int. Ed. 44: 7852–7872.

Babu, M.Y., V.J. Devi, C.M. Ramakritinan, R. Umarani, N. Taredahalli and A.K. Kumaraguru. 2014. Application of biosynthesized silver nanoparticles in agricultural and marine pest control. Curr. Nanosci. 10: 374–381.

Baeza, A., R.R. Castillo, A. Torres-Pardo, J.M. González-Calbet and M. Vallet-Regíal. 2017. Electron microscopy for inorganic-type drug delivery nanocarriers for antitumoral applications: What does it reveal? J. Mater. Chem. B5: 2714–2725.

Bala, M. and V. Arya. 2013. Biological synthesis of silver nanoparticles from aqueous extract of endophytic fungus *Aspergillus fumigatus* and its antibacterial action. Int. J. Nanomater. Bios. 3(2): 37–41.

BCC Research. 2017. Nanotechnology sees big growth in products and applications, reports BCC research, Massachusetts. https://globenewswire.com/news-release/2017/01/17/906164/0/en/

Nanotechnology-Sees-Big-Growth-in-Products-and-Applications-Reports-BCC-Research.html. Accessed 23 Oct 2017.

Bhattacharyya, A., R. Chandrasekar, A.K. Chandra, T. Epidi and R. Shetty Praksham. 2014. Application of nanoparticles in sustainable agriculture: Its current status. pp. 429–448. *In*: R. Chandrasekar, B.K. Tyagi, Z.Z. Gui and G.R. Reeck (eds.). Short views on insect biochemistry and molecular biology, Vol 2. 1st edn. International Book Mission, USA.

Bodaiah, B., U.M. Kiranmayi, P. Sudhakar Ravi, A. Varma and K. Bhushanam. 2016. Insecticidal activity of green synthesized silver nanoparticles. Int. J. Recent Scientific Res. 7(4): 10652–10656.

Bulbul, G., A. Hayat and S. Andreescu. 2015. Portable nanoparticle-based sensors for food safety assessment. Sensors 15: 30736–30758.

Calderón-Jiménez, B., M.E. Johnson, A.R. Montoro Bustos, K.E. Murphy, M.R. Winchester and J.R. Vega Baudrit. 2017. Silver nanoparticles: Technological advances, societalimpacts, and metrological challenges. Front. Chem. 5: 6.

Campos, E.V.R., J.L. Oliveira, C.M. Gonçalves da Silva, M. Pascoli, T. Pasquoto, R. Lima, P.C. Abhilash and L.F. Fraceto. 2015. Polymeric and solid lipid nanoparticles for sustained release of Carbendazim and Tebuconazole in agricultural applications. Sci. Rep. 5: 13809.

Cao, L., H. Zhang, C. Cao, J. Zhang, F. Li and Q. Huang. 2016. Quaternized chitosan-capped mesoporous silica nanoparticles as nanocarriers for controlled pesticide release. Nanomaterials (Basel) 6(7): 126.

Catley-Carlson, M. 2017. The emptying well. Nature 542: 412–413.

Cecchin, I., K.R. Reddy, A. Thome, E.F. Tessaro and F. Schnaid. 2017. Nanobioremediation: Integration of nanoparticles and bioremediation for sustainable remediation of chlorinated organic contaminants in soils. Int. Biodeterior. Biodegradation 119: 419–428.

Chang, Y.C., Y.S. Lin, G.T. Xiao, T.C. Chiu and C.C. Hu. 2016. A highly selective and sensitive nanosensor for the detection of glyphosate. Talanta 161: 94–98.

Chauhan, N., D.M. Gopal, R. Kumar, K.-H. Kim and S. Kumar. 2016. Development of chitosan nanocapsules for the controlled release of hexaconazole. Int. J. Biol. Macromol. 97: 616–624.

Cheng, H.N., K.T. Klasson, T. Asakura and Q. Wu. 2016. Nanotechnology in agriculture, *Nanotechnology: Delivering on the promise*. pp. 233–242. *In*: H.N. Cheng, L. Doemeny, C.L. Geraci and D.G. Schmidt (eds.). *ACS Symposium Series*, Vol. 1224. American Chemical Society, Washington, DC.

Chong, M.N., B. Jin, C.W.K. Chow and C. Saint. 2010. Recent developments in photocatalytic water treatment technology: A review. Water Res. 44: 2997–3027.

Conesa, H.M., M.W.H. Evangelou, B.H. Robinson and R. Schulin. 2012. A critical view of current state of phytotechnologies to remediate soils: Still a promising tool? Sci. World J. 2012: 173829.

Cortesi, R., G. Valacchi, X.M. Muresan, M. Drechsler, C. Contado, E. Esposito, A. Grandini, A. Guerrini, G. Forlani and G. Sacchetti. 2017. Nanostructured lipid carriers (NLC) for the delivery of natural molecules with antimicrobial activity: Production, characterization and *in vitro* studies. J. Microencapsul. 34(1): 63–72.

Deepa, S., K. Kanimozhi and A. Panneerselvam. 2013. Antimicrobial activity of extracellularly synthesized silver nanoparticles from marine derived actinomycetes. Int. J. Curr. Microbiol. App. Sci. 2(9): 223–230.

Devanathan, R. 2017. Energy penalty for excess baggage. Nat. Nanotechnol. 12: 500–501.

Dhami, N.K., M.S. Reddy and A. Mukherjee. 2013. Biomineralization of calcium carbonates and their engineered applications: A review. Front. Microbiol. 4: 314.

Dhewa, T. 2015. Nanotechnology applications in agriculture: An update. Oct J. Environ. Res. 3(2): 204–211.

Du, J., H. Singh and T.-H. Yi. 2017. Biosynthesis of silver nanoparticles by *Novosphingobium* sp. THG-C3 and their antimicrobial potential. Artif. Cells Nanomed. Biotechnol. 45(2): 211–217.

Duhan, J.S., R. Kumar, N. Kumar, P. Kaur, K. Nehra and S. Duhan. 2017. Nanotechnology: The new perspective in precision agriculture. Biotechnol. Rep. 15: 11–23.

Duncan, T.V. 2011. Applications of nanotechnology in food packaging and food safety: Barrier materials, antimicrobials and sensors. J. Colloid Interface. Sci. 363:1–24.

Fabrega, J., S.R. Fawcett, J.C. Renshaw and J.R. Lead. 2009. Silver nanoparticle impact on bacterial growth: Effect of pH, concentration, and organic matter. Environ. Sci. Technol. 43: 7285–7290.

FAO. 2017. Cereal Supply and Demand Brief-2017. http://www.fao.org/worldfoodsituation/csdb/en/.

FAO. 2017. Food Outlook-2017, http://www.fao.org/giews/reports/food-outlook/en/.

FAO. 2017. Food Price Index-2017, http://www.fao.org/worldfoodsituation/foodpricesindex/en/.

FAO. 2017. The future of food and agriculture—Trends and challenges, Rome.

Fita, A., A. Rodríguez-Burruezo, M. Boscaiu, J. Prohens and O. Vicente. 2015. Breeding and domesticating crops adapted to drought and salinity: A new paradigm for increasing food production. Front. Plant Sci. 6: 978.

Formoso, P., E. Pantuso, G.D. Filpo and F.P. Nicoletta. 2017. Electro-conductive membranes for permeation enhancement and fouling mitigation: A short review. Membranes 7: 39.

Faraceto, L.F., R. Grillo, G.A. de Medeiros, V. Scognamiglio, G. Rea and C. Bartolucci. 2016. Nanotechnology in agriculture: Which innovation potential does it have? Front. Environ. Sci. 4: 1–5.

Garmasheva, I., N. Kovalenko, S. Voychuk, A. Ostapchuk, O. Livinska and L. Oleschenko. 2016. *Lactobacillus* species mediated synthesis of silver nanoparticles and their antibacterial activity against opportunistic pathogens *in vitro*. BioImpacts 6(4): 219–223.

Gehrke, I., A. Geiser and A. Somborn-Schulz. 2015. Innovations in nanotechnology for water treatment. Nanotechnol. Sci. Appl. 8: 1–17.

Gericke, M. and A. Pinches. 2006. Biological synthesis of metal nanoparticles. Hydrometallurgy 83: 132–140.

Ghosh, P.R., D. Fawcett, S.B. Sharma and G.E.J. Poinern. 2017. Production of high-value nanoparticles via biogenic processes using aquacultural and horticultural food waste. Materials 10: 852.

Giroto, A.S., G.G.F. Guimaraes, M. Foschini and C. Ribeiro. 2017. Role of slow-release nanocomposite fertilizers on nitrogen and phosphate availability in soil. Sci. Rep.7: 46032.

Guilger, M., N. Bilesky-Jose, T. Pasquoto, R. Grillo, P.C. Abhilash, L.F. Fraceto and R. Lima. 2017. Biogenic silver nanoparticles based on *Trichoderma harzianum*: Synthesis, characterization, toxicity evaluation and biological activity. Sci. Rep. 7: 44421.

Gonzalez-Fuenzalida, R.A., Y. Moliner-Martínez, H. Prima-Garcia, A. Ribera, P. Campins-Falcó and R.J. Zaragozá. 2014. Evaluation of superparamagnetic silica nanoparticles for extraction of triazines in magnetic in-tube solid phase microextraction coupled to capillary liquid chromatography. Nanomater. 4: 242–255.

Grillo, R., N.Z.P. Santos, C.R. Maruyama, A.H. Rosa, R. de Lima and L.F. Fraceto. 2012. Poly(ε-caprolactone) nanocapsules as carrier systems for herbicides: Physico-chemical characterization and genotoxicity evaluation. J. Hazard. Mater. 231-232: 1–9.

Grillo, R., P.C. Abhilash and L.F. Fraceto. 2016. Nanotechnology applied to bio-encapsulation of pesticides. J. Nanosci. Nanotechnol. 16: 1231–1234.

Hosoya, H., A.S. Dobroff, W.H.P. Driessen, V. Cristini, L.M. Brinker, F.I. Staquicini, M. Cardó-Vila, S. D'Angelo, F. Ferrara, B. Proneth, Y.S. Lin, D.R. Dunphy, P. Dogra, M.P. Melancon, R.J. Stafford, K. Miyazono, J.G. Gelovani, K. Kataoka, C.J. Brinker, R.L. Sidman, W. Arap and R. Pasqualini. 2016. Integrated nanotechnology platform for tumor-targeted multimodal imaging and therapeutic cargo release. Proc. Natl. Acad. Sci. USA 113: 1877–1882.

Hsu, C.W., Z.Y. Lin, T.Y. Chan, T.C. Chiu and C.C. Hu. 2017. Oxidized multiwalled carbon nanotubes decorated with silver nanoparticles for fluorometric detection of dimethoate. Food Chem. 224: 353–358.

Hsueh, Y.-H., K.-S. Lin, W.-J. Ke, C.-T. Hsieh, C.-L. Chiang, D.-Y. Tzou and S.-T. Liu. 2015. The antimicrobial properties of silver nanoparticles in *Bacillus subtilis* are mediated by released Ag$^+$ions. PLoS ONE 10(12): e0144306.

Huo, L., X. Zeng, S. Su, L. Bai and Y. Wang. 2017. Enhanced removal of As (V) from aqueous solution using modified hydrous ferric oxide nanoparticles. Sci. Rep. 7: 40765.

Hussain, I., N.B. Singh, A. Singh, H. Singh and S.C. Singh. 2016. Green synthesis of nanoparticles and its potential application. Biotechnol. Lett. 38: 545–560.

Ingle, A., A. Gade, S. Pierrat, C. Sonnichsen and M. Rai. 2008. Mycosynthesis of silver nanoparticles using the fungus *Fusarium acuminatum* and its activity against some human pathogenic bacteria. Curr. Nanosci. 4: 141–144.

International Monetary Fund (IMF) 2017. World Economic Outlook.https://www.imf.org/en/Publications/WEO/Issues/2017/09/19/world-economic-outlook-october-2017. Accessed 22 Oct 2017.

Jamir, A. and M. Mahato. 2016. A review on protein based nanobiocomposite for biosensor application. Rev. Adv. Sci. Eng. 5: 109–122.

Jiang, W., K. Yang, R.W. Vachet and B. Xing. 2010. Interaction between oxide nanoparticles and biomolecules of the bacterial cell envelope as examined by infrared spectroscopy. Langmuir 26(23): 18071–18077.

Joint FAO/IAEA Division of Nuclear Techniques in Food and Agriculture. 2017a.Soil and water. http://www-naweb.iaea.org/nafa/swmn/soils-water.html.

Joint FAO/IAEA Division of Nuclear Techniques in Food and Agriculture. 2017b. Agricultural water management. https://www.iaea.org/topics/agricultural-water-management.

Kalidasan, K., S.K. Sahu, K. Kayalvizhi and K. Kathiresan. 2015. Polyunsaturated fatty acid-producing marine thraustochytrids: A potential source for antimicrobials. J. Coast. Life Med. 3(11): 848–851.

Karthik, L., G. Kumar, T. Keswani, A. Bhattacharya, P. Reddy and B. Rao. 2013. Marine actinobacterial mediated gold nanoparticles synthesis and their antimalarial activity. Nanomedicine 9(7): 951–960

Kaushal, M. and S.P. Wani. 2017. Nanosensors: Frontiers in precision agriculture. pp. 279–291. *In*: R. Prasad, M. Kumar and V. Kumar (eds.). Nanotechnology, Springer, Singapore.

Kavitha, K.S., S. Baker, D. Rakshith, H.U. Kavitha, H.C. Yashwantha Rao, B.P. Harini and S. Satish. 2013. Plants as green source towards synthesis of nanoparticles. Int. Res. J. Biol. Sci. 2(6): 66–76.

Kawata, K., M. Osawa and S. Okabe. 2009. *In vitro* toxicity of silver nanoparticles at noncytotoxic doses to HepG2 human hepatoma cells. Environ. Sci. Technol. 43: 6046–6051.

Kiran, V.M. and S. Murugesan. 2014. Biological synthesis of silver nanoparticles from marine alga *Colpomenia sinuosa* and its *in vitro* anti-diabetic activity. Am. J. Bio-pharmacol. Biochem. LifeSci. 3(1): 1–7.

Klaus, T., R. Joerger, E. Olsson and C.-G. Granqvist. 1999. Silver-based crystalline nanoparticles, microbially fabricated. Proc. Natl. Acad. Sci. USA 96(24): 13611–13614.

Krajina, B.A., A.C. Proctor, A.P. Schoen, A.J. Spakowitz and S.C. Heilshorn. 2018. Biotemplated synthesis of inorganic materials: An emerging paradigm for nanomaterial synthesis inspired by nature. Prog. Mater. Sci. 91: 1–23.

Kumar, S., W. Ahlawat, G. Bhanjana, S. Heydarifard, M.M. Nazhad and N. Dilbaghi. 2014. Nanotechnology-based water treatment strategies. J. Nanosci. Nanotechnol. 14: 1838–1858.

Kumar, S.R. and P. Gopinath. 2016. Nano-bioremediation applications of nanotechnology for bioremediation. pp. 27–48. *In*: J.P. Chen, L.K. Wang, M.-H.S. Wang, Y.-T.H. Hung and N.K. Shammas (eds.). Remediation of Heavy Metals in the Environment, Taylor & Francis, New York.

Kuswandi, B., D. Futra and L.Y. Heng. 2017. Nanosensors for the detection of food contaminants. pp. 307–333. *In*: A. Grumezescu and A. Oprea (eds.). Nanotechnology Applications in Food: Flavor, stability, nutrition and safety, Elsevier, Atlanta.

Lamsal, K., S.W. Kim, J.H. Jung, Y.S. Kim, K.S. Kim and Y.S. Lee. 2011. Application of silver nanoparticles for the control of *Colletotrichum* species *in vitro* and pepper anthracnose disease in field. Mycobiology 39(3): 194–199.

Lau, H.Y., H. Wu, E.J.H. Wee, M. Trau, Y. Wang and J.R. Botella. 2017. Specific and sensitive isothermal electrochemical biosensor for plant pathogen DNA detection with colloidal gold nanoparticles as probes. Sci. Rep. 7: 38896.

Le, N.L. and S.P. Nunes. 2016. Materials and membrane technologies for water and energy sustainability. Sustainable Mater. Technol. 7: 1–28.

Li, G., D. He, Y. Qian, B. Guan, S. Gao, Y. Cui, K. Yokoyama and L. Wang. 2012. Fungus-mediated green synthesis of silver nanoparticles using *Aspergillus terreus*. Int. J. Mol. Sci. 13: 466–476.

Lin, W., Y. Pan and D.A. Bazylinski. 2017. Diversity and ecology of and biomineralization by magnetotactic bacteria. Environ. Microbiol. Rep. 9: 345–356.

Liu, Y., Z. Zhang, Y. Wang, Y. Zhao, Y. Lu, X. Xu, J. Yan and Y. Pan. 2015. A highly sensitive and flexible magnetic nanoprobe labelled immunochromatographic assay platform for pathogen *Vibrio parahaemolyticus*. Int. J. Food Microbiol. 211: 109–116.

Malarkodi, C., S. Rajeshkumar, K. Paulkumar, G. Gnanajobitha, M. Vanaja and G. Annadurai. 2013. Bacterial synthesis of silver nanoparticles by using optimized biomass growth of *Bacillus* sp. J. Nanosci. Nanotechnol. 3(2): 26–32.

Malhotra, A., K. Dolma, N. Kaur, Y.S. Rathore, Ashish, S. Mayilraj and A.R. Choudhury. 2013. Biosynthesis of gold and silver nanoparticles using a novel marine strain of *Stenotrophomonas*. *Bioresour. Technol.* 142: 727–731.

Mamun, M.A.A, S. Bhattacharjee, D. Pernitsky and M. Sadrzadeh. 2017. Colloidal fouling of nanofiltration membranes: Development of a standard operating procedure. Membranes 7:4.

Manimaran, M. and K. Kannabiran. 2017. Actinomycetes-mediated biogenic synthesis of metal and metal oxide nanoparticles: progress and challenges. Lett. Appl. Microbiol. 64: 401–408.

Manivasagan, P., S.Y. Nam and J. Oh. 2016. Marine microorganisms as potential biofactories for synthesis of metallic nanoparticles. Crit. Rev. Microbiol. 42(6): 1007–1019.

Maruyama, C.R., M. Guilger, M. Pascoli, N. Bileshy-José, P.C. Abhilash, L.F. Fraceto and R. de Lima. 2016. Nanoparticles based on chitosan as carriers for the combined herbicides Imazapic and Imazapyr. Sci. Rep. 6: 19768.

Meador, M.A., B. Files, J. Li, A.H. Manohara, D. Powell and E.J. Siochi. 2010. DRAFT nanotechnology roadmap:Technology area 10, National Aeronautics and Space Administration (NASA). https://www.nasa.gov. Accessed 20 Oct 2017.

Mishra, S., C. Keswani, P.C. Abhilash, L.F. Fraceto and H.B. Singh. 2017. Integrated approach of agri-nanotechnology: Challenges and future trends. Front. Plant Sci. 8: 471.

Moghaddam, A.B., F. Namvar, M. Moniri, P. Tahir, S. Azizi and R. Mohamad. 2015. Nanoparticles biosynthesized by fungi and yeast: A review of their preparation, properties, and medical applications.Mol. 20: 16540–16565.

Mohammadi, F., F. Ahmadi and M.A. Andi. 2015. Effect of zinc oxide nanoparticles on carcass parameters, relative weight of digestive and lymphoid organs of broiler fed wet diet during the starter period. Int. J. Biosci. 6: 389–394.

Moreira, F.K.V., L.A.D. Camargo, J.M. Marconcini and L.H.C. Mattoso. 2013. Nutraceutically inspired pectin-Mg(OH)$_2$ nanocomposites for bioactive packaging applications. J. Agric. Food. Chem. 61: 7110−7119.

Nabawy, G.A., A.A. Hassan, R.H.S. El-Ahl and M.K. Refai. 2014. Effect of metal nanoparticles in comparison with commercial antifungal feed additives on the growth of *Aspergillus flavus* and aflatoxin B1 production. J. Global Biosci. 3(6): 954–971.

Nagy, L., M. Magyar, T. Szabo, K. Hajdu, L. Giotta, M. Dorogi and F. Milano. 2014. Photosynthetic machineries in nano-systems. Curr. Protein Pept. Sci. 15(4): 363–373.

Nakasato, D.Y., A.E.S. Pereira, J.L. Oliveira, H.C. Oliveira and L.F. Fraceto. 2017. Evaluation of the effects of polymeric chitosan/tripolyphosphate and solid lipid nanoparticles on germination of *Zea mays*, *Brassica rapa* and *Pisum sativum*. Ecotoxicol. Environ. Saf. 142: 369–374.

Narasimha, G., A. Janardhan, M. Alzohairy, H. Khadri and K. Mallikarjuna. 2013. Extracellular synthesis, characterization and antibacterial activity of silver nanoparticles by *Actinomycetes* isolative. Int. J. Nano Dimens. 4(1): 77–83.

Navarro, E., A. Baun, R. Behra, N.B. Hartmann, J. Filser, A.J. Miao, A. Quigg, P.H. Santschi and L. Sigg. 2008. Environmental behavior and ecotoxicity of engineered nanoparticles to algae, plants and fungi. Ecotoxicology 17: 372–386.

Naz, S., H. Shahzad, A. Ali and M. Zia. 2017. Nanomaterials as nanocarriers: A critical assessment why these are multi-chore vanquisher in breast cancer treatment. Artif. Cells Nanomed. Biotechnol. 46(5): 899–916.

Netala, V.R., S.V. Kotakadi, P. Bobbu, S.A. Gaddam and V. Tartte. 2016. Endophytic fungal isolate mediated biosynthesis of silver nanoparticles and their free radical scavenging activity and anti-microbial studies. 3 Biotech 6:132.

Nhan, L.V., C. Ma and Y. Rui. 2015. The effects of Fe$_2$O$_3$ nanoparticles on physiology and insecticide activity in non-transgenic and Bt-transgenic cotton. Front. Plant Sci. 6: 1263.

Noah, N.M., M. Omole, S. Stern, S. Zhang, O.A. Sadik, E.H. Hess, J. Martinovic, P.G.L. Baker and E.I. Iwuoha. 2012. Conducting polyamic acid membranes for sensing and site-directed immobilization of proteins. Anal. Biochem. 428: 54–63.

Nuruzzaman, M., M.M. Rahman, Y. Liu and R. Naidu. 2016. Nanoencapsulation, nano-guard for pesticides: A new window for safe application. J. Agric. Food Chem. 64: 1447–1483.

Nzila, A., S.A. Razzak and J. Zhu. 2016. Bioaugmentation: An emerging strategy of industrial wastewater treatment for reuse and discharge. Int. J. Environ. Res. Public Health 13: 846.

Oliveira, H.C., R. Stolf-Moreira, C.B.R. Martinez, R. Grillo, M.B. de Jesus and L.F. Fraceto. 2015a. Nanoencapsulation enhances the post-emergence herbicidal activity of atrazine against mustard plants. PloS One 10: e0132971.

Oliveira, H.C., R. Stolf-Moreira, C.B. Martinez, G.F. Sousa, R. Grillo, M.B. de Jesus and L.F. Fraceto. 2015b. Evaluation of the side effects of poly(epsilon-caprolactone) nanocapsules containing atrazine toward maize plants. Front. Chem. 3: 61.

Oliveira, J.L., E.V. Campos, C.M.G. da Silva, T. Pasquoto, R. Lima and L.F. Fraceto. 2015c. Solid lipid nanoparticles co-loaded with simazine and atrazine: Preparation, characterization, and evaluation of herbicidal activity. J. Agric. Food Chem. 63: 422–432.

Omanovic-Miklicanin, E. and M. Maksimovic. 2016. Nanosensors applications in agriculture and food industry. Glas. Hem. Tehnol. Bosne. Herceg. 47: 59–70.

Pantidos, N. and L.E. Horsfall. 2014. Biological synthesis of metallic nanoparticles by bacteria, fungi and plants. J. Nanomed. Nanotechnol. 5:233.

Park, B. and S.-J. Choi. 2017. Sensitive immunoassay-based detection of *Vibrio parahaemolyticus* using capture and labelling particles in a stationary liquid phase lab-on-a-chip. Biosens. Bioelectron. 90: 269–275.

Peng, B., G. Li, D. Li, S. Dodson, Q. Zhang, J. Zhang, Y.H. Lee, H.V. Demir, X.Y. Ling and Q. Xiong. 2013. Vertically aligned gold nanorod monolayer on arbitrary substrates: Self-assembly and femtomolar detection of food contaminants, ACS Nano. 7: 5993–6000.

Pereira, A.E.S., P.M.M. da Silva, J.L. de Oliveira, H.C. Oliveira and L.F. Fraceto. 2017a. Chitosan nanoparticles as carrier systems for the plant growth hormone gibberellic acid. Colloids Surf. B. Biointerfaces 150: 141–152.

Pereira, A.E.S., I.E. Sandoval-Herrera, S.A.Z. Betancourt, H.C. Oliveira, A.S. Ledezma-Pérez, J. Romero and L.F. Fraceto. 2017b. γ-Polyglutamic acid/chitosan nanoparticles for the plant growth regulator gibberellic acid: Characterization and evaluation of biological activity. Carbohydr. Polym. 157: 1862–1873.

Pereira, M.M., L. Mouton, C. Yepremian, A. Couté, J. Lo, J.M. Marconcini, L.O. Ladeira, N.R.B. Raposo, H.M. Brandão and R. Brayner. 2014. Ecotoxicological effects of carbon nanotubes and cellulose nanofibers in *Chlorella vulgaris*. J. Nanobiotechnol. 12: 15.

Peters, R.J.B., G. Bemmel, Z. Herrera-Rivera, H.P. Helsper, H.J. Marvin, S. Weigel, P.C. Tromp, A.G. Oomen, A.G. Rietveld and H. Bouwmeester. 2014. Characterization of titanium dioxide nanoparticles in food products: Analytical methods to define nanoparticles. J. Agric. Food Chem. 62(27): 6285–6293.

Poscic, F., A. Mattiello, G. Fellet, F. Miceli and L. Marchiol. 2016. Effects of cerium and titanium oxide nanoparticles in soil on the nutrient composition of barley (*Hordeum vulgare* L.) kernels. Int. J. Environ. Res. Public Health 13(6): 577.

Prabhawathi, V., P.M. Sivakumar and M. Doble. 2012. Green synthesis of protein stabilized silver nanoparticles using *Pseudomonas fluorescens*, a marine bacterium, and its biomedical applications when coated on polycaprolactam. Ind. Eng. Chem. Res. 51(14): 5230–5239.

Prasad, R., A. Bhattacharyya and Q.D. Nguyen. 2017. Nanotechnology in sustainable agriculture: Recent developments, challenges, and perspectives. Front. Microbiol. 8: 1014.

Priya, S. and T. Sivakumar. 2015. Synthesis of silver nanoparticles from fungi isolated from marine ecosystem. Int. J. Curr. Res. Chem. Pharm. Sci. 2(12): 71–73.

Qu, X., P.J.J. Alvarez and Q. Li. 2013. Applications of nanotechnology in water and wastewater treatment. Water Res. 47: 3931–3946.

Quester, K., M. Avalos-Borja and E. Castro-Longoria. 2016. Controllable biosynthesis of small silver nanoparticles using fungal extract. J. Biomater. Nanobiotechnol. 7: 118–125.

Rai, M., S. Bansod, M. Bawaskar, A.T. Gade, C.A. dos Santos and A.B. Seabra. 2015. Nanoparticles-based delivery systems in plant genetic transformation. pp. 209–239. In: M. Rai, C. Ribeiro, L. Mattoso and N. Duran (eds.). Nanotechnologies in food and agriculture. Springer, New York.

Rai, V., S. Acharya and N. Dey. 2012. Implications of nanobiosensors in agriculture. J. Biomater. Nanobiotechnol. 3: 315–324.

Rizwan, M.D., M. Singh, C.K. Mitra and R.K. Morve. 2014. Ecofriendly application of nanomaterials: Nanobioremediation. J. Nanopart. Article ID 431787.

Rodrigues, S.M., P. Demokritou, N. Dokoozlian, C.O. Hendren, B. Karn, M.S. Mauter, O.A. Sadik, M. Safarpour, J.M. Unrine, J. Viers, P. Welle, J.C. White, M.R. Wiesnerde and G.V. Lowry.

2017. Nanotechnology for sustainable food production: Promising opportunities and scientific challenges. Environ. Sci.: Nano. 4: 767.

Rui, M., C. Ma, Y. Hao, J. Guo, Y. Rui, X. Tang, Q. Zhao, X. Fan, Z. Zhang, T. Hou and S. Zhu. 2016. Iron oxide nanoparticles as a potential iron fertilizer for peanut (*Arachis hypogaea*). Front. Plant Sci. 7: 815

Ruttkay−Nedecky, B., O. Krystofova, L. Nejdl and L. Adam. 2017. Nanoparticles based on essential metals and their phytotoxicity. J. Nanobiotechnol. 15: 33.

Sabir, S., A. Arshad and S.K. Chaudhari. 2014. Zinc oxide nanoparticles for revolutionizing agriculture: Synthesis and applications. Sci. World J. Article ID 925494.

Saibil, H.R. 2000. Macromolecular structure determination by cryo-electron microscopy. Acta Crystallogr. D. Biol. Crystallogr. 56: 1215–1222.

Sakaguchi, T., J.I.B. Janairo, M. Lussier-Price, J. Wada, J.G. Omichinski and K. Sakaguchi. 2017. Oligomerization enhances the binding affinity of a silver biomineralization peptide and catalyzes nanostructure formation. Sci. Rep.7: 1400.

Saxena, J., M.M.V. Sharma, S.V. Gupta and A.V. Singh.2014. Emerging role of fungi in nanoparticle synthesis and their applications. World J. Pharm. Pharma. Sci. 3(9): 1586–1613.

Schrofel, A., G. Kratosova, M. Krautova, E. Dobrocka and I. Vavra. 2011. Biosynthesis of gold nanoparticles using diatoms–silica-gold and EPS-gold bionanocomposite formation. J. Nanopart. Res. 13(8): 3207–3216.

Seabra, A.B., M. Rai and N. Duran. 2015. Emerging role of nanocarriers in delivery of nitric oxide for sustainable agriculture. pp. 183–207. *In*: M. Rai, C. Ribeiro, L. Mattoso and N. Duran (eds.). Nanotechnologies in food and agriculture. Springer, New York.

Sekhon, B.S. 2014. Nanotechnology in agri-food production: An overview. Nanotechnol. Sci. Appl. 7: 31–53.

Seshadri, S., K. Saranya and M. Kowshik. 2011. Green synthesis of lead sulfide nanoparticles by the lead resistant marine yeast *Rhodosporidium diobovatum*. Biotechnol. Prog. 27(5): 1464–1469.

Sharma, N., A.K. Pinnaka, M. Raje, F.N.U. Ashish, M.S. Bhattacharyya and A.R. Choudhury. 2012. Exploitation of marine bacteria for production of gold nanoparticles. Microb. Cell Fact. 11(86): 1–6.

Shivakrishna, P., M.R. Prasad, G. Krishna and M.A.S. Charya. 2013. Synthesis of silver nanoparticles from marine bacteria *Pseudomonas aeruginosa*. Octa J. Biosci. 1(2): 108–114.

Siafaka, P.I., N.U. Okur, E. Karavas and D.N. Bikiaris. 2016. Surface modified multifunctional and stimuli responsive nanoparticles for drug targeting: Current status and uses. Int. J. Mol. Sci. 17(9): 1440.

Simonin, M., J.P. Guyonnet, J.M. Martins, M.V. Ginot and A. Richaume. 2015. Influence of soil properties on the toxicity of TiO_2 nanoparticles oncarbon mineralization and bacterial abundance. J. Hazard. Mater. 283: 529–535.

Simonin, M. and A. Richaume. 2015. Impact of engineered nanoparticles onthe activity, abundance, and diversity of soil microbial communities: A review. Environ. Sci. Pollut. Res. 22: 13710–13723.

Singh, C.R., K. Kathiresan and S. Anandhan. 2015.A review on marine based nanoparticles and their potential applications. Afr. J. Biotechnol. 14(18): 1525–1532.

Singh, P., H. Singh, Y.J. Kim, R. Mathiyalagan, C. Wang and D.C. Yang. 2016. Extracellular synthesis of silver and gold nanoparticles by *Sporosarcina koreensis* DC4 and their biological applications. Enzyme Microb. Technol. 86: 75–83.

Singh, P. and S. Ikram. 2017. Nanotechnology in food packaging: An overview. J. Adv. Mater. 1(1): 19–22.

Srinath, B.S., K. Namratha and K. Byrappa. 2017. Eco-friendly synthesis of gold nanoparticles by gold mine bacteria *Brevibacillus formosus* and their antibacterial and biocompatible studies. IOSR J. Pharm. 7(8): 53–60.

Stoicea, N., J. Fiorda-Diaz, N. Joseph, M. Shabsigh, C. Arias-Morales, A.A. Gonzalez-Zacarias, A. Mavarez-Martinez, S. Marjoribanks and S.D. Bergese. 2017. Advanced analgesic drug delivery and nanobiotechnology. Drugs 77: 1069–1076.

Strojny, B., M. Grodzik, E. Sawosz, A. Winnicka, N. Kurantowicz, S. Jaworski, M. Kutwin, K. Urbańska, A. Hotowy, M. Wierzbicki and A. Chwalibog. 2016. Diamond nanoparticles modify curcumin activity: *In vitro* studies on cancer and normal cells and *inovo* studies on chicken embryo model. PLoS ONE 11(10): e0164637.

Subramanian, M., N.M. Alikunhi and K. Kandasamy. 2010. *In vitro* synthesis of silver nanoparticle by marine yeasts from coastal mangrove sediment. Adv. Sci. Lett. 3(4): 428–433.

Suresh, J., R. Yuvakkumar, M. Sundrarajan and S.I. Hong. 2014. Green synthesis of magnesium oxide nanoparticles. Adv. Mat. Res. 952: 141–144.

Sutovsky, P. and C.E. Kennedy. 2013. Biomarker-based nanotechnology for the improvement of reproductive performance in beef and dairy cattle. Ind. Biotechnol. 9: 24–30.

Thatoi, P., R.G. Kerry, S. Gouda, G. Das, K. Pramanik, H. Thatoi and J.K. Patra. 2016. Photo-mediated green synthesis of silver and zinc oxide nanoparticles using aqueous extracts of two mangrove plant species, *Heritiera fomes* and *Sonneratia apetala* and investigation of their biomedical applications. J. Photochem. Photobiol. 163: 311–318.

Tripathi, D.K., A. Tripathi, Shweta, S. Singh, Y. Singh, K. Vishwakarma, G. Yadav, S. Sharma, V.K. Singh, R.K. Mishra, R.G. Upadhyay, N.K. Dubey, Y. Lee and D.K. Chauhan. 2017. Uptake, accumulation and toxicity of silver nanoparticle in autotrophic plants, and heterotrophic microbes: A concentric review. Front. Microbiol. 8: 7.

Umashankari, J., D. Inbakandan, T.T. Ajithkumar and T. Balasubramanian. 2012. Mangrove plant, *Rhizophora mucronata* (Lamk, 1804) mediated one pot green synthesis of silver nanoparticles and its antibacterial activity against aquatic pathogens. Aquat. Biosyst. 8(1): 11.

Underwood, C. and A.W. van Eps. 2012. Nanomedicine and veterinary science: The reality and the practicality. Vet. J. 193: 12–23.

Vardanyan, Z., V. Gevorkyan, M. Ananyan, H. Vardapetyan and A. Trchounian. 2015. Effects of various heavy metal nanoparticles on *Enterococcus hirae* and *Escherichia coli* growth and proton-coupled membrane transport. J. Nanobiotechnol. 13: 69.

Vardhana, J. and G. Kathiravan. 2014. Biosynthesis of silver nanoparticles by endophytic fungi *Pestaloptiopsis pauciseta* isolated from the leaves of *Psidium guajava* Linn. Int. J. Pharma. Sci. Rev. Res. 31(1): 29–31.

Vijayan, S.R., P. Santhiyagu, M. Singamuthu, N.K. Ahila, R. Jayaraman and K. Ethiraj. 2014. Synthesis and characterization of silver and gold nanoparticles using aqueous extract of seaweed, *Turbinaria conoides*, and their antimicrofouling activity. Sci. World J. Article ID 938272.

Vijayaraghavan, R., V.K. Prabha and S. Rajendran. 2012. Biosynthesis of silver nanoparticles by a marine bacterium *Bacillus subtilis* strain and its antifungal effect. World J. Sci. Technol. 2(9): 1–3.

Wang, C., Y.J. Kim, P. Singh, R. Mathiyalagan, Y. Jin and D.C. Yang. 2015. Green synthesis of silver nanoparticles by *Bacillus methylotrophicus*, and their antimicrobial activity. Artif. Cells Nanomed. Biotechnol. 44(4): 1127–1132.

Wang, L., C. Hu and L. Shao. 2017. The antimicrobial activity of nanoparticles: Present situation and prospects for the future. Int. J. Nanomed. 12: 1227–1249.

Wang, Y., C. Sun, X. Zhao, B. Cui, Z. Zeng, A. Wang, G. Liu and H. Cui. 2016. The application of nano-TiO$_2$ photo semiconductors in agriculture. Nanoscale Res. Lett. 11: 529.

World Health Organization (WHO). 2015. WHO estimates of the global burden of foodborne diseases: Foodborne diseases burden epidemiology reference group 2007–2015. Geneva, Switzerland.

Xie, J., K. Chen and X. Chen. 2009. Production, modification and bio-applications of magnetic nanoparticles gestated by magnetotactic bacteria. Nano Res. 2(4): 261–278.

Yadav, K.K., J.K. Singh, N. Gupta and V. Kumar. 2017. A review of nanobioremediation technologies for environmental cleanup: A novel biological approach. JMES 8(2): 740–757.

Yamanaka, M., K. Hara and J. Kudo. 2005. Bactericidal actions of a silver ion solution on *Escherichia coli*, studied by energy-filtering transmission electron microscopy and proteomic analysis. Appl. Environ. Microbiol. 71: 7589–7593.

Yassin, M.A., A.E.-R.M.A. El-Samawaty, T.M. Dawoud, O.H. Abd-Elkader, K.S. Al Maary, A.A. Hatamleh and A.M. Elgorban. 2017. Characterization and anti-*Aspergillus flavus* impact of nanoparticles synthesized by *Penicillium citrinum*. Saudi J. Biol. Sci. 24: 1243–1248.

Yazgan, I., N.M. Noah, O. Toure, S. Zhang and O.A. Sadik. 2014. Biosensor for selective detection of *E. coli* in spinach using the strong affinity of derivatized mannose with fimbrial lectin. Biosens. Bioelectron. 61: 266–273.

Yearla, S.R. and K. Padmashree. 2016. Exploitation of subabul stem lignin as a matrix in controlled release agrochemical nanoformulations: A case study with herbicide diuron. Environ. Sci. Pollut. Res. 23(18): 18085–18098

Zhang, C., M. Li, X. Xu and N. Liu. 2015. Effects of carbon nanotubes on atrazine biodegradation by *Arthrobacter* sp. J. Hazard. Mater. 287: 1–6.

Zhang, Y., B. Wu, H. Xu, H. Liu, M. Wang, Y. He and B. Pan. 2016. Nanomaterials-enabled water and wastewater treatment. NanoImpact 3-4: 22–39.

Zhao, L., Y. Huang, C. Hannah-Bick, A.N. Fulton and A.A. Keller. 2016. Application of metabolomics to assess the impact of Cu(OH)$_2$ nanopesticide on the nutritional value of lettuce (*Lactuca sativa*): Enhanced Cu intake and reduced antioxidants. NanoImpact 3-4: 58–66

Zhao, L., B. Peng, J.A. Hernandez-Viezcas, C. Rico, Y. Sun, J.R. Peralta-Videa, X. Tang, G. Niu, L. Jin, A. Varela-Ramirez, J.Y. Zhang and J.L. Gardea-Torresdey. 2012. Stress response and tolerance of *Zea mays* to CeO$_2$ nanoparticles: Cross talk among H$_2$O$_2$, heat shock protein and lipid peroxidation. ACS Nano 6(11): 9615–9622.

Index

Printed and bound by CPI Group (UK) Ltd, Croydon, CR0 4YY

24/10/2024

01778307-0004